The Maker of Pedigrees

INFORMATION CULTURES

Series Editors

This book series examines how information has been produced, circulated, received, and preserved in the historical past. It concentrates principally, though not exclusively, on textual evidence and welcomes investigation of historical information both in its material forms and in its cultural contexts. Themes of special interest include the history of scholarly discourses and practices of learned communication, documentary management and information systems—including libraries, archives, and networks of exchange—as well as mechanisms of paperwork and record-making. This series engages with vital discussions among academics, librarians, and digital humanists, presenting a vivid historical dimension that uncovers the roles played by cultures of information in times and places distant from our own.

Also in the Series

Sailing School:
Navigating Science and Skill, 1550–1800 (2019)
Margaret Schotte

Leibniz Discovers Asia:
Social Networking in the Republic of Letters (2019)
Michael Carhart

A Centaur in London:
Reading and Observation in Early Modern Science (2023)
Fabian Kraemer

The Maker of Pedigrees

Jakob Wilhelm Imhoff and the Meanings of Genealogy in Early Modern Europe

MARKUS FRIEDRICH

Johns Hopkins University Press
Baltimore

Printed in the United States of America on acid-free paper
2 4 6 8 9 7 5 3 1

Johns Hopkins University Press
2715 North Charles Street
Baltimore, Maryland 21218
www.press.jhu.edu

Cataloging-in-Publication Data is available from the Library of Congress.
A catalog record for this book is available from the British Library.

ISBN: 978-1-4214-4579-3 (hardcover)
ISBN: 978-1-4214-4580-9 (ebook)

CONTENTS

ACKNOWLEDGMENTS

The origins of this book lie over ten years in the past. While working on the history of archives in Europe, I came to appreciate the importance of genealogy in the early modern period. After searching for new ways to explore the history of archives after the appearance of the original German version of *The Birth of the Archive* in 2013, I eventually decided to focus on the knowledge cultures of genealogy more comprehensively than I had done before. My initial impetus to do so was stimulated—yet again—by the Max Planck Institute for the History of Science in Berlin, which resulted in a paper published under the aegis of the institute. Many thanks again to Lorraine Daston and Christine von Oertzen for hosting me at that magnificent institution.

When I subsequently became aware of Jakob Wilhelm Imhoff and his rich paper trail, I decided to explore genealogy still further—ultimately leading to this book. My archival research on Imhoff began in earnest in early 2016 with a brief but highly memorable visit to Coburg. I still recall very fondly the company of my father and especially of my daughter Frieda, then only twenty months old, on her first archival trip. The bulk of the archival work was carried out in Nuremberg in the following years—thankfully, most of it was finished before the coronavirus pandemic hit in early 2020. I am grateful for the professional help provided to me by the staff of Nuremberg's state archive, city archive, and especially the Germanisches Nationalmuseum. Dr. Christine Sauer, the director of Nuremberg's city library, also lent me valuable support by helping me identify Imhoff's surviving volumes. Many other institutions throughout Europe facilitated my research in various ways, not least by providing copies of documents and books in difficult times.

During the process of researching and writing, I had the chance to present preliminary results and thoughts—many of them so preliminary that they barely survived the moment they were enunciated—at various venues, including conferences, seminars, and lectures in Dunedin, Vienna, Padua, Trier, Leipzig, Hannover, Münster, Graz, Hamburg, and Rome. Many people in attendance helped me sharpen my ideas, and quite a few of their observations made it more or less directly into the book. Several colleagues and experts helped shape my thinking on early modern genealogy; I wish to thank, in alphabetical order, Friedrich Beiderbeck, Stefan Benz, Sven Erdner, Nora Gaedicke, Michael Hecht, Paola Molina, Elisabeth Timm, Stefan Waldhoff, and Thomas Wallnig. Holger Zaunstöck was instrumental in clarifying Imhoff's connection to Halle, Germany. I also wish to acknowledge Federico Del Tredici's help in clarifying details of the history of the Visconti family. Robert Bizzocchi, doyen of all historians of genealogy, graciously shared his time one night in Pisa for an extended discussion over wine and pizza. Tom Tölle read an early draft of the entire text and made me rethink almost all of it. I greatly appreciate his help with that and with much else. Volker Bauer, leading expert on early modern genealogy and a great source of inspiration for my own research, kindly read an earlier version as well and answered many subsequent questions in the later process of revisions. John Noël Dillon, language wizard, worked his magic once again. After translating two of my previous books, he agreed to polish my own version of "English" and turned this book into a readable text—many thanks, John, for this and so much more.

Earl Havens, Tony Grafton, and especially Ann Blair, the editors of the Information Cultures series for Johns Hopkins University Press, supported me from the beginning and shared vital ideas at a later step. I am very grateful that my work may now appear in their great collection of monographs. An anonymous reviewer asked several tremendously important, critical questions and encouraged me to elaborate on some of the broader topics of my work more explicitly. I can only hope I have lived up to his or her expectations. Kyle Kretzer, Beth Gianfagna, and the entire editorial team at the press did an outstanding job in transforming my rough manuscript into a beautiful book.

Funding for the research and publication of this book came from various sources. The University of Hamburg provided ongoing support in several ways. A significant portion of the text was written during an Alexander von Humboldt Fellowship at the University of Otago in Dunedin, New Zealand, in early 2020, a stay that was unfortunately cut short by the onset of the pandemic. There I enjoyed early morning writing sessions in Twizel, with spectacular views of Mount Cook, and up on Highgate, with stunning vistas over Dunedin Harbor as the sun rose over the peninsula. I am also happy to acknowledge the University of Otago's support, mediated in particular by Will

Sweetman—many thanks for a great time and a wonderful experience. The project received further funding from the Deutsche Forschungsgemeinschaft (German Research Foundation) under Germany's Excellence Strategy–EXC 2176 "Understanding Written Artefacts: Material, Interaction and Transmission in Manuscript Cultures," project no. 390893796. Many thanks to Hamburg's Centre for the Study of Manuscript Cultures and all my colleagues there for never-ending discussion of manuscripts, their features, and their relevance. I am sure that more than a few of my ideas here reflect insights from our engaging exchanges.

As I grow older, family becomes ever more important in my life. My writing on family history perhaps mirrors this everyday experience. I am fortunate in that I frequently could combine family and research over the past few years, as my wife and daughters have accompanied me to many conferences and on research and writing trips to destinations from Bavaria to New Zealand. For all their support, but above all for the simple joy of their company, I dedicate this book to all the branches of my own family.

Introduction

In late 1700, the European crisis of the Spanish succession was heating up. For years, the great powers of Europe had anticipated the death of Charles II (1661–1700), king of Spain, who had failed to produce a male heir. With his death, the Spanish branch of the house of Habsburg would come to an end; Spain would need a new king from outside. As the monarch neared his demise, his rivals in Europe looked on with growing interest. Several potential successors had emerged over the previous years. Johann Ferdinand of Bavaria had been a compromise candidate with broad support. After his premature death in 1697, however, only two serious contenders remained: Philip (1683–1746), Duke of Anjou and grandson of Louis XIV of France, and Archduke Charles (1685–1740), son of the Holy Roman emperor. Both princes had married a daughter of the late Spanish king; hence they both felt entitled to claim the throne. Under pressure from France, Charles II of Spain eventually bequeathed the throne to Philip in his will, dated October 2, 1700. After Charles's death on November 15, the French duke was proclaimed Philip V, king of Spain, on November 16. France had carried the day, but the other European powers refused to accept French dominance in western Europe. The fourteen-year War of the Spanish Succession broke out shortly thereafter.

A month and a half after Philip's accession in Spain, Jakob Wilhelm Imhoff (1651–1728) shared his perspective on these recent events in a private letter.[1] Imhoff, the protagonist of this study, was a patrician from the free imperial city of Nuremberg in southern Germany and by then a widely known expert of dynastic history and genealogy. It was no wonder that he had his own view on the Spanish succession. For the moment, he agreed, the French had gained

Jakob Wilhelm Imhoff, drawn by P. Stroebel and engraved by Jan de Leeuw, most likely shortly before 1700. StA N, Mss 286.

the upper hand in European politics. The accession of Philip of Anjou to the Spanish throne was a victory for the aging Sun King and the kingdom of France. Imhoff knew well how this looked to many of his contemporaries. The union of Spain and France under a single house created an "imbalance" of power. Potentially, he mused, "Europe's freedom [was] over" (es seye um Europae freyheit geschehen). Perhaps even a new "universal monarchy" now loomed. Imhoff's assessment echoed popular anti-French sentiment. Modern historians have often used similar categories and concepts to analyze the international status quo at the beginning of the War of the Spanish Succession.

But Imhoff was not content with such a mundane perspective. First, being a pious Lutheran, he added a brief religious analysis: perhaps, he argued, God was using Bourbon supremacy to chastise the Continent for its lack of religious integrity. Even more strikingly, however, Imhoff gave yet another, seemingly counterintuitive reading of the events, one that he elaborated in much greater detail and with much more gusto than he devoted to the familiar lamentations and religious interpretations of French supremacy. Over one and a half pages, Imhoff situated the Spanish succession of 1700 in a broader genealogical panorama.

Genealogical history suggested to Imhoff a different understanding of the French takeover of Spain. In his eyes, France's short-term political success was a clear sign of future catastrophe, a "fatale prognosticon." Could the French really afford to send away one of the potential direct heirs to the French crown? Imhoff acknowledged that the French monarchy had a long line of potential successors to the Sun King. Yet he knew of several parallel cases in French history that had resulted in dynastic disaster. Had not the Capetian and the Valois dynasties died out shortly after acquiring unprecedented power? Imhoff saw significant parallels here, and he anticipated the collapse of the Bourbon dynasty in the near future. His judgment was, of course, almost prophetic. By 1712, almost all of the Sun King's many potential heirs had died, while the Duke of Anjou, the only surviving adult male heir, was now ineligible for the French throne because he was the king of Spain. The nation's hopes now rested on a sickly infant, the future Louis XV, great-grandson to the still-reigning Louis XIV. Imhoff's genealogical premonition turned out to be eerily accurate.[2]

Genealogy in Early Modern Culture and Politics

The crisis of the Spanish succession is perhaps the most obvious example of the enormous political, social, and cultural relevance of genealogy in early modern Europe. Its people looked at their world with a "genealogical gaze."[3] Genealogy was a widely respected and efficient way for contemporaries to understand both their immediate social surroundings—family, kinship, marriage—and the basic structures of society at large—political alliances, social hierarchies, and legal frameworks. Knowing the genealogy of one's own family and of numerous others, both noble and common, was a key to comprehending and successfully navigating everyday life. This fundamental role of genealogy also makes it an especially fruitful subject for historical study. As this book highlights, Europeans' interest in and utilization of genealogical knowledge gives modern historians great insights into the social praxis of a deeply hierarchical society. Many practices of power and distinction and many of the fundamental political conflicts of this period were expressed in genealogical language through concepts such as "descent" and "family," "house" and "branch."

The impact of genealogy on the lives of early modern people was profound. Wars were fought in the name of family relations, and genealogical facts of descent and marriage often decided whose claims to power were considered credible. A small group of closely related and interconnected princely families governed the many polities of early modern Europe, and intrafamilial developments in one part of the Continent easily affected events in faraway places. The numerous marital alliances in Europe's "society of princes" constituted a dense network of overlapping biological relationships that harbored the potential for significant genealogical competition among sovereign families.[4] Knowledge about family ties, marriages, and the current state of dynastic affairs, including knowledge about the number, age, sex, and health of future heirs or potential marriage partners, was of utmost importance for one's assessment of contemporary politics. Genealogical information was a vital political resource, and so it was both therefore consciously generated and jealously guarded. Many ruling dynasties regarded details of family history, along with the documentation that underpinned them, as state secrets and were extremely reluctant to grant outsiders—and in some cases even insiders—access to their genealogical *arcanum*. Imhoff knew full well that he was trading in highly sensitive knowledge, and not only in the obvious case of the Spanish succession.

Since genealogy was a guiding principle of international relations, it was little wonder that its study was hotly contested. Political rivals often cultivated alternative versions of individual genealogies. Challenging an opponent's understanding of family history was an important tactic of international politics and diplomacy. In many cases, genealogical publications were media of self-serving dynastic propaganda, produced more or less with the obvious goal of bolstering one's own social and political agenda. Throughout his career, Imhoff had to cope with the frequently one-sided and biased nature of genealogical knowledge and on more than one occasion had to contend with alternative, sometimes even contradictory versions of the same family's history.

In addition to its immediate political relevance, genealogy also conferred cultural legitimacy on early modern social stratification. Pedigree and lineage underpinned the social hierarchies of premodern Europe. The European nobility increasingly viewed "pure" and "venerable" ancestry as defining criteria of their social superiority, especially as aristocratic dominance faced growing challenges from new bureaucratic and economic elites that appeared in the late Middle Ages. Since a high cultural value was attached to antiquity generally, identification as "old" nobility conferred particular prestige. Noble families vied for ancient glory and contemporary power with elaborate and often largely fictional origin stories, a practice that did not come under critical scrutiny until Imhoff's own lifetime. Genealogical thinking stabilized social hierarchies by anchoring noble families in the remote past.[5]

Early modern genealogies thus need to be understood as "sites around which power relations were negotiated."[6] Scholars have traditionally focused on the political, military, and economic power bases of the nobility. In recent years, additional sources of social prestige have received increased attention. Studies now emphasize the *cultural* dimensions of noble hegemony, prominently citing genealogy as a tool for defending social stratification. Recent studies on genealogical media, including lavishly illustrated books and manuscripts, paintings, tapestry, and epitaphs, among many other items, have highlighted the ceremonial and performative contribution that genealogy made to upholding the nobility's claim to social privilege.[7] Careful scrutiny of Imhoff's work reveals in detail how nobles used genealogies and family histories to lend cultural legitimacy to their social status.

More broadly, genealogy also must be considered as a formative influence on Europe's social imaginary.[8] By Imhoff's lifetime, basic notions of family and kinship were habitually understood in terms of genealogy. Even today, families are mostly regarded as social groups resulting from biological procreation and marriage alliances. As institutions, they appear to transcend time and are characterized by relatively strong internal cohesion, which is thought to result from the biological ties between the related members. Traditionally, genealogies were understood simply as representations of such "natural" familial connections. Over the past few decades, however, scholars have critically reconsidered and radically de-essentialized the concepts of family and kinship—and, consequently, genealogy. As it turns out, there is very little that is "natural" about these social categories, notwithstanding widespread reliance on a naturalizing language that includes terms such as "branches" or "blood."[9]

Accordingly, instead of simply reflecting natural facts, genealogies are now regarded as active forces in promoting culture-specific notions of "family." They underpin an understanding of families as intergenerational institutions weathering the vicissitudes of birth and death. Genealogical thinking, by insisting on continuous biological descent as the defining feature of a family, establishes transtemporal stability and continuity. Insofar as it makes "family" an explicit object of research, discussion, and representation, genealogy also has a self-reflexive function for living relatives. Genealogy enables—and requires—one to take stock of who did and does, and who did and does not, belong to any given "family." It provides a framework for conceptualizing what a "family" is and, at the same time, introduces the many members of individual families. Taken together, these functions and effects of genealogy are rightly considered essential means of negotiating family identities within the nobility and beyond.[10]

It has become increasingly clear just how malleable families actually were (and are). Scholars point out, for instance, the ambivalence of the place of women in early modern understandings of the family. The question of whether one's paternal and maternal ancestors should count equally as family received contradictory answers in Europe at different times and for different purposes.[11] Moreover, the new scholarly approach to understanding genealogy has also revealed how conflict-ridden these ongoing processes of constituting and defining families actually were. Kinship groups were "communities of strife," and one of the most important arenas in which intrafamilial disputes played out in early modern Europe was genealogy.[12] Unloved family members, sometimes even entire branches, might disappear from a family's genealogy, condemned to a kind of familial *damnatio memoriae*. The exclusionary power of genealogy in particular has recently been highlighted.[13] Little wonder then that genealogy was often a highly contested practice, and Imhoff's case likewise encourages us to appreciate the selective nature of genealogical memory, since he quite frequently encountered the delicate problem of "lost," overlooked, and excluded individuals and branches.

The practice of genealogy thus was a mainstay of private and public affairs, yet it operated in highly creative and contested ways. Genealogy comprised a flexible and dynamic body of knowledge about who belonged to any given noble house or dynasty. Vice versa, by focusing on genealogies, we may better grasp that families themselves exist as communities of knowledge—more specifically knowledge of "who's who." Genealogies were, first and foremost, vast collections of individual biographies. They allowed readers to understand who was family. Their extensive biographical detail fleshed out abstract notions of the family in socially specific, tangible terms. Contemporary families not only needed a general framework to know who they were; they also required information about specific members. As communities of knowledge, families depended on the more or less consensual sharing of information about people whom they considered to be relatives. This is exactly what early modern genealogies helped to achieve.

By undertaking an in-depth study of Jakob Wilhelm Imhoff, one of the leading genealogists of his time, this book helps develop this revisionist understanding of family as a social structure that is constituted by practices of sharing knowledge. It stresses the role of genealogy as a cultural tool for safeguarding the nobility's social privilege. It investigates the massive part that genealogy continued to play in elite self-representation at the dawn of modernity, especially by discussing the continuing relevance of historical thinking for ceremonial and performative behaviors of social distinction. Finally, it contributes to a better understanding of early modern cultures of political decision making, insofar as it studies how genealogical knowledge was embedded in contemporary political analysis across Europe.

Studying the History of Genealogical Knowledge

Few bodies of (historical) knowledge were as immediately relevant to social and political life as genealogy. The field, therefore, is particularly well-suited to assess how important changes in the knowledge culture of Europe intersected long-standing social, political, and economic structures. Genealogy helps us reconsider broader questions of how developments in epistemology, research practices, and the history of knowledge at large affected basic parameters of the social, political, and intellectual status quo. While genealogy played a consistently important part in structuring social and political life and in expressing (elite) identity throughout most of the second millennium of European history, it also underwent important changes. Its core functions may have remained relatively stable, but genealogy nevertheless continued to evolve and diversify, often in tandem with broader cultural and epistemic changes, as Europe moved from the Renaissance through the Scientific Revolution toward the early Enlightenment. Genealogy evolved as the *sites* and settings of knowledge production multiplied, as information *markets* and new genres of publications emerged, as new *practices* of researching and managing information evolved, and as knowledge gained new *functions*. Sites, practices, markets, and functions of knowledge: genealogy witnessed its share of change on all four of these fronts.

Sites of Genealogical Knowledge

In early modern Europe, sites of knowledge production proliferated. Alongside traditional settings, such as monasteries and universities, new "places of knowledge" emerged, including state-sponsored academies, administrative offices, and private households, among many others.[14] The proliferation of knowledge sites allowed for the pursuit of alternative epistemic agendas and the cultivation of various types of knowledge, often of conflicting natures. Places of genealogical knowledge production also burgeoned, and each of them afforded different opportunities and imposed different limits. Monasteries and princely courts were among the oldest venues for genealogical knowledge production, while state bureaucracies, courts of law, and academic institutions became increasingly prominent after 1500 or so.

Of the many emerging differences between various sites of genealogical knowledge, one is especially important in the following chapters, namely, the difference between sites of "auto-genealogical" and sites of "hetero-genealogical" activity. These concepts, first articulated by Olivier Bouquet in 2011 with respect to twentieth-century Turkey, distinguish between genealogies produced *by* (noble) families themselves from those written *about* them from an outside perspective.[15] There are many occasions throughout this book to observe both the intimate connections and important differences between internal and external approaches to family history. As sites of hetero-genealogical investigation multiplied in the early modern period, the role of the nobility gradually shifted from that of the subject to that of the object of genealogical activity, a process that had important consequences.

Throughout this book, I argue that Imhoff productively engaged with most emerging genealogical contexts. He relied on information from many different sources in his work, integrating the various settings of genealogy into a coherent whole. Imhoff's ability to work with multiple forms of genealogy derived in part from his personal circumstances. His own genealogical activity was likewise situated in a particular place of knowledge, namely, the early modern city. Imhoff lived, worked, and published primarily in Nuremberg, a prominent free imperial city in southern Germany. This urban context afforded him a range of unique opportunities, although it also imposed a few specific limitations on his work. The fact that Imhoff wrote genealogy as a wealthy patrician in an independent urban center, and not as a bureaucrat or courtier at a royal court, had significant consequences for his genealogical output.

The plurality of sites of genealogical knowledge, as it emerges from this study of Imhoff, invites us to consider the fact that genealogy was hardly a unified enterprise in early modern Europe. To some extent, the very concept of "genealogy" had become an umbrella term that covered different projects and approaches. While many other topics, such as natural phenomena and politics, were also discussed in numerous settings, genealogy pertained to an exceptionally wide range of contexts, from ecclesiastical to academic, courtly, urban, and bureaucratic sites. In different sites of knowledge, different types of knowledge and alternative criteria of truth were often preferred. One obvious example that features prominently in the following chapters is the ambivalent status of hearsay, gossip, and other forms of oral knowledge. Scholars have highlighted the increasingly contested status of orally transmitted information in early modern Europe, and that was also true for early modern genealogists.[16] What counted as legitimate family knowledge depended on its context and genealogical setting. Yet despite all the diversification of early modern genealogy, Imhoff's case also demonstrates that experts from different backgrounds continued to cooperate fruitfully to a significant extent.

Markets and Genres

Scholars today largely agree that an increasingly commercialized market for books, manuscripts, and information in general emerged around 1700.[17] Knowledge became commodified, that is, it circulated primarily because people intended to sell it at a profit. Writers and publishers worked in ever more competitive economic contexts. Books now needed to be advertised, and they were especially prominent goods in the nascent advertising business.[18] Successful authors' names were increasingly used as marketing tools to promote printed products to a growing audience. Genealogy also played a prominent part in the emerging trend of establishing literary brands.

Many scholars and writers started to take economic considerations more explicitly into account, increasingly drawing their livelihood from selling knowledge-bearing media. As a result, information started to move in the form of "public circulation," most obviously through print media addressed to unspecified paying audiences, but also through public manuscripts.[19] This often led to the blurring of boundaries between "serious" scholarly circles, on the one hand, and more entertaining and journalistic material on the other, intended for a wider audience. Genealogy also opened up quickly to the new world of journalistic information-mongering.

Imhoff allows us to see the consequences of this transformation of the field in detail. Genealogical knowledge gradually escaped the confines of aristocratic control, as people put it in circulation for profit. The nobility was losing its near-monopoly on the production and dissemination of family-related information, tilting the field in general from aristocratic auto-genealogy to commercialized hetero-genealogy. More and more genealogists no longer wrote primarily *for* the nobility, but rather increasingly *about* them from an outside perspective. They began to report *on* the nobility just as journalists reported on events. In short, genealogical knowledge burst onto the public stage, granting access to everyone willing to participate in the growing genealogical information market.

The accelerating commodification of genealogical knowledge resulted in the rise of new literary genres, which publishers considered especially marketable. New types of genealogical literature emerged around 1700, including genealogical calendars and cheaply produced compendia, that catered to a paying audience of anonymous readers.[20] Imhoff himself contributed significantly to the rise of the market-focused genre of "universal genealogy," or the genealogical encyclopedia.[21] Encyclopedism, of course, was not limited to genealogy at the time. A growing demand for handy reference works that provided exhaustive and authoritative knowledge characterized many other contemporary fields as well. Imhoff's easy success in publishing his works resulted at least in part from choosing such an attractive genre. Genealogy's growing commercialization thus allows us to gauge effectively the impact of this vibrant new information market on Europe's culture of knowledge at large.

Practices of Obtaining and Managing Knowledge

The early modern period saw momentous shifts in the working practices of scholars and authors. Scholars employed a new set of "little tools of knowledge" to acquire, organize, evaluate, and process raw data.[22] They devised ever more sophisticated forms for storing information and invented numerous types of written artifacts in the process, from notebooks to paper slips filled with excerpts, to "fact sheets" that contained open-ended lists of simple data recorded in minimalist, straightforward notation. Genealogy likewise relied on many such material practices of writing to process relevant information, especially personal data about numerous individuals—a phenomenon that has not yet been studied widely.[23] Imhoff's toolbox of research methods was well-stocked with such tools, which he had encountered in his family, at the schools and university of Nuremberg, in its municipal government, and on his travels throughout Europe.

Two skill sets were particularly important to Imhoff: he became an expert manager of correspondence and proved to be a masterful compiler. His ability to facilitate the circulation of genealogical data and sources by letter, especially outside of their natural habitats, was one of his greatest assets. He gathered numerous incoming genealogical dossiers from all over Europe, today filling almost two dozen volumes of several hundred pages each. By applying sophisticated processes of bricolage, many involving the complex manipulation of physical piles of written artifacts, Imhoff assembled these incoming materials into his genealogical encyclopedias. He employed a variety of methods, including various strategies of visualization, to perform this feat. We will later have the opportunity to witness up close his never-ending routine of drawing preliminary genealogical diagrams, one of his preferred ways of sorting and organizing genealogical facts.

Closely connected with material practices of acquiring and managing information, broader epistemic developments had reshaped the methods and standards of the historiographical study of texts. By Imhoff's lifetime, textual criticism and antiquarianism had transformed the knowledge culture of Europe. Scholars such as Gottfried Wilhelm Leibniz and Antonio Ludovico Murotari, both of whom corresponded with Imhoff, exemplified the new, sophisticated methods of scrutinizing sources. They relied, among other things, on innovations in the rapidly evolving fields of paleography, diplomatics, and sphragistics to interpret historical documents, building on the methodological and empirical achievements of the secular humanists since the fifteenth century and, more recently, of ecclesiastic scholars such as the Jesuit Bollandists and the Benedictine Maurists.

What drove these scholars was an urgent desire to distinguish historical truth and fact from historical fiction and forgeries. Critical scholarship, including philological and antiquarian approaches, served as a kind of epistemic safeguard against the many attacks on the reliability of historical knowledge that characterized the later seventeenth century.[24] Genealogy also eventually came under the sway of skeptical source-criticism, and understandably so. Earlier genealogists had frequently created fictitious pedigrees, uncritically reveled in flattering anecdotes, and occasionally even forged documents to "prove" what the nobility wanted to hear. Late-seventeenth-century genealogists created a powerful rhetoric of methodological progress, distinguishing themselves from their predecessors, whose research was allegedly much sloppier and less vigorous. Careful and critical scrutiny of traditional stories and historical sources was now also state of the art in genealogy, which as a discipline had become obsessed with debunking fictions and forgeries.

Yet Imhoff made only selective use of the contemporary toolbox of critical methods. He kept his distance from some of the most prominent scholarly trends of his day. While he was aware of recent advances in philology and historiography, there is little evidence that Imhoff himself ever engaged seriously in discussions about paleography or diplomatics. There is also little evidence that Imhoff ever visited an archive for his genealogical projects. It is hard to miss the degrees of methodological rigor and technical precision that set Imhoff and other eminent scholars apart. The Nuremberg patrician may have kept abreast of the latest developments in early modern scholarship, but he actively cultivated only a personalized subset of contemporary scholarly methods. Imhoff's occasionally lax methodology may ultimately have derived from his primary understanding of the function of genealogical knowledge.

Functions and Goals of Genealogical Knowledge

As the sites, genres, and practices of genealogical activities multiplied, so did the functions of genealogical knowledge. On the one hand, genealogy continued to fulfill its traditional role as legitimating ideology. Despite advances in critical methodology, genealogy's legitimating function was hardly challenged. Imhoff himself was comfortable with the use of genealogy to buttress social hierarchy, aristocratic identity, and political power. In fact, he often prioritized celebratory considerations over alternative perspectives. This allowed him to cooperate with the nobility very closely, albeit, as we will see, in carefully calibrated ways.

It would have been difficult to write genealogy entirely without, let alone in opposition to, the nobility, so a modicum of accommodation to their vision of genealogy was advisable for most practitioners. Nevertheless, collaboration with the nobility was far more than a mere necessity to Imhoff: he evidently enjoyed maintaining positive relationships with aristocrats. His good relations guaranteed an extraordinary flow of information, although they also shaped the content of his works, for instance, necessitating a certain amount of restraint with respect to overt criticism. Imhoff's positive attitude toward a genteel genealogical perspective may be reflected in the high artisanal quality of his works as well. However, there were limits to cooperation, and these too must be highlighted. Imhoff was never any one family's house genealogist, and he let them know it. He retained a remarkable sense of independence, despite his endorsement of the traditional ideological functions of genealogy.

Genealogy's shifting position in the broader culture of knowledge, however, gradually brought to the fore several additional functions. Some of these Imhoff openly endorsed, while others are only implicitly reflected in his work. Perhaps most important, genealogy gradually acquired a significant role as a guide to *contemporary* politics. "The contemporary" was discovered as a category and ultimately challenged the established cultural primacy of "tradition" in Imhoff's lifetime. While ancient ideas remained influential reference points, the relevance of *recent* developments and standpoints nevertheless grew.[25] "News" was invented as a category, and the social role of the "nouvellist" or provider of news emerged.[26] Specific media, including journals and newspapers, were created to satisfy the growing thirst for news.[27] One of the consequences of this movement for genealogists was a growing tension between contemporary and historical outlooks.[28] A new imperative to provide up-to-date, contemporary information about the nobility emerged, and it brought significant changes to how genealogists worked, how they acquired data, and how publishers conceived of genealogical publications. A certain amount of journalistic flair came to shape contemporary genealogy, and Imhoff himself combined serious Latinate scholarly erudition with a fascination with the *au courant*, occasionally even indulging in sensationalism.

Eventually, genealogy also functioned in further unanticipated ways, potentially even contributing to the "invention" of what is traditionally known as the "ancien régime."[29] Books like Imhoff's paraded, for better or worse, the nobility's enormous pride in its own achievements, past and present. Genealogies, moreover, embodied the elitism of the aristocracy like few other literary genres; indeed they openly celebrated the self-professed exceptionalism of their aristocratic subjects. Family histories, especially those of a celebratory nature, continued to portray the nobility as the driving force of social and political life. By the same token, genealogical literature also provided ample empirical information for anyone who wanted to describe the nobility from a (critical) outside perspective.

By compiling dozens, possibly even hundreds of normalized family and royal genealogies in a single framework, encyclopedic genealogy also paved the way toward a reexamination of the subject. Works like Imhoff's made "the nobility" palpable as a social group or class. As they learned about thousands of nobles, who belonged to hundreds of noble families, Imhoff's readers may have gradually shifted their attention from individual noble privileges and the idiosyncrasies of individual families to "the" nobility as such. Encyclopedic genealogy homogenized the heterogeneous plurality of individual noble families; it fashioned a singular collective, a unified social class or estate, fleshing out abstract notions of social theory with a plethora of colorful detail.

To highlight the ongoing dynamics of change in the field of genealogy, the following chapters frequently refer to genealogical *knowledge* instead of simply saying "genealogy" or "genealogies." This terminological decision connects the history of early modern genealogy to current research perspectives often summarized as the "history of knowledge."[30] This approach entails a greater focus on the constantly evolving nature of cultural attitudes toward knowledge and knowledge practices. In the case of genealogy, this perspective gives us a new appreciation of change and dynamics in a field that is all too often considered stable and static. Indeed, the power of genealogy as a means of structuring social and political life remained enormous and went largely unquestioned in premodern Europe right up to the fall of the ancien régime.[31] There were few fundamental attacks on genealogy and genealogical thinking, as such. Even modern ideas of kinship, family, pedigree, and ancestry often remain closely connected to traditional genealogical thinking. Yet, despite the largely unchallenged prominence of genealogy in Europe, the way in which people actually went about using this all-important mental matrix changed significantly. This study of genealogical *knowledge*, therefore, helps refine the scholarly inquiry into continuity and change in the social imaginary of Europe at the crucial moment when modern forms of science and research were just beginning to gain ground.

Limits of Knowledge: The Instability and Preliminary Nature of Genealogies

A history-of-knowledge approach also focuses our attention on the little-studied everyday challenges of producing genealogical tables and narratives.[32] Imhoff's example highlights the innumerable difficulties, the daily ups and downs, the slow progress and the myriad setbacks that accompanied the preparation of genealogical works. Stressing the common vicissitudes of genealogical knowledge production allows us to appreciate how genealogical knowledge was never a fixed body of information; on the contrary, it was constantly evolving. As new dynastic events continually occurred, and new sources shed additional light on established facts, genealogical knowledge was hardly ever stable, definitive, or "finished." As political and social alliances shifted, genealogical priorities and perspectives also changed. Genealogies had to be adapted to accommodate changing preferences. For all these reasons, genealogy was characterized by "fundamental instability."[33] Genealogical work was perpetually work in progress.

The fragility of genealogical knowledge may also have magnified the already considerable economic risks associated with genealogical representation. As a rule, genealogical representation became ever more elaborate (and hence costly) in early modern Europe. Its products frequently incorporated the most exquisite contemporary architectural, pictorial, and print media. The complexities of scholarly research necessitated lengthy and expensive research campaigns. If one added to that base cost the growing awareness that genealogical works needed to be constantly updated, many families must have found the representational and epistemic standards of European genealogical culture economically prohibitive. As Christophe Schuwey recently suggested in an important article, the very expensiveness of genealogy may have put numerous families in a tight spot.[34] Economic distress may have forced them to make do with journals and other market-oriented genres, including genealogical encyclopedias, as cheap surrogates for lavish personalized genealogical representation.

For all of these reasons, the availability of genealogical knowledge and appropriate media with which to present it was by no means self-evident or common. Quite to the contrary, we will encounter numerous families whose genealogical knowledge was in less than perfect shape for one reason or another. Genealogy's legitimating potential was enormous, but it was not easy, and certainly not cheap, to tap it. While social and political life remained fixated on genealogical thinking, the task of documenting and representing individual genealogies often proved to be truly cumbersome. Genealogy could become a trap, an overwhelming constraint. Although no one denied the fundamental validity of genealogical thinking for the basic social and political structures of early modern Europe, at least some participants in this economy of knowledge found the constant need to explicate genealogical facts to be a nuisance.

Approaching the field from the vantage point of the history of knowledge thus exposes a certain discrepancy between the public and the private, familial roles played by genealogy. Lavish public displays of dynastic magnificence may have served to promote the great sovereign houses, but many nobles had mixed feelings privately. The stabilizing social function, the institutionalizing power of genealogy had a soft underbelly. Viewed from the production side, genealogical knowledge was unstable and vulnerable. Many nobles knew at least intuitively that arguments based on genealogy, despite their essentializing language and largely undisputed conceptual framework, did not necessarily provide the social stability they sought. Although genealogy claimed to be a source of power, influence, and privilege, it was just as often a source of anxiety, exhaustion, and exasperation.

The Protagonist

The case of Jakob Wilhelm Imhoff is particularly well-suited to a reassessment of early modern European genealogy along these lines. Imhoff was famous across Europe as the author of numerous acclaimed books on the European nobility. At home, he was known simply as "the erudite Imhoff."[35] He was connected to important scholars, administrators, and churchmen across Europe, and he engaged all of them in his vast project of acquiring genealogical knowledge: a European *respublica genealogica*, a republic of genealogists, emerges from his correspondence. Imhoff's letters show how genealogy was a shared concern of writers from many regions and social contexts.

Imhoff's genealogical activities are, moreover, exceptionally well documented. In addition to his many letters and genealogical publications, he left behind a huge cache of working papers. Whereas very little of Imhoff's private correspondence or personal notes survives, his research papers were well maintained and cared for after his death. Imhoff filled dozens of ledgers with genealogical dossiers. Although this enormous mass of documentation thus sheds relatively little light on Imhoff's private life in Nuremberg, his papers give us detailed insight into his genealogical activities on a European scale.[36] These materials provide the documentary basis for the following chapters.

The future master genealogist was born on March 7, 1651, in Nuremberg, a powerful southern German free imperial city; he was the oldest of four surviving children (out of six).[37] The Imhoffs were one of the city's most elite families. As was typical for a Nuremberg patrician, his education began at the local public school at age seven, before he moved on to the city university, located in nearby Altdorf, in 1667. In 1670, again typically for his class, he went on a Grand Tour of Europe. Imhoff traveled in the company of two friends, Severin Walther Schlüter (1646–1697, along with his brother Matthaeus, 1648–1719) and Johannes Fabricius (1644–1729). Both men would later become important contributors to his genealogical research. Heading west from Nuremberg, the party first visited Strasbourg and Frankfurt, where Imhoff met the eminent genealogist Philipp Jakob Spener (1635–1705) before traveling onward to the Netherlands. From September 1670, he stayed almost a year in Paris and then passed the winter of 1671–72 in northern Italy, where he undertook several extended sojourns in Venice.[38] Later destinations in Italy included Rome and Naples. Imhoff constantly met fellow travelers, famous scholars, and local dignitaries along the way. Given how important Naples was to Imhoff later in life, it is surprising that he presumably spent only fifteen days there in 1672. On his way back north, he stopped in Florence, where he befriended the famous scholar Antonio Magliabecchi (1633–1714) and joined the Accademia degli Apatisti. In early June 1673, the young patrician returned to Nuremberg. These years of schooling and traveling were formative for Imhoff. He forged long-lasting social connections and acquired important intellectual interests and habits. It is not surprising that he looked back fondly on these times later in life.[39]

After his return, Imhoff settled into adult life. For decades to come, he would divide his time and energy between just a few occupations. First and foremost, he started to work on genealogy. After several years of preparation, his first books appeared between 1683 and 1685, including his *Notitia Sancti Romani Germanici Imperii Procerum tam Ecclesiasticarum quam Saecularum* (Compendium of nobles of the [Holy] Roman Empire, both secular and ecclesiastic).[40] The *Notitia Procerum* became Imhoff's most successful book, even though he did not yet use what would become his signature format, a combination of genealogical tables and narratives. The book immediately established Imhoff's genealogical credentials. Despite the fact that it was a hefty—and constantly growing—tome of hundreds of densely printed Latin pages without illustrations or genealogical diagrams, the *Notitia Procerum* saw four subsequent editions, the last of which was posthumous.

Around the same time that Imhoff published the *Notitia*, he also joined his patrician peers in the city government. He became head of the office for almsgiving (Landalmosenamt), overseeing several of the city's wealthy caritative foundations.[41] When he became Losungsamptmann (roughly, "tax official") in 1688, a post that he held until the end of his life, he became heavily involved in managing the city's finances, although no details about this aspect of his life are currently available.[42] After 1700, as the War of the Spanish Succession strained Nuremberg's finances, Imhoff helped manage an increasingly overextended budget.[43] At least occasionally, he also took on extraordinary duties. In 1697, he served as the city's envoy on a diplomatic mission to the elector

Lothar Franz von Schönborn (1655–1729), archbishop of Mayence.[44] In late 1711, Imhoff complained to a colleague, who was out of town on a diplomatic mission, that he was overburdened with administrative work.[45]

Despite his growing administrative obligations, Imhoff continued his genealogical work. In the late 1680s, he decided not to pursue an even more distinguished civic career so that he could dedicate himself to his studies more extensively.[46] He preferred to broaden his research beyond German-speaking lands rather than pursue the *cursus honorum* higher. In 1687, his *Excellentium Familiarum in Gallia Genealogiae a Prima Earundem Origine usque ad Praesens Aevum Deductae* (Genealogies of outstanding families in France, from their origins to the present age) appeared.[47] From that point onward, Imhoff's works combined Latin text with large-scale genealogical diagrams, creating a recognizable visual formula. The high craftsmanship of his publications, including his expert combination of text and diagrammatic representation, was one of their most important selling points. Just three years later, Imhoff published a similar compendium, the *Regum Pariumque Magnae Britanniae Historia Genealogica* (Genealogical history of the kings and peers of Great Britain).[48]

By now, the idea of a genealogical encyclopedia had taken root. As Imhoff stated in a private letter, he thought of himself as in the midst of a "historico-genealogical journey" that was far from over.[49] As Imhoff pressed on with his project, he maintained his focus on genealogical affairs. His nongenealogical cultural interests, by contrast, were vague and unspecific.[50] Even activities closely related to genealogy failed to attract his attention. When Christian Franz Paullini (1643–1712), founder of the Imperial Historical College, a group of historians dedicated to the study of German history, invited Imhoff to join the enterprise, the Nuremberg patrician politely declined—he preferred to spend his time and energy on the projects he had set for himself.[51]

As he set up his life in Nuremberg, Imhoff also founded a family. In 1683, he married Anna Felicitas Holzschuher (1662–1712), the daughter of another distinguished local potentate. The marriage seems to have been a happy one, and the couple had eight children, four of whom survived. Imhoff took great care to orchestrate the social placement of his children.[52] In an ostentatious break with family tradition, he sent his only surviving son away from Nuremberg for his secondary education in 1702. Jakob Wilhelm Sr. enrolled his young namesake in the newly fashionable Paedagogium at Halle/Saale.[53]

This decision also indicates that Imhoff entertained a positive attitude toward pietism, which perhaps even extended to its more radical forms.[54] This was a reformist brand of Lutheranism that drew significant inspiration from the theology and piety of Philipp Jakob Spener—whom Imhoff had encountered as a genealogist in Frankfurt at the beginning of his grand tour. Pietism had gained a growing presence in Nuremberg since the 1690s.[55] Imhoff clearly knew (or knew about) the most prominent sympathizers of pietism in Nuremberg.[56] He expressed typically pietist criticism of the religious status quo, including a declaration that he considered most Protestants to be "Christians only in name."[57] He preferred "praying, singing, and reading the Bible" to "extensive preaching."[58] His lifestyle was, in accordance with his pietist leanings, relatively modest. He opposed "luxury" and preferred a "simple," austere existence.[59]

Pietist religiosity and genealogical scholarship often coexisted harmoniously in Germany circa 1700, despite occasional tension.[60] Spener, for instance, gradually moved away from the study of genealogy (and heraldry) as his religious commitments grew. Occasionally, Spener expressed moral reservations about his fascination with the worldly matters of family history—yet, he never gave up the topic completely. Imhoff, in contrast, kept his scholarly persona largely independent from his pious views. He combined religious dedication with a generous dose of moderate rationality along enlightened lines. He once openly declared, "I am not superstitious," insisting on his mental sobriety even in challenging times.[61] The genealogist, moreover, opposed all forms of zealotry and applauded religious moderation, openly sympathizing with the anti-rigorist standpoint advocated by his old friend Fabricius.[62] He encouraged a form of "prudent Christian caution" in managing one's life and affairs.[63] In a world of complex and morally ambivalent political decision making, he found consolation in the old and pragmatic adage that "we have to go with the lesser evil."[64]

His religiosity tempered by such attitudes, Imhoff continued to publish genealogical works. His *Notitia Procerum*, after already appearing in a second edition by 1687, went through two more editions in the 1690s (1693, 1699), each of them the result of thorough rewriting and additions. Imhoff, moreover, put great effort into an expanded republication of two recent works by Georg Lohmeier (after 1650–91), a genealogical author from Lüneburg. Imhoff had contributed to the first edition of Lohmeier's *Der Europäischen Reiche Historische und Genealogische Erläuterung* (Historical and genealogical explanation of the European monarchies, 1690),[65] and he agreed in 1695 to reedit the popular work after Lohmeier's premature death. In the same year, Imhoff also reedited, as a companion volume, Lohmeier's conventional survey of world history in tables, the *Neu-erläuterte Zeit- und Jahr-Rechnung*

(Newly explained chronology).[66] In 1701 and 1702, two further installments of Imhoff's own genealogical encyclopedia appeared, namely, books about Spain and Italy.

Thereafter, Imhoff temporarily turned to a publishing venture outside of Nuremberg. His next book—the first to appear in Amsterdam with Zacharie Châtelain—differed dramatically from his previous works. The *Recherches historiques et genealogiques des grands d'Espagne* (Historical and genealogical research on the grandees of Spain) of 1707 was Imhoff's only attempt at publishing in a vernacular language. It was a highly untypical work for Imhoff,[67] and it remained a one-off experiment, closely connected in all likelihood to the growing interest in Spanish affairs during the ongoing War of the Spanish Succession. In 1708, again in his customary style, Imhoff published a further installment of his genealogical encyclopedia, albeit a much shorter volume, covering the royal families of Portugal. Two years later, in 1710, Châtelain finally published a much anticipated sequel on Italy, again adopting Imhoff's standard format.[68]

While Imhoff thus remained extraordinarily productive during the first decade of the eighteenth century, his attitude toward life and work started to shift. The genealogist felt that the time was ripe for a fundamental change of lifestyle. In 1705, he felt drained by overwork, age, and poor health. He began to think about "finishing" his work.[69] By about 1710, Imhoff was living an increasingly secluded life. He once missed an (admittedly, brief) visit by Anton Ulrich (1633–1714), Duke of Brunswick-Wolfenbüttel, because, he wrote, on Sundays he "had no one around to tell me recent news."[70] His ties to his intellectual peers diminished as his correspondence became less frequent, less voluminous, and less consistent.[71] A fifth edition of the *Notitia Procerum* after 1710, he said, was out of the question, because most of his contacts had passed away or moved on. He would have had to find "new correspondents," but this was no longer an option.[72] When, in 1712, his beloved wife passed away, his mood darkened even further.[73] He suddenly felt that his own end was drawing near. Imhoff began to dedicate himself more exclusively to religious activities. He "retired from the world"; as he wrote to one friend, "I am closing up shop."[74]

Imhoff indeed had to outsource the publishing of the fourth and final volume of his Italian and Spanish genealogies in 1712. As the Nuremberg patrician explained in the uncharacteristically brief preface to the book, by a happy accident the Leipzig publishing entrepreneur Johann Burkhardt Mencke (1674–1732) had passed through Nuremberg and convinced Imhoff to hand over the manuscript. Unsurprisingly, given Mencke's involvement, the text appeared under the Gleditsch imprint in Leipzig. After that, only two more, very brief genealogical publications followed in the years to come. A *Genealogia Ruthenorum Comitum et Dominorum in Plauen* (Genealogy of the Counts and Lords von Reuss in Plauen) appeared in 1715, but this was just a thin pamphlet containing only a few genealogical tables.[75] In 1719–20, at the insistence of an anonymous "patron" from Italy, Imhoff produced a genealogy of the Albani family, from which the reigning Pope Clement XI (1649–1721) hailed: it was a small booklet of merely four pages.[76] This was an unassuming end to one of the most ambitious genealogical enterprises undertaken circa 1700. Nevertheless, this was not the end of Imhoff's works. Johann David Köhler (1684–1755), professor at Altdorf University, republished the best-selling *Notitia Procerum* posthumously one more time in two volumes that appeared in 1732 and 1734.[77] Jakob Wilhelm Imhoff had passed away years before, on December 20, 1728.

Taking the Nuremberg patrician as an example, this book investigates the creation, circulation, and compilation of genealogical knowledge about the European nobility circa 1700. Imhoff's research and publications enable us to situate genealogy at the crossroads of antiquarian scholarship, aristocratic self-representation, political considerations, and the emerging market for news media. Viewed through the lens of Imhoff's works and research, genealogy was neither political nor scholarly, neither celebratory nor informative—rather, it tried to be everything all at once, with often highly idiosyncratic results.

Chapter 1 illustrates the state of the discipline circa 1700. It points to the many different forms in which genealogical knowledge was socially and politically relevant and discusses different forms of presenting this knowledge to fulfill a variety of functions. Imhoff, as it becomes clear, drew on all of them. Chapter 2 investigates Imhoff's own site of genealogical knowledge production, highlighting the many ways in which his genealogical projects and research practices were a product of his background. The political and cultural place of Nuremberg in the Holy Roman Empire was a major influence on Imhoff's approach to genealogy. In chapter 3, I present an analysis of Imhoff's dealings with the nobility, primarily in Germany and Italy, which are by far the two best-documented cases. It becomes obvious, on the one hand, that Imhoff continued to support the nobility's utilization of genealogy. On the other hand, however, the chapter also demonstrates how Imhoff deliberately kept his distance from the nobility and occasionally even pressured them to cooperate. As we will see, a very complex and ambivalent dynamic was at work in these seemingly asymmetrical power relations. Chapter 4 sketches the early modern *respublica genealogica*, presenting and analyzing the many different contributors who helped Imhoff acquire material for his publications. A careful look at the identity and motives of Imhoff's informants clarifies the mechanisms that enabled genealogical knowledge to spread so widely, but it also reveals the limits of cooperation and some of the difficulties and challenges that genealogists had to overcome. Chapter 5 showcases the genealogist at work, as we observe as closely as possible his routines for assembling and managing relevant information and watch him construct narratives and diagrams from preliminary documents and dossiers. Imhoff's surviving papers allow us to closely observe the work of encyclopedic bricolage before it culminated in his splendid and authoritative genealogical accounts. This chapter highlights in particular the materiality of compilatory working routines and presents several case studies of Imhoff's ability to "fit together" pieces of evidence from different materials. Finally, chapter 6 concludes this study by exploring the position of Imhoff's works on the book market. His books were largely successful, and he cooperated with a range of well-respected international publishers. The high artisanal quality of his publications, together with the excellence of their information, guaranteed a wide reception, and this chapter also takes a closer look at some of the responses of Imhoff's readers to his work.

All told, the following chapters explore the place of genealogy in the cultural, social, and political history of Europe circa 1700. By examining the life and works of Jakob Wilhelm Imhoff, master genealogist from Nuremberg, we may observe how changes in the knowledge culture of Europe affected genealogy and how the evolving nature of the genealogical imaginary subtly affected early modern Europeans' understanding of society, politics, and scholarship. It turns out that, although genealogy was relatively stable in its primary social roles, it was nevertheless a highly diverse, and continuously diversifying, body of knowledge. By looking at the processes of creating and circulating genealogical knowledge, we will see time and again how genealogy was, simultaneously, a powerful resource that stabilized society and organized the intergenerational transmission of goods, privileges, and influence, and a sometimes burdensome obligation, a trap, and even an inhibiting framework. As the culture of genealogical knowledge evolved, people's views of traditional genealogical models of thought had to evolve, too.

CHAPTER ONE

Genealogy circa 1700

In late 1712 or very early 1713, Heinrich Hartwig Knorn (1650–post-1717), an erstwhile correspondent of the young Gottfried Wilhelm Leibniz and a self-professed admirer of Jakob Wilhelm Imhoff, was traveling through the Prussian countryside.[1] At some point, his stagecoach stopped at a local tavern or inn. Perhaps Knorn intended to spend the night there. In the tavern, he met a stranger and they struck up a conversation—and it was genealogy that brought them together. Knorn was curious about English and Scottish families, and his new acquaintance referred him to several recent publications. When Knorn found it difficult to obtain these works from his local booksellers, however, he grew suspicious and asked Imhoff, as a leading authority, whether the books really existed or whether the stranger had played a cruel prank on him.

As this anecdote vividly illustrates, by Imhoff's day, even the local tavern could be a productive setting for genealogical knowledge exchange. Genealogy had become a subject of everyday conversation, connecting strangers and providing arresting topics for an evening chat even at local inns in small country villages. While such everyday exchanges about genealogy may have been rather superficial, the mere fact of conversations about noble families, their histories, and their genealogies, even in the most remote of places, nevertheless demonstrates how profoundly these questions had penetrated European culture. In continental Europe, in particular, widening participation in genealogical discourses and their casual presence across the countryside coincided with the preeminence of genealogy as a tool for defining—and contesting—the possession and transmission of status, wealth, and power.

The ubiquity of genealogy circa 1700 was the culmination of a long and complex history. In the first part of this chapter, I survey the rise of genealogy in Europe in the centuries before Imhoff's lifetime, briefly summarizing the developments that coalesced into his generation's widespread "obsession" with family history.[2] I trace how lineage became the bedrock of noble identity and how the writing of genealogies rose as a key practice of asserting noble status. In the second and third sections, in turn, I sketch the landscape of late-seventeenth-century genealogy. These sections highlight some of the settings in which genealogical knowledge circulated after roughly 1650 and set forth some of the broad ways in which genealogists conceived of their own doings. One of the major points to emerge is that genealogy circa 1700 was a highly heterogeneous affair and by no means an orderly and coherent realm in the early modern knowledge economy. On the contrary, a set of alternative, potentially even contradictory forms of genealogy existed, situated in specific social, cultural, and political settings, that relied on specific research routines and media. As genealogy diversified, its social and cultural implications multiplied. Genealogy evolved from a tool, largely controlled by the nobility, for expressing aristocratic status into a body of knowledge that was cultivated to describe early modern society and politics in a relatively detached, analytical fashion. Although the many functions of genealogical knowledge usually overlapped and often complemented one other, the growing diversification of the field must not be overlooked. Scholars need to be very clear about which discourse they mean when writing about "genealogy."

The Rise of Genealogy

Genealogy did not always play a dominant part in European culture.[3] It is not a "natural" way of imagining kinship and family.[4] For much of the Middle Ages, genealogical literature had been only a "minor genre."[5] That is not to say that people did not identify relatives or remember ancestors. All families are communities of knowledge, acquiring social contours through the sharing of who's-who knowledge about family members. Yet, family-constituting knowledge need not be genealogical in form, and it certainly did not always circulate in formats that became prevalent in early modern Europe. On the contrary, notions about kinship and family in the early Middle Ages were much more flexible, vague, and opaque. Outside of nuclear family compounds, "links to kin . . . were usually reliable, but secondary, and subject to choice and negotiation," writes one authority.[6] While there were strong incentives for memorializing ancestors, in particular in connection with praying for the dead in purgatory, this body of family-related knowledge was not normally the subject of an explicit discourse or subjected to formal organization. Genealogical knowledge existed, but genealogy was not a carefully curated field.

The rise of genealogy as a distinct field of knowledge did not take place until after the turn of the first millennium. An increasingly nuanced and complex genealogical culture emerged across Europe after roughly 1100. Knowledge about one's ancestors, and the successful presentation of that knowledge, began to play a crucial part in the articulation of one's place in society.[7] Ever more occasions—such as marriages, entry into certain orders or church institutions, or accession to a family's seat in local diets—required, or at least invited, the presentation of skillfully created genealogies in numerous media.[8] Specific literary and visual conventions for managing and circulating relevant knowledge now emerged. Stand-alone works dedicated exclusively to familial filiation, for instance, became popular.[9]

The European nobility was the first to become interested in genealogy. The earliest examples of extensive genealogical discussions mostly stem from the highest echelons of European society, royal houses. Among the earliest genealogical works are those produced for the Carolingians and, slightly later, the Capetians.[10] In due course, some of the most important nonroyal families also acquired a taste for genealogy. In the late fifteenth and early sixteenth centuries, the obsession with genealogy reached the middle levels of the aristocracy. Around the same time, moreover, non-noble elites also began to participate in the emerging genealogical culture. The rich merchants of European cities, such as the patricians of Venice or, for that matter, Nuremberg, intensified their genealogical activities. Florence, in particular, is a well-studied example of late medieval non-noble genealogy.[11] By the sixteenth and seventeenth centuries, the trend was spreading still farther down, reaching middling urban milieus and, at least occasionally, the families of craftsmen and farmers.[12]

The rise of genealogy was not the rebirth of ancient models.[13] It was rather the product of complex social and political changes that defined European society for much of the following millennium. Scholars have long observed a progressive "nuclearization of family" after the turn of the first millennium: European concepts of kinship shifted from a horizontal model of the family, which focused predominantly on smaller and more flexible local groups of kin living, working, and acting together, toward a much more vertical understanding focused on intergenerational descent, and especially on agnatic descent from father to son.[14] Genealogy rose to prominence as a means of expressing this new descent-oriented, lineage-focused understanding of kinship.

Even though the broad thesis of a simple and unidirectional "lineage shift" (*mutation lignagère*) has been harshly criticized in recent decades, there is nevertheless no denying that, especially from the fourteenth century onward, families were frequently imagined as communities of biologically related members.[15] People thus also thought about kinship in increasingly naturalistic terms. The metaphor most often employed to convey this growing reliance on biological relatedness was "blood."[16] Belonging to a "family" thus increasingly required one to demonstrate a biological link to a common ancestor.

Insofar as families were the fundamental building blocks of society, new concepts of "family" affected social practices at large. Marriage, for instance, was increasingly subjected to genealogical analysis, as blood relatives were forbidden to intermarry and legal practices were devised to suppress illicit marriages.[17] As ever more couples had to present their marriage plans before law courts, genealogical diagrams gradually became a common feature of court documents.[18] This genealogically attuned understanding of kinship in turn contributed to the rise of new concepts of inheritance that prioritized biological criteria in more formalized and carefully regulated ways.[19] Especially in unusual situations, detailed knowledge of family relations soon became indispensable to adjudicate inheritance cases.[20] Genealogical charts that mapped out the biological and marital ties between potential competitors for an inheritance became important resources for lawyers. By Imhoff's day, legal manuals conventionally printed examples of family trees at the beginning of discussions of individual inheritance cases.[21]

The rethinking of inheritance according to a new, biological understanding of intergenerational relationships eventually resulted in increasingly complex arrangements for the transmission of possessions and titles from one generation to the next. The rise of primogeniture reflects a growing need to refine rules of inheritance.[22] Even more important, as families came to be viewed as stable social entities that transcended individuals and generations, certain portions of the family property were removed from easy circulation. Essential properties, such as ancestral castles and vital fiefdoms, now belonged to "the family" rather than to any one of its members. The European nobility developed legal remedies, such as the creation of fee tails and trusts (*fideicommissa*) to pool family property and attach it to "the family" by law.[23] Continuous attachment to individual properties in turn fostered the growth of dynastic identities. Castles and manors, for instance, became focuses of aristocratic pride, as they embodied permanence, whereas individual generations and family members came and went.[24] Genealogy, the proof of a family's contin-

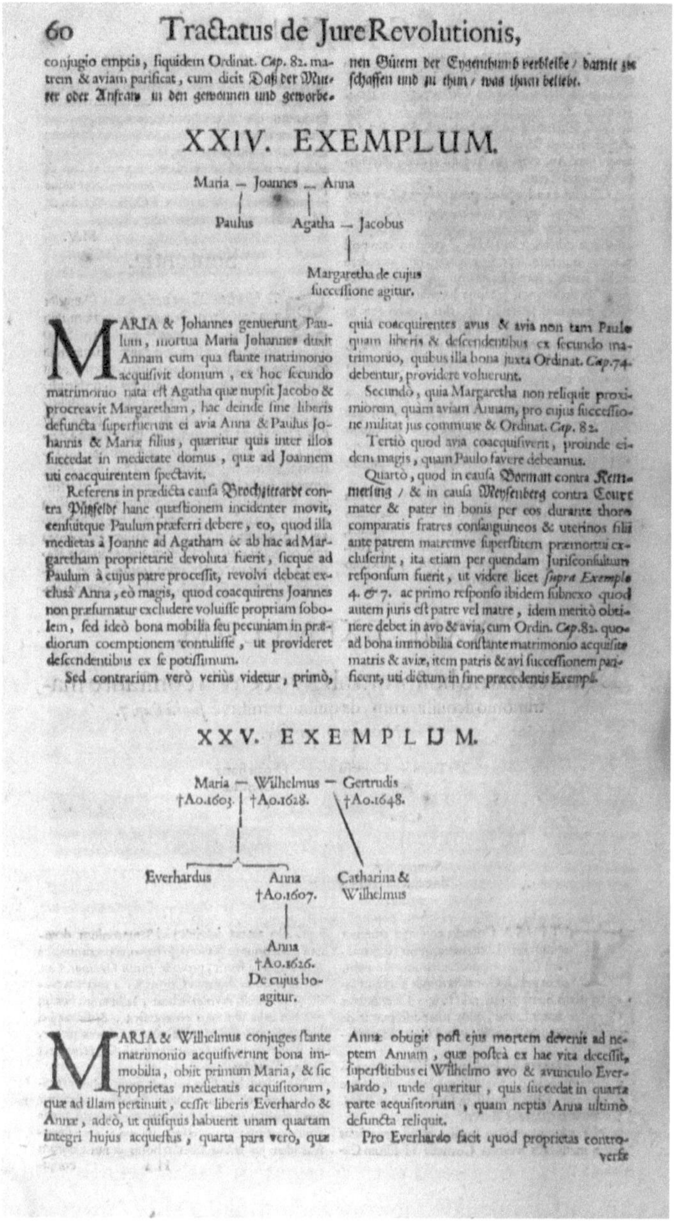

60 Tractatus de Jure Revolutionis,

conjugio emptis, siquidem Ordinat. *Cap.* 82. matrem & aviam parificat, cum dicit Daß der Mutter oder Anfrau in den gewonnen und geworbenen Gütern der Eygenthumb verbleibe / damit zu schaffen und zu thun / was ihnen beliebt.

XXIV. EXEMPLUM.

Maria — Joannes — Anna

Paulus Agatha — Jacobus

Margaretha de cujus successione agitur.

MARIA & Johannes genuerunt Paulum, mortua Maria Johannes duxit Annam cum qua stante matrimonio acquisivit domum, ex hoc secundo matrimonio nata est Agatha quæ nupsit Jacobo & procreavit Margaretham, hac deinde sine liberis defuncta superfuerunt ei avia Anna & Paulus Johannis & Mariæ filius, quæritur quis inter illos succedat in medietate domus, quæ ad Joannem uti coacquirentem spectavit.

Referens in prædicta causa Brochgitrardt contra Püttfeldt hanc quæstionem incidenter movit, censuitque Paulum præferri debere, eo, quod illa medietas à Joanne ad Agatham & ab hac ad Margaretham proprietariè devoluta fuerit, sicque ad Paulum à cujus patre processit, revolvi debeat exclusâ Anna, eò magis, quod coacquirens Joannes non præsumatur excludere voluisse propriam sobolem, sed ideò bona mobilia seu pecuniam in prædiorum coemptionem contulisse, ut provideret descendentibus ex se potissimum.

Sed contrarium verò veriùs videtur, primò, quia coacquirentes avus & avia non tam Paulo quam liberis & descendentibus ex secundo matrimonio, quibus illa bona juxta Ordinat. *Cap.* 74. debentur, providere voluerunt.

Secundò, quia Margaretha non reliquit proximiorem, quàm aviam Annam, pro cujus successione militat jus commune & Ordinat. *Cap.* 82.

Tertiò quod avia coacquisiverit, proinde eidem magis, quam Paulo favere debeamus.

Quartò, quod in causa Borman contra Kemmerling / & in causa Meysenberg contra Court mater & pater in bonis per eos durante thoro comparatis fratres consanguineos & uterinos filii ante patrem matremve superstitem præmortui excluserint, ita etiam per quendam Jurisconsultum responsum fuerit, ut videre licet *supra Exemplo* 4. *&* 7. ac primo responso ibidem subnexo quod autem juris est patre vel matre, idem meritò obtinere debet in avo & avia, cum Ordin. *Cap.* 82. quoad bona immobilia constante matrimonio acquisita matris & aviæ, item patris & avi successionem parificent, uti dictum in fine præcedentis Exempli.

XXV. EXEMPLUM.

Maria †Ao.1603. — Wilhelmus †Ao.1628. — Gertrudis †Ao.1648.

Everhardus Anna †Ao.1607. Catharina & Wilhelmus

Anna †Ao.1626. De cujus bo- agitur.

MARIA & Wilhelmus conjuges stante matrimonio acquisiverunt bona immobilia, obiit primum Maria, & sic proprietas medietatis acquisitorum, quæ ad illam pertinuit, cessit liberis Everhardo & Annæ, adeò, ut quisquis habuerit unam quartam integri hujus acquestus, quarta pars verò, quæ Annæ obtigit post ejus mortem devenit ad neptem Annam, quæ posteà ex hac vita decessit, superstitibus ei Wilhelmo avo & avunculo Everhardo, unde quæritur, quis succedat in quarta parte acquisitorum, quam neptis Anna ultimò defuncta reliquit.

Pro Everhardo facit quod proprietas controverse

Printed juridical literature of Imhoff's lifetime frequently displayed genealogical trees in order to explain details of fact in lawsuits. This example is from Melchior Voets, *Tractatus de Jure Revolutionis: Ad Lucem Ordinationis Judiciariae Cap. Beschluß von Succession 88* . . . , 2nd ed., 1694, 60. Munich, Bayerische Staatsbibliothek, 2 J.germ. 90#Beibd.3, urn:nbn:de:bvb:12-bsb10490447-0.

uous existence through time and space, of an unbroken nexus of biological descent, was the conceptual backbone behind these efforts to enhance intergenerational stability.

There also was a political dimension to the genealogical turn. Since rulership was conceived as heritable, the rules of genealogy also regulated the question of monarchic succession. Since a series of rulers was ideally a series of biological relatives or, at the very least, of relatives by marriage, an understanding of genealogical developments was indispensable to understand contemporary power structures. Genealogical uncertainties or ambiguities became major points of debate when rivals vied for political power and legitimacy.[25] The outbreak of the War of the Spanish Succession in 1701, which Imhoff witnessed so closely, is only the most spectacular early modern example. To members of Imhoff's generation, the intimate relationship between international politics and genealogy was self-evident.[26]

The tendency to think about kinship and family in genealogical terms—that is, basing social relationships on biological "facts" of descent—gradually also affected European conceptions of nobility. Since the birth of the European nobility in the Middle Ages, there had been tension between two rival understandings of the source of noble status. On the one hand, a rich tradition claimed that descent from noble parents transmitted social excellence through physical procreation; on the other, an equally prominent line of thought argued that true nobility derived from one's personal conduct. Debate whether virtue or birth was the source and sign of "true" nobility raged long before Italian and northern European humanists famously took up the question.[27] In most cases, writers supported both arguments, holding them in a precarious balance. While birth was not everything, it was nevertheless of vital importance. Debate continued into the fifteenth and early sixteenth century, coming no closer to a definitive conclusion. "In no way can one deny someone of noble birth the right to call himself noble, but it is not generally true that he is truly noble," one sixteenth-century Italian writer noted in typically ambivalent phrasing.[28]

Significant change occurred nevertheless in the sixteenth century. Over thirty years ago, Ellery Schalk famously identified a shift "from valor to pedigree," implying that "nobility" ceased to be something one *did* (and how one did it) and became something one *was*.[29] "Nobility" now consisted primarily of one's ancestry; being noble became a question of descent. Genealogies that could demonstrate a noble family's venerable and pure lineage thus became "hallmarks of nobility." Genealogical self-consciousness became intrinsic to the early modern nobility.

Recent scholarship has criticized Schalk's conclusions as simplistic. On the one hand, scholars have highlighted the abiding importance of military values for defining nobility well into the eighteenth century.[30] On the other hand, the equation of nobility and virtue endured outside of military ideology as well. One of Imhoff's acquaintances from Nuremberg, for instance, claimed that it was "vanity" (*vanitas*) to base one's social status exclusively on "the nobility of one's blood or the fame of one's family."[31] True nobility consisted of a happy combination of both noble birth and noble conduct. Moreover, a more careful rereading of Schalk's sources reveals that the new language of "blood," "race," and "birth" often communicated ideas that were surprisingly similar to those traditionally expressed by the language of virtue or "valor."[32] Virtue was never completely dropped as a hallmark of nobility.

Yet Schalk's empirical observations about the dramatic rise to prominence of genealogy within the nobility have generally withstood criticism. There is no denying that the quantity of genealogical literature exploded. In Italy, family histories started to appear in the 1560s, including works about the Orsini, the Este, and the Gonzaga families.[33] Germany witnessed a similar chronology.[34] For France, Jay M. Smith has listed dozens of specialized genealogical works for every decade of the later sixteenth and the seventeenth centuries.[35] These numbers alone indicate the newfound significance of genealogy. Even though the definition of nobility in Europe remained more ambivalent or polyvalent than Schalk allowed, the prominence of pedigree and biological frameworks in asserting noble status rose significantly. Genealogy was a major beneficiary of this development. In Imhoff's lifetime, genealogy really was everywhere.

Contexts and Settings

As genealogical thinking spread far and wide, its forms and functions also multiplied. Genealogical knowledge appeared in new places of knowledge. To a certain extent, "genealogy" became an umbrella term that integrated a range of different knowledges with different purposes and attendant practices. Imhoff's friends and acquaintances made use of their genealogical expertise in numerous contexts and accordingly produced very different varieties of genealogical texts. Imhoff's own works and working papers reflect the enormous variety of social and literary contexts in which genealogical knowledge might be needed or useful. For the sake of clarity, we can distinguish three major arenas in which Imhoff's various contacts explored genealogical knowledge: for royal and noble patrons; for institutional, political, and juridical purposes; and for the early modern market for politically relevant information.[36]

Genealogy at Court: Dynastic Identity and Family Memory

Probably the greatest amount of genealogical knowledge continued to emanate from and circulate in dynastic and courtly contexts. Genealogy rose to prominence as an expression of "dynastic identity" after the new, lineage-driven understanding of family and dynasty had taken hold.[37] Genealogy thus was often celebratory and affirmative, reflecting the preferred "self-definition" of a given house.[38] Courtly genealogy, to a very large extent, was aristocratic "auto-genealogy" or "propaganda," and it promoted social ambitions rather than laying out simple "facts." Genealogists working in courtly settings presented the histories of princely and noble families "to set before the eyes of Your Highness and the Public the origin, extent, and grandeur of a house as illustrious as yours."[39] Writing genealogy from the vantage point of the nobility was still the most natural option around 1700. Imhoff knew this world of courtly genealogy very well and profited enormously from the knowledge produced therein, but he consciously kept his distance and could be highly critical of the marked tendency of dynastic genealogies to take one-sided and biased positions.

With respect to the antiquity of a family, for a long time nobles were seldom content to recount the few generations that were actually well documented. On the contrary, early modern nobles strove to connect their families to the most remote, mythical beginnings of European civilization and even mankind. The inscrutability of this remote past conferred splendor and glory, as expressed by Michel de Montaigne (1533–92) in a well-known bon mot. True nobility, he claimed, had no obvious origin. It was "without birth, like the river Nile."[40] This idea remained popular in Imhoff's time. In 1669, a French treatise riffed on Montaigne: "The knowledge of things, which is advantageous in all other circumstances, is not at all that in this connection, because the most illustrious nobility is that whose beginning is unknown; it has been compared to the river Nile, the sources of which inquisitive persons from all ages have yet to reveal to us." True nobility, according to the writer Alexandre de Belleguise, was characterized by a "happy ignorance of the beginning of our forebears' glory."[41] Many of Imhoff's friends thought similarly.[42]

Besides its claim to unfathomable antiquity, the glorious origin of a noble family also required a link to one (or several) of the many early modern master narratives of universal history. Hence, most noble houses "found" either biblical or ancient Greek or Roman ancestors, or, after the humanists had shifted their attention to national histories, claimed descent from some Germanic or Gallic hero associated with the fight against the Romans. A tradition of fabulous family origin stories arose in the Middle Ages. Almost every dynasty fabricated such a fantastic identity narrative. The Habsburgs, for instance, claimed to descend from the Trojans and from biblical figures.[43] The Wettin family from Saxony held up Widukind, a famous, local, early medieval warrior-prince, as their earliest progenitor, while the Visconti of Milan still promoted the eighth-century Lombard king Desiderius as an early family member late in the seventeenth century.[44]

Hardly less creative were speculative dynastic connections to prominent foreign families. The dozen or so families from north of the Alps who improbably claimed direct descent from the Roman Colonna and Orsini families are especially well known.[45] Imaginary as they may have been, such origin myths followed a clear rationale: The prestige of the alleged progenitor of a house helped bolster its social prominence in the present. There also was a competitive side to such fictions: "Comparing the antiquity" of noble lineages was the implicit purpose of such constructions.[46] Moreover, by grafting itself onto this or that branch of universal history, a family could express specific political and ideological loyalties and values. Crafting an origin myth and smartly inserting it in the series of "real" historical ancestors were thus parts of an eminently strategic process that required a significant amount of planning and creativity.

In the process of inventing such illustrious pedigrees, genealogists not only made speculative assertions but also occasionally fabricated some of the evidence they needed. Genealogy was rife with forgeries, reflecting the rampant reliance on fabricated evidence in premodern European legal and historical practice. There are numerous examples, even as late as the seventeenth and eighteenth centuries. The Cartulary of Brioude, a medieval collection of documents from the southern French church of Saint-Julien in Brioude, is one such famous case. Although it consisted largely of authentic medieval documents, Cardinal Bouillon had a number of small but significant additions inserted in the codex in the late seventeenth century to raise his family's genealogical profile.[47] The Visconti in Milan also resorted to the outright forgery of documents, a task performed by a Milanese scholar named Carlo Galluzzi in 1669.[48] Carlo's son Giacomo Antonio Galluzzi was publicly executed in 1685 for a later series of forgeries.

In addition to the antiquity of their lineage, nobles also wanted their family histories to be populated by heroes who excelled in military valor, moral virtue, and political acumen. Deeds of the past shined glory and splendor on nobles of the present. Naturally, each family aspired to show how its members were "singular" in terms of the "multitude and brilliance of great events."[49] Genealogists active in courtly contexts, including Imhoff's Neapolitan acquaintance Biagio Aldimari, who wrote these words about the Carafa family, worked hard to find shining examples of dynastic greatness. They searched for signs of virtue, achievements, distinguished service, and pious conduct, often twisting and tweaking historical facts along the way. Stories about "great" family members were regularly told at length.

Showcasing the glorious history of a family was only one function: celebratory biographies also served an edifying purpose. Stories of ancestral bravery and piety helped the living family members tread the "path that you and your brothers now follow so worthily," thereby orienting and guiding young nobles, as Horatius Flacchio, chief genealogist of the de la Tour et Taxis family, wrote.[50] Genealogists' recounting of family history usually provided moral lessons for current and future generations. Nobles looked to genealogy as a collection of pedagogical *exempla* that could be used to inculcate morally sound behavior into newcomers. Genealogy exhorted contemporary nobles to live up to the standard set by previous generations, thus inserting family identity into the process of individual self-fashioning.[51] Genealogical knowledge helped to "guide future generations to the highest and true glory."[52] The nobility wanted to have genealogical knowledge of their ancestors readily available so they might have "veins filled with their blood and a soul filled with their virtues."[53]

Genealogy also helped to integrate individual families into broader cultural and political contexts. By Imhoff's day, genealogical identity-fashioning had long transcended the narrow confines of the family proper. Whereas Olivier de Laborderie has argued that late medieval English genealogies contributed significantly to the rise of a "national" identity, Richard Helgerson has emphasized the importance of genealogy to sixteenth-century forms of nationhood, and Jan Broadway has explained in detail how genealogy and family history helped give rise to new regional identities in the seventeenth century.[54] Writers now often embedded their genealogical works in a broader framework dedicated to celebrating their "patria" or homeland. By that time, a family's distinguished ancestors regularly featured as the "best men" of a certain region.[55] Genealogists working for individual families associated them with historically established regional identities. Girolamo Maria di Sant'Anna (fl. seventeenth century), for instance, a Discalced Carmelite from Naples and one of Imhoff's most important contacts, seamlessly integrated genealogies into a research agenda that also included a study of Saint Gennaro, patron saint of Naples, and the investigation of southern Italian "antiquities" generally. All of these projects, from genealogy to sacred history, were part of a coherent project to glorify his native city and *patria*.[56]

As genealogy evolved into a very powerful, versatile, and sophisticated tool to express social aspirations and to cultivate dynastic identities, it also produced unintended side effects and ran into unanticipated difficulties. Genealogy may have increasingly felt like a trap to at least some members of the nobility itself. There was little one could do to escape the dominance of genealogy. At best, one could hope for mastery of the discourse, including the aptitude to employ mythical genealogies when advantageous and to reject them when expedient.[57] Engaging in mythical genealogy could also prove to be surprisingly risky. Overplaying one's hand could be self-destructive; too outrageous a genealogical conceit was counterproductive. The powerful French family de la Tour-Bouillon offers an instructive example. Cardinal de Bouillon, the head of the family around 1700, had dared to promote a genealogy that would have overshadowed even that of the royal house of Bourbon itself.[58] What was the result for him? An utter fall from royal grace. It took decades for his family to reestablish itself at the heart of the French nobility. The cardinal had completely misjudged the social implications of genealogy in his attempt to outdo his peers.

Even without incurring such dramatic consequences, the constant need to update one's genealogy in light of the most recent social and political developments often forced nobles to undertake the tiresome task of rewriting existing bodies of knowledge. Although individual nobles here and there could potentially afford to ignore genealogical matters, especially when they felt secure in their social status and could tap additional resources to push their agenda, no dynasty could simply recuse itself from the general genealogical tug-of-war. Yet the very open-endedness of genealogical discourse itself might sap the symbolic power of family history. The constant rewriting of once-powerful origin-myths shattered revered stories into myriad alternative versions, dissipating their celebratory power. Horatius Flacchio (a bête noire to many of Imhoff's friends) labored to find a rhetorical way out of this conundrum. His solution was not terribly convincing: "The different ideas that authors have about the origin of the illustrious House de la Tour cannot diminish the opinion of its magnificence that has been passed down to the present day; and even if they do not accept the same premise, they nonetheless agree completely with respect to its antiquity and nobility."[59] This was a rather feeble opening to a laudatory book dedicated to the dynasty, yet it was a typical way of dealing with the sobering experience that generations of legitimating genealogies had produced little more than an unconvincing mass of contradictory claims.[60]

Court Genealogists as "Artisans of Glory"

The "authors" to whom Flacchio alluded were the numerous genealogists in the employ of the de la Tour family. The nobility could rarely satisfy its need for genealogical knowledge on its own. By 1700, it had become conventional to outsource genealogical projects to closely monitored specialists. If many early modern historians were "artisans of glory," genealogists working for particular noble houses were so a fortiori.[61] Courtly genealogists routinely monitored their employers to ensure that they liked the results. Giovanni Battista Pigna, for instance, a prominent genealogist working for the Este family in sixteenth-century Italy (whose book was valuable source for Imhoff) implored his employer, the Duke of Ferrara, to read drafts of the work he was writing. "It is altogether not only good, but even necessary that Your Excellence take the time to read it bit by bit," he declared.[62] Imhoff himself occasionally forwarded preliminary versions of his writings to princes and aristocrats for advance approval.[63]

This was a fine line for genealogists to walk, however. Total submission to princely wishes verged dangerously close to the abandonment of veracity. Genealogists openly lamented this dilemma: "I will never knowingly and willingly write something false, fabricated, or spurious; the highest necessity, however, demands that I sometimes hide the truth," Caspar Sagittarius, another correspondent of Imhoff with close ties to the ruling elite of central Germany, once noted.[64] Biagio Aldimari, the genealogist of the house of Carafa, couched the same dilemma of the court-genealogist in even more dramatic words: "Writing about families is as difficult as it is dangerous, if one wants to write the truth. Seldom are people satisfied who are always eager to seem to be more than what they really are. And if one accommodates their desires, it is necessary to compose more myth than history, bringing little esteem to the author and the greatest harm to the public and to the learned."[65]

Court genealogists clearly appreciated that they were caught between a rock and a hard place, given their considerable dependence on princely favor. Although, by Imhoff's time, many courtly genealogists thus portrayed themselves as "victims of power," it is hard to overlook the many advantages such patronage conferred.[66] Entering the service of a potent patron was, of course, an important social and economic practice for scholars in all disciplines.[67] In the field of genealogy, though, this broader pattern acquired a special poignancy: serious genealogical research was almost impossible without special access to archives, which was, in turn, contingent on noble patronage. By entering the service of a particular noble, a genealogist opened up otherwise inaccessible troves of historical sources. Constraining as it may have been in the short term, working under such conditions ultimately helped scholars expand their knowledge significantly. The need for discretion about archival findings on one occasion did not preclude one from utilizing them later. As Sagittarius said, "If there is anything displeasing to the masters, I omit it until another time and place comes in which I can bring it to light."[68] Current work on the genealogy of a family could easily become the starting point for a new project in the future.[69]

Courtly genealogists, biased as they may have been, often made very good use of their special archival access. Dedicated exclusively to one family, and usually with plenty of time to produce their works, these genealogists often published luxurious tomes designed to convey power and pride with their material refinement. These lavishly produced monographs frequently integrated long narratives in sophisticated, rhetorical prose and first-rate typesetting with magnificent illustrations, detailed archival references, and verbatim quotations from numerous original sources. The documentary apparatus of such works was often astonishing. In France, the renowned genealogist André Duchesne (1584–1640) developed a new standard format that combined monographic dynastic histories with additional volumes of endless transcriptions of medieval documents that served as "proof" of the facts reported in the narrative.[70] Imhoff owned several such French works and would have been familiar with this documentary approach.[71] Aldimari in Naples also based his genealogy of the Carafa on extensive documentary research and proudly listed the many archives he had visited.[72] In addition to numerous archival documents, he also searched for antiquities and ancient epigraphical monuments, eventually compiling at least "300 volumes," both handwritten and printed, on genealogical matters.[73] Entering the service of a powerful family as its genealogist, therefore, was an attractive option for historians in search of sources, knowledge, and expertise. No wonder, then, that Imhoff cultivated friendship with many court genealogists. Despite their bias, they were often Imhoff's most knowledgeable informants.

The nobility's constant need to obtain, present, and disseminate genealogical knowledge also presented very attractive economic opportunities. Noble patrons often richly remunerated their genealogists. By the latter half of the seventeenth century at the latest, the growing number of men who wrote for a living would have known about the potential rewards for producing genealogical treatises. Cultivating a reputation as a reliable and productive genealogical author could pay handsomely. One of Imhoff's (older) contemporaries in Nuremberg (and a potential acquaintance), Sigmund von Birken (1626–81), made a career this way. Birken became famous as a major German baroque poet but wrote commissioned genealogical works throughout his career.[74] It all began in 1660, when Michael Endter, a prominent Nuremberg publisher who printed many of Imhoff's works received a commission from Vienna to publish an updated genealogy of the Habsburgs.[75] Endter needed a man who could do this "in a contemporary style" and eventually subcontracted Birken, who was known for his lively and readable writing.[76] Their surviving correspondence reveals not only that Birken's writing was closely supervised by the Viennese authorities but also that he received valuable information from the Habsburg court and that the economic arrangements played a prominent part in their relationship. Further genealogical works followed, as Birken continued to market his talents to other houses in need of genealogical literature.[77]

Imhoff, in contrast, wrote neither for money nor in the service of a particular noble house. Although he frequently conformed to some of the typical behavior of court genealogists, such as proactively submitting to censorship, he differed both from men like Birken, who worked as genealogists for hire, and from men like Aldimari, who dedicated long stretches of their lives to a single house, visiting numerous archives and reading countless original sources. Imhoff's avoidance of direct employment by a noble house meant that he lacked direct access to and familiarity with the archival holdings of any one dynasty. Imhoff could never turn noble archives inside out as Aldimari did for his research on the house of Carafa. Large-scale searches for original sources were out of the question for Imhoff. Judging by the scarcity of handwritten underlining and manuscript marginalia in the French books of documentary evidence that he possessed, he did not even make very extensive use of these rich documentary appendixes himself, but rather relied primarily on narratives produced by court genealogists.[78] He appreciated, and perhaps admired, their work routines, utilizing a highly sophisticated analysis of archival sources, but he did not follow their example. Imhoff had long opted for a different literary genre, for which readers neither would nor could reasonably expect extensive archival research.

Public Genealogy: Into the Early Modern Information Market

A man named Nicolaus Rittershausen (1597–1670) was potentially Jakob Wilhelm Imhoff's greatest inspiration. Rittershausen had served for many years as a professor of law at the University of Altdorf, located very near Nuremberg. His claims to fame, however, were his significant achievements in the fields of geography, cartography, and genealogy.[79] From 1653 to 1664, he published three constantly growing editions of his *Genealogiae Imperatorum, Regum, Ducum, Praecipuorumque Aliorum Procerum Orbis Christiani* (Genealogies of emperors, kings, dukes, and other distinguished nobles of the Christian world). The final version of this book, which consisted exclusively of high-quality genealogical tables, was a massive success.

What is particularly noteworthy about Rittershausen's work is that it was *not* produced for one particular family.[80] Rittershausen abandoned the traditions of courtly genealogy to a significant extent. Universities emerged as sites of genealogical knowledge production, fostering an alternative approach to the field. Rittershausen first experienced the transformative potential of an academic setting for genealogy while studying at the University of Helmstedt. One of his teachers there, Rudolph Diephold (1572–1626), taught him to approach genealogy from the vantage point of a scholar of public law.[81] Diephold and, later, Rittershausen were interested in genealogy not as a means of promoting dynastic legitimacy but rather as a window on the dynastic background of contemporary European politics. They both viewed genealogy as an analytical tool for understanding the alliances and conflicts within Europe's "society of princes."[82] As Diephold claimed and Rittershausen echoed, genealogy had potentially vast implications on politics. "I would be preaching to the converted if I were to recommend genealogies: their usefulness recommends itself, and a subject that encompasses the affairs, rights, claims, marriages, successions, and treaties, and other such affairs of kings and princes between one another has no need for commendation," the Altdorf professor noted. Ignorance of genealogy, he continued, "has caused squabbles, disputes, and even major wars."[83] Administrators and politicians needed to take into account how princes and aristocrats were related to one another in their assessment of both international and domestic politics.[84] Indeed, *every* person interested in contemporary affairs had to understand the intricate genealogical framework that lurked behind many events of European history.

In order to refashion genealogy into a tool for contemporary political analysis, the aspiring genealogist had to cover not one, but many—and ideally *all*—of the reigning families of Europe. Since the international society of princes consisted of multiple intermarried dynasties, the diligent observer had to know all of them. By shifting focus from discussion of a single family to the presentation of multiple genealogies, Rittershausen almost single-handedly pioneered a new genre of genealogical literature, that of "universal genealogies" or genealogical encyclopedias.[85] His work ultimately evolved into a collection of standardized genealogical diagrams that devoted roughly equal space to every reigning house. Genealogies (and families) were normalized and thus now could be compared.

Genealogical encyclopedism, born outside the courts, required scholars like Rittershausen to maintain objective distance from individual families and their genealogical agendas. Rittershausen and Imhoff after him wrote from the standpoint of impartial experts on not just one, but all families. Their works, dedicated to elucidating the relationships that bound together Europe's ruling elite, helped reimagine "the nobility" as a collective entity, thus bringing into focus a vision of the second estate as a coherent social class that transcended individual families. The newly fashionable trend of encyclopedism thus played a major part in redefining the field of genealogical inquiry. Encyclopedic genealogy did more than simply compile preexisting knowledge for convenience; it gave new contours to the conversation about the nobility, politics, and the social order at large.

The rise of genealogical encyclopedism had profound consequences for the status and prestige of genealogy circa 1700. First and foremost, it changed the basic parameters of research. It was impossible for compilers like Rittershausen and Imhoff to investigate every family tree from scratch. In-depth archival research was out of the question. The vast collections they produced necessarily relied on secondary knowledge. Rittershausen appears to have contacted the (regional) nobility directly to ask for material.[86] Several of his friends, including well-known scholars like Hiob Ludolph (1624–1704) and the eminent genealogist Philipp Jakob Spener, also supplied information and advice.[87] Imhoff later created his own system of secondary information-acquisition, which I will consider in detail later. His system laid the foundation for his own compilatory work.

The second major consequence of the rise of encyclopedic genealogy was a shift in genealogy's "regime of historicity."[88] Genealogy was profoundly reshaped by the discovery of contemporaneity in Europe circa 1700.[89] "Contemporary affairs" and "news" became buzzwords.[90] Despite a still widely shared fascination with the remote past of dynastic history, including (mythical) origin stories, genealogists attached ever greater importance to *current* genealogy. Insofar as universal genealogies were intended to orient readers in the contemporary political and social landscape, they took on a distinctively presentist bias.[91] In Rittershausen's paradigmatic view, the primary object of genealogy was not the heroes of the past, but rather reigning heads of state. He conspicuously declined to discuss genealogy prior to the year 1400 in his book, thus highlighting the growing presentism of encyclopedic genealogy.

A generation later, when Imhoff was at the zenith of his career circa 1700, the fascination with contemporaneity had become even more urgent. It was then, as many historians of journalism and media have noted, that a sophisticated information market had emerged. News was commercialized; information became a commodity.[92] Providing up-to-date information could be lucrative, and numerous publishers and other information-mongers tried their hand at establishing profitable businesses. Journals and newspapers are the most obvious examples of this phenomenon, but many other genres were also profoundly influenced by the publication strategies of the incipient news cycle. Encyclopedic genealogy also left the confines of academic discussion and burst out onto the public market. Universal genealogies were quickly associated with commercialized news reporting about current events.[93]

Unsurprisingly, Imhoff personally corresponded with some of the most important figures active in the news business.[94] In nearby Franconia, he could also observe several spectacular examples of how journalistic authors created a profitable business model by compiling genealogical reference works for the public market. The Nuremberg career writer Johann Christoph Beer (1638–1712), for instance, churned out a series of biographical and genealogical surveys of the royal dynasties of various European states.[95] The success of his 1668 work *Der Könige in Franckreich Leben Regierung und Absterben* (The life, reign, and death of the kings of France) induced Beer to publish numerous sequels from 1684 to 1695, covering the royals of Spain, Bohemia, Denmark, Sweden, Hungary, and Italy, as well as the archdukes of Austria.[96] The fact that Beer cleverly combined his contemporary genealogies with additional forms of journalism is yet another illustration of how quickly and thoroughly genealogy was integrated into a dynamic news market.[97]

Johann Seifert (1655–1733), a soldier's son and a contemporary of Imhoff from nearby Regensburg, also exploited the new genealogical information economy.[98] In 1696, he published the first of a large number of genealogical tables on noble and patrician families of Franconia. This series included, in 1723, a set of twenty-one genealogical tables dedicated to the Imhoff family itself.[99] Although he sometimes merely reprinted existing tables, Seifert usually solicited fresh information through correspondence.[100] Seifert skillfully engaged elite families, occasionally inducing individuals to leak controversial documents. In case of the Imhoff family, for instance, he sweet-talked the unsuspecting Anton Ignaz Imhoff (1697–1775) of Augsburg into giving him access to private information.[101] Seifert produced his tables proactively at his own risk, in the hope that the families would honor his efforts *post factum*.[102] While this was often a successful strategy, it could also backfire. The Imhoffs, for example, were not at all pleased by Seifert's publication (for unspecified reasons) and even contemplated a public rebuttal.[103] Whether this resulted in any actions against Seifert himself is unclear; the incident nevertheless shows how risky such freelance genealogizing could potentially be. It was not easy to anticipate the movement of the early modern genealogical information market. Exploiting the economic opportunities of encyclopedic genealogy proved to be a challenging way to make a living.

Beer and Seifert's market-driven, German-language projects to compile up-to-date, encyclopedic information on the nobility put Imhoff's works into perspective. While Imhoff also encouraged the public circulation of genealogical knowledge in novel ways, his books nevertheless differed significantly from these authors' publications. In contrast to Beer and Seifert's predominantly market-oriented works, Imhoff's books were much more carefully produced, more luxurious, and—most important—written in Latin; they thus targeted a much narrower, even elite, audience. Although Imhoff's genealogical writing flourished in an age in which journalistic forms of rapidly produced information were becoming regular, he remained faithful to an earlier type of genealogical book, the deluxe folio volume. Hence, although he was anchored in a rapidly developing publishing economy, Imhoff did not fully participate in the most radical changes to the media ecology of genealogy circa 1700. His oeuvre remained beholden to traditional conceptions of universal genealogy.

Bureaucratic Judgments: Institutional and Juridical Genealogy

Of all the newly emerging sites of genealogical knowledge production, the important part played by administrative offices may be the least known today. Imhoff, however, was acutely aware of the fact that bureaucrats and state-sponsored genealogical officials were amassing large amounts of high-quality information about the nobility. Among Imhoff's many contacts, no one exemplifies institutional genealogy better than Charles-René d'Hozier (1640–1732), one of his longtime correspondents.

Following in the footsteps of his late father and his older brother, d'Hozier served the French monarchy for decades as *juge d'armes*.[104] This was a heritable royal office, established in 1615 to supervise and monitor the nobility's use of coats of arms. In addition to heraldic matters, d'Hozier and his office also became the monarchy's primary clearinghouse of genealogical information on the French nobility. In particular, the *juge d'armes* was responsible for scrutinizing the pedigree of young nobles who aspired to enter the royal service as Pages of the Royal Stables.[105] D'Hozier's lifelong, everyday work on heraldic and genealogical questions made him one of the most renowned experts on the subject, as had been true of his father and brother and would eventually also be true of his successor, his nephew Louis-Philippe. During his years of service, d'Hozier filled hundreds of volumes with notes, papers, and documents relating to the French nobility, their coats of arms, and their genealogies. He sold this massive collection to the king on leaving office in 1717.

D'Hozier and his office valorized genealogy in a very specific, institutional way. Genealogical administrators like himself guarded access to elite offices and corporations that gave noblemen both economic opportunities and social prestige. Membership in noble corporations such as the Pages of the Royal Stables, and also in numerous ecclesiastical bodies, had long been a defining privilege of the nobility. Beginning in the mid-sixteenth century, however, these institutions increasingly narrowed their criteria of admission and demanded ever more elaborate proof of noble status.[106] Proof of nobility centered predominantly on genealogy. Corporations required nobles to submit solid evidence that three, four, or five generations of the applicant's ancestors, both male and female, were of pure noble stock. Although oral testimony remained very important, "proof" tended more regularly to consist of archival records that documented the noble status of one's ancestors. Candidates had to submit dossiers of such documents, which were carefully scrutinized. This is what d'Hozier did for the Royal Pages on the king's behalf.

From the perspective of social history, this development was at least partially a result of growing status-insecurity among early modern nobles. In marked contrast to their publicly exhibited image as the bedrock of tradition and long-term continuity, the European nobility was in fact a fast-changing social group.[107] The sixteenth and seventeenth centuries in particular witnessed significant new social dynamics, as new elites emerged (bureaucrats, merchants, financiers), and a combination of civil war and religious upheaval decimated the established elites of many places. It was becoming increasingly clear to impartial observers that much of the now-living nobility was neither particularly old nor particularly pure. At least some of these upstarts were excluded from full membership in the nobility on account of hardening genealogical criteria. Documentary proof of nobility drew visible boundaries between noble and non-noble, as well as between old and new nobility, in an otherwise increasingly chaotic world in which social hierarchies were being upended. Tellingly, the nobility itself had initially asked Louis XIII in 1614 to appoint a *juge d'armes* whose main task would be to create a "register" of true nobles, thus producing a fixed and reliable index of who belonged.[108] The French nobility thus voluntarily ceded at least some of its monopoly on genealogical knowledge and the definition of noble status to the bureaucratic state in order to reinforce the social status quo. To some extent, the nobility actively supported the rise of hetero-genealogy.

Submitting noble status to bureaucratic control, however, quickly proved to be a two-edged sword, even for many well-established aristocrats. The obligatory proof of nobility exposed time and again how far noble families fell short of documentary standards. Administrative genealogy occasionally put enormous archival pressure on the nobility. For French nobles, dealing with d'Hozier and his office often entailed complicated (and potentially degrading) negotiations, as many families did not have the necessary dossiers available. Only very gradually did aristocrats upgrade their archival infrastructure to improve the evidentiary basis of their genealogies. Mandatory evidence of nobility drove home the point that families needed to keep their documentation in order. In some cases, the needs of institutional genealogy may have significantly contributed to the transformation of family archives into focal points of aristocratic identity, but it was a slow and unsteady process.[109] For a long time to come, many nobles still anticipated with dread the moment when they would be called on to present their genealogical dossiers.

D'Hozier's files attest to the nobility's difficulties in this regard in ample detail. Tellingly, many families were forced to make elaborate excuses as to why they could not provide adequate documentation for their claims. One aristocrat cited a lack of funds and asked d'Hozier to consider only the absolute minimum of generations, perhaps hiding other, more serious problems. Others told a different story to explain the fragmentary status of their papers: "My original titles [to prove my pedigree] are currently unavailable, because the recent death of several family members has entirely dispersed our papers," one eighteenth-century applicant explained, potentially dissimulating inconsistencies in the historical record.[110] In Italy, too, nobles cited "wars" and "invasions" that had afflicted "every kingdom, province, city, and village" to explain their inability to provide appropriate evidence.[111] Embarrassment over these and similar deficits often compelled individual nobles to scramble for evidence. In 1729, for instance, Adriana Carolina von Syberg zu Aprath, a member of the Westphalian lower nobility, wrote frantically to a cousin, Johann Abraham von Syberg zu Aprath, to ask for information for a genealogical dossier that she was preparing for one of her sons.[112] Many other noble parents are on record similarly pleading with relatives and outside experts for help in assembling a convincing genealogical dossier. Nobles not only had to negotiate the strengths and weaknesses of their own pedigree; they also had to assess the strengths and weaknesses of their own *knowledge about* their pedigree.

We should not overestimate the bureaucratic rigor of such protocols, though. France was an extreme case, where evidence was scrutinized by royal officers; but even there, we find ample evidence of local laxity and manipulation.[113] Elsewhere, it was the noble corporations themselves who conducted evaluations, meaning that nobles were judged by their peers. Local connections often trumped detailed documentation in such contexts. In Westphalia, a recent study has found that the corporation of Free Imperial Knights in the Holy Roman Empire (Ritterschaft, Reichsritter) utilized both the occasional investigative scrutiny of applicants' documents and a more traditional reliance on hearsay and peer acclamation in the later seventeenth century.[114] While nobles were prudent to worry about the rigors of institutional genealogy, they could also rely to some extent on their family's public reputation among their peers.

The whole procedure was least problematic for the small group at the very top of the European aristocracies. The nobility of reigning dynasties, in particular, was usually considered noble beyond doubt. Yet, even though no contemporary would have seriously questioned the claim of such a family to impeccable noble status, some of the most elite families nevertheless routinely submitted to formal bureaucratic genealogical scrutiny. In 1669, for instance, Duke Teobaldo Visconti presented his eldest son Ercole (1646–1712) to the Milanese Corporation of Jurists. Membership in this corporation hinged on noble status; hence, Teobaldo was obliged to produce his son's proof of nobility. He willingly did so, and the Visconti easily assembled the required dossier and had young Ercole duly approved.

The real point for the Visconti here, however, was to instrumentalize the procedure to gain public approval of a new genealogical vision—highlighting the authenticating power of institutional genealogy. The Visconti not only presented the specific documents needed for Ercole's approval; on the contrary, they submitted a vast dossier that included some of the Galluzzi forgeries mentioned above. The family thus used the occasion to have a spectacular new pedigree officially approved, one that "overthrows most genealogies of the historians."[115] The dossier also appeared as a book that fascinated scholars for decades.[116] In 1679, the Visconti submitted the same lineage again, once more seeking official approval.[117] Although they clearly were not the primary object of institutional genealogy, the highest echelons of the European nobility still endorsed the practice, if only to exploit it for their own purposes.

Proof of nobility was part of a broader trend toward stricter legal control of noble status and life. Although traditional forms of aristocratic dispute-resolution, including brigandage, dueling, and armed confrontation, had not yet died out in Imhoff's lifetime, nobles increasingly turned to legal procedures to settle internal conflicts. Formal litigation became an everyday occurrence among the nobility in the seventeenth and eighteenth centuries, and questions of genealogy often played a key part, for instance, in disputes over inheritance, succession, and ceremonial precedence. As one legal authority from Imhoff's era noted: "Both in our civil and canon law, it is very important to know the status and condition of birth, to what parents one is born, from whence the rights of legitimacy arise or originate. The power and usefulness of genealogies stand out especially in cases concerning succession and marriage, in contracts, guardianship, the possession of an estate, and testimony in court."[118] Another contemporary genealogical authority, discussing the importance of genealogy as a scholarly field, had only one thing to say: "the legal claims of great lords cannot firmly be decided without the help of genealogists."[119] The quotation is remarkable as much for what it says as for what it does not. In the eighteenth century, the legal implications of genealogy occasionally trumped its honorific qualities, at least among authors who were surveying the impact of genealogy for the general public.[120]

The legal relevance of genealogy necessitated new forms of genealogical knowledge and inspired new media and genres. The propagandistic grandeur of courtly genealogy counted for little in the sober world of legal argumentation. Elaborate narratives about individual bravery, moral conduct, and other edifying lore were no help in a court of law. Instead, genealogical facts needed to appear in such a way that supported legal arguments. For that purpose, genealogical arguments had to be based on hard evidence in the form of legally authoritative, preferably archival documentation. A new literary genre accommodated these needs: the legal brief, one of the most significant new genealogical media that emerged in the seventeenth and eighteenth centuries.

Imhoff and his German peers knew such writings as *deductiones*. They were, in the words of a younger citizen of Nuremberg, "writings intended to explain the nature of a legal dispute." They "present the matter under discussion to the public in the most advantageous way in order to convince readers of the legality of a given case."[121] Such texts, which existed all over Europe, presented both formal legal arguments *and* relevant (genealogical) evidence. Although written in an ostensibly objective legal style, briefs usually promoted the perspective of only one party, presenting genealogical facts and legal arguments in such a way that they supported their sponsor's case.[122] What mattered most in these genealogical texts was not the symbolic power of family lore but rather the case-specific presentation of (seemingly) well-documented genealogical facts. To that end, they contained a wealth of genealogical documentation, which was eagerly appropriated by genealogists such as Imhoff.

Deductiones and similar genres addressed a wide range of topics, from the minutest conflicts between local noblemen to grave matters of international politics. Since inter-state relations in the early modern society of princes were still, to a great extent, inter-nobility relations, genealogy became a decisive factor in European politics. A good share of international conflicts were conceptualized in genealogical terms, such as disputes over the rightful succession to or inheritance of princely crowns. The War of the Spanish Succession is only the most obvious example, in which different contenders claimed the Spanish throne on the basis of their relationship to the late Spanish king. Whenever successions proved contentious, every contender usually tried to back up his claim to power by producing, and often also publishing, legal briefs or *deductiones* full of genealogical detail.[123] Genealogy thus became relevant to politicians and administrators, and politically sensitive genealogical information became a state or dynastic secret.

Imhoff was fully in tune with these developments. He did his utmost to access governmental genealogical arcana wherever he could; and as we will see, numerous bureaucrats and administrators opened their secret dossiers and provided information for him. Even more important, however, Imhoff was deeply curious about the daily legal drama surrounding many families of the German and non-German aristocracies, as families and family members fought over inheritances, precedence, and successions. By far the largest individual entry in the *Index rerum* of Imhoff's *Notitia Procerum* concerned "famous controversies" (*controversiae illustres*) between noble houses: such disputes appeared in his book because of they were fought with genealogical information. Even more so than his printed books, however, Imhoff's numerous letters show just how much news about such legal battles and lawsuits flowed through the genealogical information conduits of the Holy Roman Empire. Imhoff, posing as a disinterested genealogical observer, carefully noted how legal positions shifted, new conflicts arose, and what future debates might be anticipated in light of the most recent dynastic marriages, births, and deaths. Although he was rarely personally involved in any of these struggles, Imhoff nevertheless watched in perpetual fascination as nobles struggled to resolve their conflicts in a highly legalistic fashion. Genealogy often played a key part, and the genealogist profited greatly from the knowledge that was marshaled for such cases.

New Scholarly Methods

As forms and sites of genealogy multiplied, a common language nevertheless connected many actors in the field across all divides. Most genealogists after 1650 claimed, and probably believed, that their generation was much more discerning and critical than earlier ones when it came to making genealogical assertions. Imhoff and his peers claimed to follow the new methodological standards of critical historiography. Beginning in the fifteenth and sixteenth centuries, Renaissance humanist scholars had demonstrated that traditional historiography was full of false and fictitious stories.[124] Rigorous new philological techniques now made it possible to date sources more precisely and to distinguish forgeries from authentic pieces.[125] Humanist critics from Lorenzo Valla to Hermann Conring demolished fairy tales like the Donation of Constantine or the Translation of Empire (*translatio imperii*). Protestant critics attacked many of the comforting exempla of saints' lives. Catholic scholars, especially the Jesuit Bollandists and the Benedictines of St. Maur, also began to clear away unfounded and spurious hagiographical narratives, in the process setting influential examples for critical-source work. Lawyers like Hugues Doneau and François Hotman demonstrated the historicity of legal systems that, they showed, did not harken back directly to ancient Roman times. All of these developments eventually coalesced into a broad consensus that the past required a thorough cleansing of fictitious stories and "myths," whether they originated from ignorance or outright fraud. Slowly, a "revolution of mentalities" took place in historiography.[126] Genealogists of Imhoff's time regularly claimed to be part of this development.

A Rhetoric of Progress . . .

Imhoff's friends saw an epistemological divide between their own practice of genealogy and that of previous generations. The field had advanced, they insisted, and Imhoff's lifetime represented a new era of "mature" genealogy.[127] Genealogists presented a powerful narrative of change, highlighting their field's supposed progress from "myths" to facts.[128] We may take as an example Andrea Gittio's harsh criticism of the three-volume *Teatro genealogico delle famiglie nobili* by Filadelfio Mugnos (1607–75) of Sicily.[129] These volumes were full of "myths and inventions," Gittio claimed; hence, "no one should put any trust in these."[130] Similarly, Andrea Pusterla, head librarian of the Ambrosiana in Milan and an early Italian supporter of Imhoff, accused Giovanni Pietro Crescenzi (1615–50) of producing nothing but "myths" in an effort to discredit his work in 1693.[131] In Paris, d'Hozier similarly denounced a recent publication on the genealogy of the Grimaldi family of Monaco as full of "myths." He "wondered" how a "smart man" like the author could have produced such a book that predictably would not be trusted by anyone.[132]

Such intellectual concoctions, d'Hozier complained more generally, indicated the "ridiculous" competition among the nobility. The invention of evermore ancient origins for their houses was, in his view, "vanity" pure and simple.[133] He similarly observed a general "folly of men who all want to descend from some royal house."[134] The "myths" that early modern dynasties paid genealogists to produce drew scorn and ridicule from Imhoff's friends, who considered them nothing but "a figment of the imagination" (*ens rationis*, humorously).[135] A divide between old and new genealogy thus seemed to open up. Traditional genealogy, Imhoff's friends claimed, relied all too often not on historical evidence, but on wishful thinking. They contrasted genealogical "fables" with the "pure light of history" (*pura historiae lux*).[136]

The impact of this new style of attested genealogy was felt most strongly in treatments of the nobility's earliest history, especially their origin stories. In effect, Imhoff's friends promoted an alternative understanding of noble "historical origins" and the "antiquity" of nobility that diverged significantly from the earlier notions discussed above.[137] Unknowability ceased to be a virtue. Since historical narratives needed to rely on documentary evidence, a family could only claim to be ancient to the extent that its history was securely documented. One had to remain conspicuously silent about everything else. Against the tendency to embellish and enlarge family trees through speculation, genealogists circa 1700 promoted epistemic economy, cutting away everything that they considered unsubstantiated. At the very least, genealogists needed to determine what was documented and what was not. Imhoff once distinguished incontrovertible genealogical "demonstration," by means of explicit sources, from error-prone and speculative "conjecture."[138]

Gottfried Wilhelm Leibniz agreed. All genealogical claims, he wrote, must rest on solid evidence and reliable sources.[139] The "lack of proper proof" in traditional genealogies made it "difficult to trust these authors, even if they may be saying truthful things." Although unsubstantiated statements may indeed be true in and of themselves, truth could be *ascertained* only by means of proof. Yet genealogies did not need to be merely true; they needed to be *demonstrably* true.[140] Gone were the days when genealogists could gloss over a lack of documentation with rhetorical sleight of hand.[141] What the sources did not explicitly record could not be known. As Heinrich Meibom Jr. (1638–1700), an older contemporary of Imhoff, explained, genealogy discovers the origin of families only "insofar as it may be determined with certainty."[142]

These ideas led to a rhetoric of conspicuous epistemic submission. Historians lived in a state of epistemic "necessity," as Benedetto Bacchini (1651–1721), another of Imhoff's acquaintances, claimed, insofar as they were driven by evidence, nothing else.[143] Genealogists were likewise "unfree," insofar as they were resigned to source dependency. Only poets, by contrast, were free to "choose" how they presented the past. Genealogists and historians were not "creative"; they merely "drew forth" what was contained in the sources, Leibniz concurred. Whereas earlier authors had "imperfectly reported their proofs," presumably because such proofs were often simply lacking, the generation of scholars "today" relied on "so many monuments."[144] In short, Leibniz thought, genealogy had made a quantum leap in epistemological reliability, because it now, finally, lived up to the necessary standard of "caution" (*autant de soin*).[145] All this made genealogy much more "exact," Leibniz insisted.

Imhoff enthusiastically agreed, occasionally with a rhetorical flourish. Notwithstanding the fact that, as an encyclopedic genealogist, he was not heavily involved in working directly with original sources, he still embraced the rhetoric of modern evidence-based historiography. He participated in the early modern trend of celebrating "the archive" as *the* locus of historiographically relevant evidence par excellence. As long as an author relied on archival documents, his conclusions could not be doubted, Imhoff declared.[146] Genealogical publications that "report nothing from old and unedited charters," in contrast, were more or less useless.[147] The failure to consult archives met with Imhoff's "bewilderment."[148] Without archival information, it was impossible to investigate family history and constitutional law.[149] Judging from these statements, all made casually by Imhoff in his letters, archives as symbols of reliable evidence had become a fixed trope of the new scholarly rhetoric.

With such impressive declarations, the new genealogists fashioned their personae as supremely rigorous scholars. Debunking erroneous stories and falsehoods became something of a habit. On one occasion, Imhoff (or one of his friends) declared a certain "Sigismund," allegedly bishop of Meissen in the fourteenth century and a key figure in the Wettin family's traditional genealogy, to be purely fictitious. The reason: no original source from the period mentions this "Sigismund." For this claim to stick, one argued that the total silence of the written record was unlikely if the person had really existed: "he [the fictitious Sigismund] lived in a century, the fourteenth, which saw so many writers of documents in Saxony and elsewhere"; hence one could reasonably expect that "they would have mentioned him at least in passing." The relevant point here is that the new critical attitude not only looked for positive evidence; it also placed individual pieces of evidence (or the lack thereof) in a broader perspective that integrated careful assessments of the methods and forms of source transmission.[150] Genealogical "facts" that did not pass this test had to be discarded from family history, no matter how well they fit a family's self-perception.

Jakob Friedrich Reimmann summarized the new critical attitude in 1710 with a contemporary catchword: he asked, "Who is so bleary-eyed as not to see that skepticism is highly important in the study of genealogy?"[151] Reimmann does not explain what "skepticism" meant to him in this context. Yet the term was popular in contemporary historical methodology, and it generally implied a mindset of reservation and restraint with respect to historical claims.[152] Leaving unanswerable questions unanswered was a key provision. "Unless further old, correct documents are found, it will probably remain unclear who the landless margrave Heinrich really was," Imhoff once stated matter-of-factly.[153] At the very least, what was "certain" (*certa*) about a family had to be clearly distinguished from what was merely "probable" (*probabilis*).[154]

Acceptance of the partial inscrutability of the past became a cardinal virtue of the genealogist's mindset. Johann Christoph Beckmann (1641–1717), a German historian and genealogist in the duchy of Anhalt, accordingly divided historical time into three eras. There was a "*Tempus Obscurum*, i.e. times in which there is no certainty at all; next, *Tempus Mythico-Heroicum*, in which some things can be made out, but in highly uncertain conditions; and last, *Tempus Historicum*, for such things that can be reported with greater certainty."[155] In practice, as many scholars agreed, the boundaries of documentable genealogy fell in most cases somewhere between the years 1100 and 1400.[156] Besides a few exceptional houses, the poor survival rate of older historical evidence made proper genealogical research next to impossible before that time. Genealogists did well to steer clear of speculating about this "dark age." In effect, periods that lacked documentary evidence had become ahistorical, or so the new genealogists liked to claim.

People accepted this emerging consensus of critical and evidence-based genealogy for a variety of reasons. Scholars were attracted to genealogical skepticism by notions about philological rigor, which expressed a distinctive erudite habitus. Kings and royal ministers, in contrast, found more stringent criteria for nobility helpful because it benefited the royal purse. Rigorous scholarship tended to reduce the number of tax-exempt families—a major motive behind the review of all claims to nobility in France under Louis XIV. As these new methodological standards became firmly ensconced in scholarly communities and governmental institutions, the new rhetoric of philological criticism eventually also became a reference point for the nobility. All too implausible origin stories now might become sources of embarrassment. Toppling fictitious claims to (ancient) nobility and erecting proper genealogies translated into protecting the social status quo. Aristocrats accordingly endorsed the scholarly criticism of family history as "protection" against the social fallout of genealogical naïveté.[157] The new critical attitude toward genealogical knowledge thus served as a rhetorical platform that could accommodate multiple supporters with otherwise different interests and standpoints.

. . . Resulting in Ambivalence

For all the rhetoric, however, this was definitely not a story of indisputable epistemic progress in the field of genealogy. Imhoff and his peers may have preferred evidentiary proof in the form of solid documentation as much as they liked; they nevertheless neither could nor indeed did avoid recourse to myth, speculation, and conjecture. If better information was unavailable, a friend of Imhoff's sighed, "we must depend either on myths or on the conjectures of others."[158] Often enough, "conjecture" (*conjecturare*) remained the only way forward. These educated guesses sometimes proved to be correct, sometimes not.[159] If one of Imhoff's peers found new evidence, he would simply corroborate or falsify his preliminary intuition.[160] This was not ideal, but unavoidable. It was best practice.

Most contemporary protagonists also found that historical evidence and outrageous genealogical claims were by no means inevitably incompatible. Biagio Aldimari, for instance, an exemplar of the new courtly genealogy, did not hesitate to fabricate a self-serving and purely invented ancestry for his own family.[161] Fabulous genealogies remained popular well into the eighteenth century and beyond.[162] No genealogist could simply ignore the ideological allure of such myths. Even institutional genealogists, working with the support of the early modern state or other public authorities, were flexible in practice when it came to accepting myths and forgeries.[163] Seventeenth- and eighteenth-century genealogy was often characterized by such a "double soul."[164] The results of thorough empirical research were not infrequently kept hidden so as maintain appearances.[165] As we will see later, Imhoff also followed this tradition of "genealogical doublethink."[166]

By Imhoff's lifetime, genealogy had expanded far beyond the confines of noble courts and aristocratic manors. To be sure, it continued to fulfill its important purpose as an expression and legitimation of noble status. European elites continued to treat ancestry and family history as important sources of dynastic identity. One's forebears and their glorious achievements still defined a noble family's place in society and still held the exemplary power to entice young nobles into behaving like "true" scions of such family traditions. Genealogical media, both two-dimensional and three-dimensional, displayed the pride that nobles felt about their ancestors and invited readers and viewers to share in that feeling. Genealogies still expressed and explained where an individual stood in the world, and why.

Although the fundamental function of genealogy was hardly questioned, the ways in which one might utilize this framework of ancestral pride and the reasons why families felt obliged to publicize their knowledge about the past subtly changed. The status of genealogy shifted, first, as genealogy came under the purview of humanistic philology and documentary antiquarianism and, second, as it entered the era of news-reporting, popular encyclopedism, and commercialized information markets. The sites for producing and assessing genealogical knowledge multiplied, as outside observers became interested in noble genealogies for various reasons. States and bureaucrats scrutinized the nobility out of fiscal interests or in the hope of controlling social mobility. Elite corporations, meanwhile, attempted to arrest the tide of social change, jealously safeguarding their exclusive status. Scholars, trained in philology and new historiographical methods, in contrast, could hardly avoid bringing their new critical tools to bear on genealogical sources, even if they were often inclined to look the other way when uncompromising scholarly rigor would have undermined their career prospects—many scholars were still embedded in patronage networks and needed to keep their patrons' favor. For all the basic assumptions about the social and cultural importance of genealogy that these various actors shared, the field became clearly divided into a growing number of specialized genres and practices.

Jakob Wilhelm Imhoff was well connected to most of the different forms of genealogy in vogue circa 1700. His own vision of the field drew on all of them, yet was not beholden exclusively to anyone. Imhoff's works drew on all of these contexts and combined them in creative ways. He followed Rittershausen, Spener, and the emerging tradition of encyclopedic genealogy in compiling second-hand genealogical information. Producing an encyclopedia entailed a strong commitment to constantly evolving contemporary genealogy, yet Imhoff's presentist bias was less pronounced than that of others. Imhoff was also more inclined to condone the celebratory and propagandistic rationale of aristocratic genealogy. He certainly did not distinguish himself with a particularly aggressive opposition to origin stories. As the following chapter demonstrates, much of Imhoff's peculiar position between old and new genealogy (albeit with a very large dose of the latter), between dynastic propaganda and encyclopedic information-mongering, was a consequence of his social background as a proud and well-to-do citizen of one of the most important German imperial cities: Nuremberg.

CHAPTER TWO

A Patrician Genealogist and His City

In 1726, a few years before Jakob Wilhelm Imhoff's death, his near-contemporary cousin Christoph Jakob Imhoff (1654–1726) passed away. A local scholarly journal, one of the many that had recently sprung up all over Germany, published a brief obituary that portrayed the late patrician as an embodiment of the city's civic spirit and cultural sophistication.[1] Christoph Jakob had been a prominent scholar, working in several fields, and he contributed significantly to Nuremberg's reputation as a city of books: he had collected an unusually large library that impressed many of his contemporaries. Christoph Jakob's range of learning matched the riches of his collection. As the author noted, the deceased had often been called a "walking museum and archive" (*Museum & Archivum obambulans*), thus exemplifying the intimate relationship that the citizens of Nuremberg had with their city's past. The obituary also insisted that Christoph Jakob had descended from impeccable origins. The Nuremberg elite, it suggested, was hardly less distinguished than the surrounding Franconian nobility. "It would be easy to recount and list not only sixteen but even many more ancestors of this very old and very noble family, in unbroken series on both his maternal and paternal side," the text proclaimed, appropriating the aristocracy's language of proofs of nobility.[2] Christoph Jakob, the obituary continued, had been very conscious of his "noble birth," yet his pride about his family was tempered by devotion to religion and commitment to academic training and education. The remaining pages of the obituary reviewed his family life and highlighted his dedication to his native city, where he served in different official positions, including that of city archivist. According to this brief biography, Christoph Jakob Imhoff was thus paradigmatic of the potent mixture of scholarly learning, civic responsibility, aristocratic identity, and Christian piety that supposedly characterized early modern Nuremberg patricians at large. Attending to genealogical detail was a key element of this mindset.

For all its panegyrical and hagiographic exaggeration, the obituary was not far off the mark. In Nuremberg and many other free imperial cities of early modern Germany, urban infrastructure and patrician culture converged in fruitful ways, making these municipalities into important centers of early modern German knowledge culture. European cities in general were major hubs for scientists and scholars in various disciplines, including genealogy.[3] Contemporary and encyclopedic genealogy circa 1700 in particular relied on Europe's network of cities as their preferred knowledge sites. The field's growing affiliation with journalism, encyclopedism, and the printing industry made genealogy an urban phenomenon.

Imhoff's case will allow us to explore the distinctively urban trajectories of early modern genealogy in greater detail, thereby building on recent initiatives to assess the lasting importance of early modern cities as places of knowledge in the later seventeenth and in the eighteenth century.[4] Long after the heyday of Renaissance urban culture, cities remained vitally important to Europe's knowledge culture, even in the age of the territorial state. Despite the ascendancy of courtly centers of learning and state-sponsored cultural projects, the cities of Europe retained many of their traditional infrastructural advantages, social opportunities, and economic appeal for scholars, publishers, and other information entrepreneurs. This chapter explores how one major early modern city, the southern German metropolis of Nuremberg, became the site of major genealogical research and publishing projects circa 1700.

In order to understand both Nuremberg's broad profile as a center of knowledge and its specific role for Imhoff's project of encyclopedic genealogy, we must take into account the city's peculiar social structure and political status in the Holy Roman Empire. Before Napoleon reorganized Germany around 1800, Nuremberg was one of several dozen imperial free cities. This status meant that the city enjoyed "imperial immediacy"; in other words, it answered to no higher authority than the Holy Roman emperor himself. It was not subject to a territorial state. Free imperial cities were fiercely independent and proud of their autonomy. In the Middle Ages and as late as the sixteenth century, they remained important political players in the Holy Roman Empire. They enjoyed the right to vote in the Imperial Diet (the Reichstag), at least collectively, and significantly influenced the politics of the empire for much of the early modern period.

In addition to being politically independent and powerful, free imperial cities had also been powerhouses of Germany's cultural Renaissance and economic expansion. Cities like Nuremberg and Augsburg were centers of artistic innovation in the decades around 1500, as they opened their gates to influences from Italian art and culture. They had, moreover, wielded considerable economic clout. For a while, the important merchant firms—and families—of urban southern Germany had bankrolled many of the activities undertaken by the Holy Roman emperor and by other European princes, who repeatedly took out urban credit to finance their wars and political schemes.

Most free imperial cities retained their formal autonomy even after the Thirty Years' War (1618–48), at the time of Jakob Wilhelm Imhoff's birth in 1651. Although the balance of power tilted ever more in favor of emerging territorial states, traditional scholarly narratives about the decline of free imperial cities now appear to be vastly overblown. As the first section of this chapter demonstrates, these places remained first-rate cultural centers, even as their overall economic and political influence shrank in the seventeenth and eighteenth centuries.[5] When the obituary for Christoph Jakob Imhoff praised Nuremberg's preeminence as a site of learning and knowledge production, it was by no means nostalgically reviving outdated *topoi*. On the contrary, Nuremberg remained a dynamic urban cultural environment circa 1700, providing important institutional, logistical, and social infrastructure for a wide variety of intellectual projects. Jakob Wilhelm Imhoff's patrician peers, as will become clear, contributed decisively to this status. Many of them developed into veritable "families of literary producers."[6]

The free imperial cities of early modern Germany were governed by a proud and sophisticated oligarchy of international standing: the patriciate. Initially, elite urbanites had been proudly non-noble, representing an alternative claim to social excellence. Despite the fact that many patricians eventually imitated noble behavior and openly aspired to ennoblement, urban elites maintained significant distance from the nobility. The second section of this chapter further explores this patrician identity. As will emerge, Imhoff's specifically urban, elite background not only shaped his private life in many ways, but it also colored his understanding of society in general and of the nobility in particular. As we will see on numerous occasions, his special perspective on many genealogical issues can be productively linked to his status as a Nuremberg patrician.

As the cultural and political importance of free imperial cities grew after the fourteenth century, their governing families likewise became ever more self-conscious and sought to project compelling images of their elite status.[7] Urban elites, rich, cultured, and influential as they were in their peculiar ways, became as obsessed with displaying and legitimizing their high social status as the nobility was. The patricians accordingly devised a highly effective culture of celebratory self-representation. In Nuremberg and elsewhere, architecture, artwork, and material culture at large played a fundamental part in the propagation of patrician urban identities. There were "Tucher altars" and "Imhoff chapels" that embedded prominent dynasties in the cityscape itself, thereby also inscribing them in the public memory.[8] Patrician families continued to celebrate their contributions, architectural and otherwise, to the grandeur of their city.[9] Historiography and genealogy were significant vehicles for creating and displaying patrician identity.[10] Unsurprisingly, the basic structures and approaches of patrician genealogy had much in common with noble or princely auto-genealogy.[11] Patricians and nobles shared a basic understanding of how and why genealogy and family history were valuable resources for legitimizing, expressing, and defending their individual and class identity. The third section of this chapter highlights Nuremberg's patrician genealogical culture in which Jakob Wilhelm Imhoff was enmeshed. Among the many local patrician lovers of genealogy, none was more influential to the genealogist than his father, who initiated Jakob Wilhelm into the world of encyclopedic genealogy.

A Prosperous Place of Learning

Nuremberg exemplifies the enormous power of early modern cities as sites of knowledge even in an age of growing princely dominance and emerging territorial states. Its genealogical culture, out of which Jakob Wilhelm Imhoff emerged, was one facet of a broader civic commitment to learning and scholarship. The urban patriciate in particular provided a dynamic and affluent milieu for various intellectual and scientific pursuits. Patricians made Nuremberg home to several important academic institutions and created opportunities for spectacular scientific research. Imhoff's genealogical projects were embedded in a lively environment of scholarly activity. When Imhoff decided to dedicate his life to intellectual endeavors, he would draw much inspiration from his fellow patricians. There would have been little need to explain or justify his decision to put scholarship foremost in his life. Few people would have wondered at seeing such wide-ranging and intense scholarly activities flourish in an elite urban family.

Patrician (and other) cultural pursuits thrived in Nuremberg not least because the city boasted excellent infrastructure. Scholars there could find numerous venues for their activities, academic and ecclesiastical, scholarly and literary. The city facilitated communication and connections with many places across Europe and beyond. As a population center, Nuremberg attracted travelers and migrants, who contributed significantly, if sometimes only fleetingly, to the spread of ideas and information. Imhoff's genealogical research drew heavily on the many resources—institutional, communicative, logistical, and social—that his home city afforded him.

A Patrician Culture of Learning

Nuremberg had been a first-rate center of learning since at least the Renaissance. Numerous important German humanists around the year 1500, including Hartmann Schedel and Conrad Celtis either came from Nuremberg or had important contacts there.[12] Several patricians were also deeply involved in humanistic learning, most famously Willibald Pirckheimer (1470–1530). Nuremberg boasted a highly developed educational infrastructure.[13] In 1538, the city council ordered the creation of a public library, which held more than eight thousand titles by 1600.[14] The city also founded the Egidiengymnasium, although it first existed for only a few years, from 1526 to 1535. After it was reestablished in 1633, it became a progressive elite educational institution. From 1575, the city also had its own university in the small neighboring town of Altdorf.[15] The university was famous for its adventurous cultural milieu and frequently flirted with heterodoxy.[16] Steeped in humanist tradition, it attracted a considerable number of innovative intellectuals.[17]

Meanwhile, the city itself became a prominent hotbed for religious reform. Nuremberg was quick to open up to Reformation thought. By the mid-1520s, Lutheranism had put down deep roots.[18] Later, in the seventeenth century, Nuremberg again became an important center of religious reform. Spearheaded by influential city pastors such as Johann Saubert (1592–1646) and Johann Michael Dilherr (1604–69), a special brand of reformist, open-minded Lutheran orthodoxy emerged that focused on popular piety, affective religiosity, and Christian education.[19] Both churchmen advocated the promotion of an emotional Christian piety that could touch people's hearts, not least by focusing on images and music as bearers of religious messages. Dilherr in particular put great emphasis on updating the city's pedagogical institutions. Many of Nuremberg's intellectual and cultural luminaries embraced this reformist religious climate. Imhoff's own pietism would later build on these traditions.

Nuremberg's rich literary scene was aligned especially closely with this reformist variant of Protestantism.[20] Although Nuremberg's famous tradition of *Meistersang* ended in 1613, the city remained a major hub for literary activities in the following decades.[21] Nuremberg was the home of several literary associations, most famously the Pegnesischer Blumenorden.[22] This was an influential group of poets, first gathered by the patrician Georg Philipp Harsdörffer (1607–58) in 1644 and, after a period of stagnation, reinvigorated, in the 1660s by Sigmund von Birken, the genealogical writer encountered in the previous chapter, who was also a famous baroque poet. The first city theater (Stadttheater) of the Holy Roman Empire, the so-called Fechthaus, opened in Nuremberg in 1628.[23] In addition to literature and theater, the city also boasted a lively musical scene, including an early opera house (founded in 1668) and several music academies.[24] Quite a few members of the patriciate, including several Imhoffs, had close personal ties to these institutions.[25]

In addition to its vibrant religious and literary scene, Nuremberg also was home to considerable activity in the natural sciences. After 1650, the city broadly endorsed the new culture of scientific investigation.[26] An important observatory opened its doors in Imhoff's lifetime.[27] Several patricians were also prominent members of the imperial German Academy for Natural Research Leopoldina, founded in 1652.[28] One of Imhoff's peers, Johann Georg Volckamer the elder (1616–93), became the academy's third president in 1686 and made Nuremberg the academy's de facto headquarters.[29] Another example of naturalist activity in the city was the growing importance of horticulture. Building on a well-established local tradition, one of the sons of the president of the Leopoldina, Johann Christoph Volckamer (1644–1720), transformed Nuremberg into a center for the cultivation of exotic citrus fruits.[30] In 1708, he published the lavishly illustrated *Nürnbergische Hesperides*, a milestone in scholarly literature on citrus fruits.[31]

Imhoff thus grew up with and lived among many peers who had dedicated themselves to science, the arts, literature, and other scholarly pursuits. In all likelihood, he knew many of the people discussed here personally. Imhoff's observations and personal exchanges with these men would have taught him how to achieve success in such endeavors and how to comport himself among scholars and collectors. His embrace of the life of the mind would have been a largely noncontroversial, potentially even natural, life-choice.

The City as Information Hub

This rich ensemble of intellectual activities relied on Nuremberg's extraordinary logistical infrastructure. Contemporaries counted Nuremberg among the most prominent information hubs in central Europe, a point generally corroborated by modern research.[32] Its long-standing tradition as a center of book culture was particularly important. The city hosted an important publishing industry. Nuremberg printers, including the leading firms of Endter and Sonntag, produced large numbers of books each year. Not only could Imhoff publish many of his own books with the city's highly professional printers; Nuremberg's status as a publishing center also meant that numerous books could be found circulating in the city and were thus more easily accessible to Imhoff there than elsewhere. Even rare books from outside of Germany could be found in Nuremberg's bookstores. In 1692, for instance, Imhoff was able to buy locally Paolo Mini's book on Florentine nobles, written in Italian and published in Florence in 1593.[33]

Numerous postal lines, key arteries of the early modern news circuit also ran through Nuremberg, allowing scholars like Imhoff to receive incoming news and letters as speedily and conveniently as possible.[34] Nuremberg's prominent role in German politics also ensured that the city stayed well stocked with political news. Regensburg, the permanent seat of the Imperial Diet after 1663, was not too far away, and contact between the two cities was close and regular. As a patrician and leading member of the city government, Imhoff would have had regular contact with the city's envoys to Regensburg (and elsewhere). Given his prominent position, diplomats could hardly refuse to help him find genealogical material. Marriage ties to some of these diplomats could only have increased their willingness to help.[35]

Nuremberg also attracted large numbers of visitors as a major urban center. Most travelers traversing the western and southwestern territories of the Holy Roman Empire, including numerous aristocrats and diplomats, would have passed through Nuremberg (and Altdorf). Imhoff frequently took advantage of such opportunities and interviewed travelers about genealogical issues.[36] In November 1689, for instance, William Paget (1637–1713), the new English envoy to Vienna, came through town on his way south. When Imhoff let slip that he was working on English genealogy, Paget immediately offered his help and support. There was, moreover, a man with many diverse interests, including meteorology and other natural phenomena, traveling in Paget's entourage: St George Ashe (1658–1718).[37] Most important for our context, Ashe was a herald, and as such highly interested in genealogical questions.[38] Ashe in turn became an important source of information on English genealogy.[39]

While these connections with England were accidental, Nuremberg also served as an excellent base for more stable and reliable contacts across the Channel. Imhoff published his work on English genealogies (his *Regum Par-*

iumque Magnae Britanniae Historia Genealogica of 1690) well before a wave of Anglophilia swept the Continent in the early Enlightenment; his strong interest in the British Isles was somewhat exceptional for his lifetime.[40] Yet the genealogist followed an established local tradition. Nuremberg had a fair share of Anglophiles in the late seventeenth century, particularly among the members of the Blumenorden, the aforementioned influential literary association.[41] Christoph Wegleiter (1659–1705), a prolific author of spiritual poetry, was one of them.[42] Wegleiter lived in England in 1686 and 1687 and moved in an elite circle of scholars, whom he linked back to Imhoff.[43]

Two other pioneering Nuremberg Anglophiles were perhaps even more important as facilitators of Imhoff's research on English genealogy: Christoph Arnold (1627–85) and his son Andreas (1656–94), who were also associated with the Blumenorden.[44] Christoph Arnold was a "darling" of erudite Germany and a key figure in the European republic of letters.[45] As early as the late 1640s, Christoph had established contacts with several leading scholars of the Netherlands and England, including John Milton. His son Andreas relied on his father's network for five years during his own travels in France, the Netherlands, and England after 1680.[46] Through the two Arnolds, Imhoff established direct correspondence with Elias Ashmole in Oxford, an expert genealogist.[47] Jakob Wilhelm's work on England, and indeed his broad interest in the country, clearly benefited from local tradition and Nuremberg's unique intellectual and cultural links across the Channel.

The Arnold case is also noteworthy, as it highlights how transregional communication often depended on local networks of patronage, which were inevitably grounded in local social relations. The Arnold family hailed from low origins. Although precise details are lacking, it seems that the Arnolds rose to prominence under the protection of several patrician families, including the Imhoffs. Andreas Arnold repeatedly wrote letters to Wilhelm Imhoff, the genealogist's father, while Jakob Wilhelm wrote in the younger Arnold's *Album amicorum*.[48] The cooperation of the Blumenorden in general was probably also at least in part a result of the Imhoff family's ongoing patronage. Again, Imhoff Jr. seems to have benefited from his father's sustained support of and interest in the activities of this association.[49]

Nuremberg's role as a center of patronage also influenced the flow of genealogical information in other cases. The city not only attracted newcomers and migrants, but it also sent many of its citizens out into the wider world, sometimes only for limited periods of time, sometimes for good. Few former citizens or residents of Nuremberg would have forgotten Imhoff's prominent background, and many may therefore have felt obliged to support his research, even from afar. Among many loyal Nuremberg informants are Johann Wilhelm Pfeiffer (1645–95), who had found employment in Curland and sent Imhoff genealogical material on his employers, the Wallenroth family,[50] and Tobias Hardesheim (or Herdesianus, d. 1681), who sent information about the counts of Isenburg-Büdingen, for whom he worked as a tutor.[51] Although the precise motives of Pfeiffer or Hardesheim are hard to discern, the opportunity to maintain a positive relationship with a high-profile citizen of their native city may have been an incentive.

Imhoff's Nuremberg-based, yet transregional patronage network also relied on the city university in nearby Altdorf. During his studies there, Imhoff forged personal connections with several other students, including Severin Walther Schlüter (1646–97), scion of an equally elite family from Hamburg (his father was mayor, his brother a prominent senator of the city). Schlüter, who joined Imhoff on the first leg of their grand tour through Europe, became superintendent in the small duchy of Saxony-Lauenburg in 1684. From there, he not only sent genealogical information himself (about northern Europe); more important, he also established contact on Imhoff's behalf with the Danish royal secretary and archivist Johannes Moth (1639–1705) in Copenhagen, who contributed to the project as well.[52] The cooperation of Nuremberg's academic community would have been reinforced by the fact that Imhoff's father had supervised the university for the city government. It may have been out of reverence for his father that several Altdorf professors helped Imhoff procure information.[53]

The support that Imhoff received from the city's merchant communities nicely rounds out the picture of the infrastructural and logistical opportunities that Nuremberg afforded scholars and writers. The city's far-flung European commercial networks were famous, and they facilitated not only the exchange of goods and money. On the contrary, the rich portfolio of local contacts that Nuremberg merchants had in numerous major European cities also translated into crucial support for cultural and intellectual pursuits, including Imhoff's genealogical research. On many occasions, for instance, Imhoff could count on the city's commercial contacts to arrange book transfers. Local merchants' excellent connections to Venice and the strong German community of the Fondaco dei Tedeschi in Venice itself proved to be highly efficient. Members of the Pommer family provided logistics, and the physician Johannes Böhm (ca. 1640–1731) helped Imhoff procure rare books in the city of St. Mark.[54] On other occasions, Imhoff leaned heavily on printers and booksellers to receive or forward messages and books. The vigorous native publishing industry of Nuremberg would have presumably helped him establish these contacts.

Overall, Imhoff's experience would have corroborated modern ideas about the status of Nuremberg and its academic satellite Altdorf as a major hub of information in central Europe. This was a very advantageous place for compiling knowledge. Cities like Nuremberg inevitably attracted and hosted a potent crowd of information mongers, many of whom Imhoff managed to involve in his projects. They also opened up a range of communication channels that connected the city to numerous other urban and courtly sites of knowledge. Clearly, Imhoff's genealogical research depended intrinsically on the infrastructure of his native city. His genealogies were, to a very high degree, an urban project.

The Life of a Patrician: Urban Elites and the Nobility

Although much of Nuremberg's cultural life and intellectual infrastructure would have been open and accessible to scholars and authors of many kinds, Imhoff's social position as a patrician enabled him to take full advantage of the city's abundant opportunities. Hailing from the tiny yet powerful and ambitious group of families at the top of Nuremberg society profoundly benefited Imhoff's prospects as a genealogist. His background, moreover, shaped his very perspective on what genealogy was and could be.

Nuremberg's Patrician Elite

Like most of the free imperial cities, Nuremberg claimed to be a "republic." Such cities prided themselves on the fact that they were not governed in monarchical fashion. In reality, however, most early modern city republics were ruled by closed oligarchies. In Nuremberg, this oligarchic elite consisted of no more than forty-three families, who formed the "patriciate."[55] Members of this urban elite, including the Imhoffs, controlled every political and administrative position of any significance. Through a complex apparatus of councils and offices, the patricians held power over the city and its territory. The actual governing body was the "college of septemviri," led by three senior members, including the Vorderer Losunger. The Vorderer Losunger was de facto the most powerful politician in town and constituted the apex of the city's administration.

Whereas Nuremberg's constitutional framework was conciliar in nature, a large number of executive offices assumed responsibility for particular areas of policy and administration, from justice to finance to infrastructure to education. There were, at one point in the early modern history of Nuremberg, about 120 specialized offices or standing committees dedicated to the various tasks of managing daily life in the city—all of them headed and staffed by a small number of patricians. The result was a bewildering accumulation of responsibilities and titles among politically active citizens. Since some of these administrative posts were often held for long periods or for life and were highly lucrative, this system was also economically important to the patriciate.[56] By maintaining this institutional framework, the patricians reinforced their social prestige and wielded far-reaching political influence both within the city and beyond throughout the early modern period.

The patricians had originally derived their power from economic success.[57] Most had been merchants and energetically participated in the commercial boom Europe witnessed around 1500. A dense network of long-distance trading relations connected Nuremberg and its merchant families to many regions in Europe and the early colonial world.[58] As the focus of European trading gradually shifted away from the Mediterranean in the later sixteenth century, however, the merchants of Nuremberg found it increasingly difficult to adapt. The patricians withdrew from commerce en masse during these decades. The Imhoffs were relatively late in closing their firm in 1635.[59] A new class of merchants took over local commerce, yet the patriciate retained its firm grip on all aspects of political power, shutting out the mercantile newcomers.[60] Although the new economic elite's dissatisfaction with patrician supremacy grew after about 1700, Nuremberg rarely saw outright social violence in the early modern period. In this respect, the Franconian city differed greatly from other urban centers, such as Frankfurt or Hamburg, which were repeatedly shaken by fierce domestic conflicts during these decades. Nuremberg's ancient constitution was not abandoned, and the oligarchic patriciate's monopoly on power did not topple until 1794. Imhoff lived and worked in a largely stable social environment of undisputed privilege.

A Well-Established Family

The broad stability enjoyed by Nuremberg patricians also applied to the Imhoff family, who had come to Nuremberg in 1340 from their ancestral home in Lauingen.[61] On July 4 of that year, Konrad Indemhof became a citizen; eleven years later, his stepbrother Hans (II) joined him. The stepbrothers quickly established themselves. Their widowed (step-)mother, Anna, remarried and became the wife of the first known ancestor of the Pirckheimer family, thus becoming the ancestress of another important patrician family. Hans II already held important offices in the city government. Trading in a wide variety of goods, including spices, precious metals, silk, and other luxury products, the family consolidated their growing wealth.[62] A number of important donations to the major churches are on record.

The family ascended quickly. In the early sixteenth century, the Imhoffs truly joined the ranks of the small number of elite families. By now, they counted among the richest of the rich.[63] The Imhoffs regularly sat on the city council and held some of the highest administrative offices. For most of the subsequent three centuries two family members sat on the executive council (Kleine Rat), the city's primary governing body, and four members of the family even held the supreme office of Vorderster Losunger in the early modern period. The family also continued to expand. An important branch of the dynasty relocated to Augsburg, only to divide further as the century wore on. Some of the Augsburg Imhoffs remained Catholic after the Reformation, whereas the Nuremberg lines—and the majority of other branches—converted to Protestantism. During the Reformation, Andreas I (or Endres) Imhoff (1491–1579), the first to serve as Vorderster Losunger, steered his city and family through the rough waters of rapid change. Andreas I is considered one of the most important members of the family.[64] He was regularly involved in the high politics of the Holy Roman Empire. He, his son, and grandson (all named Andreas) frequently hosted the emperor when he came to town.

Andreas I also initiated the family's gradual transition from trade to finance. One of his grandsons, Wilhelm I (1558–1630), great-grandfather of Jakob Wilhelm Imhoff, was the last Imhoff to manage the family firm personally. In 1635, five years after his death, the company went out of business. By Jakob Wilhelm's birth in 1651, the Imhoff merchant family had evolved into wealthy and influential cultural and political powerbrokers. The Imhoffs continued to invest heavily in the city government. Two more Imhoffs rose to the supreme office of Vorderster Losunger in the seventeenth century: Georg (1601–59; in office as Losunger from 1658) and Georg Paulus (1603–89; Losunger from 1675). Many more family members held other important administrative positions. One last Imhoff, Johann Christoph I (1659–1736), served as Vorderster Losunger in the eighteenth century. Although not every family member rose to such heights, many of them, including Jakob Wilhelm, reached positions of importance. His personal and family experiences would have taught the genealogist to look at the world from a confident position of power and influence.

This picture of safe eminence did not change even as a few family members attracted public scorn. After a brief career in Nuremberg's administration, Johann Hieronymus Imhoff II (1627–1705) converted to Catholicism and entered the service of several Catholic princes. With the spiteful fervor of a renegade, he wrote numerous polemical treatises against Nuremberg's Lutheran preachers. Although his aggressive stance against Nuremberg's official religion caused concern, Johann Hieronymus's behavior did not greatly affect the family's social standing.[65] His earlier accomplishments as city archivist seem to have trumped his newfound antagonism.[66] The family's public standing also emerged largely intact from the one major scandal that it faced in Jakob Wilhelm's lifetime. In 1672, the obscure Jeremias Imhoff IV (1630–72) was secretly executed for "thievery."[67] He had misappropriated large sums of money from the city's public pawnshop (Leihhaus), the finances of which he oversaw. While this affair led to a minor setback for the family in terms of its political influence, the Imhoffs by and large weathered the storm largely unscathed.[68] There are few signs that Jakob Wilhelm or his relatives felt that their social standing was threatened.

The Imhoffs also confidently shared their class's intense commitment to culture. By far the family's best-known collector and connoisseur of art was Willibald Imhoff (1519–80), a self-declared "lover of medals and antiquities."[69] Thanks to Willibald's efforts, the family owned—and was famous for owning—one of the most significant collections of Albrecht Dürer's works until 1633, when Willibald Imhoff's grandson Johann Hieronymus I began to sell most of it at a time of economic hardship.[70]

In addition to artworks, the Imhoffs owned a significant number of books. Easily the most important Imhoff library was that of Christoph Jakob (1654–1727), with whose obituary this chapter began.[71] He owned thousands of books and took special care to collect and preserve small and ephemeral printed works. His collection also included several autographs by Martin Luther, and it was no wonder that prominent scholars from afar contacted Christoph Jakob when they needed an expert opinion on Luther's handwriting.[72] Following this family tradition, albeit in comparatively minuscule format, Jakob Wilhelm also collected books. Between 1675 and roughly 1700, he amassed many international works on genealogy, around seventy of which still survive today, identified by his distinctive ex libris. In the 1720s, a manuscript catalog of his collection was produced and circulated regionally among Imhoff's friends and acquaintances.[73] Shortly before his death, he donated numerous genealogical works to the city library of Nuremberg. The donation must have consisted of enough titles to warrant designated storage: the "Imhoff bookcase" could be found in the "second aisle" of the library.[74]

Imhoff's ex libris. Wolfenbüttel, Herzog August Bibliothek, Berlepsch Exlibris T. I, p. 106, no. 102.

Some of the women of the Imhoff family also wielded significant cultural influence in Jakob Wilhelm's lifetime. Today Regina Clara Imhoff, perhaps the mother of Johann Hieronymus II, is prominent among them. She has a place in the history of music on account of her *Tabulaturbuch* of 1649.[75] Another Imhoff lady, likewise named Clara Regina (1664–1740), was one of a group of adolescent girls who gathered around the famous botanist and naturalist Maria Sibylla Merian (1647–1717) while the latter was living in Nuremberg between 1670 and 1682.[76] Merian also befriended other members of the Imhoff family, including Christoph Friedrich, a renowned "investigator of nature."[77] He owned a remarkable collection of naturalia, including stones and shells, that he displayed in "elegant" rooms to handpicked friends and visitors. We may imagine a well-stocked curiosity cabinet, containing a combination of natural objects and paintings, statues, and other antiques.[78] Jakob Wilhelm, the genealogist, thus grew up in a family of considerable cultural refinement. The arts, scholarship, and publishing were part of his direct personal and familial experiences.

Imhoff's patrician family not only instilled in him stable elite status, social pride, and wide-ranging cultural expertise; it also provided him with a versatile network of family contacts outside of Nuremberg. There were in fact so many influential Imhoffs in sundry places that it became necessary to maintain lists to keep track of living family members.[79] By 1500, the family had established bases across Europe and beyond. Frequent visits and prolonged stays by individual Imhoffs connected the family to France. One branch of the family even moved to Bari in southern Italy.[80] Numerous family members also worked in Lisbon around 1500, often in collaboration with other southern German merchants.[81] From Portugal, these German merchants quickly reached the newly discovered world outside Europe.[82] In subsequent generations, several Imhoffs had successful careers overseas. Gustaaf Willem (1705–50), who became governor of Dutch Batavia on behalf of the Dutch East India Company, hailed from a reformed branch of the family located in Emden. Toward the end of the eighteenth century, moreover, Christoph Adam Imhoff (1734–88) made his way to British India in the company of Warren Hastings (1732–1818).[83] By the mid-eighteenth century, the Imhoff family had become intimately involved in colonial power and global expansion.

It was no wonder, then, that the patrician families of Nuremberg also functioned as conduits of knowledge. While they still engaged in trade, these families hoarded knowledge about goods, markets, moneys, and trading opportunities, and Jakob Wilhelm Imhoff could still rely on the family network even after the Imhoffs' withdrawal from commerce.[84] He repeatedly contacted several family members when working on his genealogies, depending on them to procure and forward valuable information. His distant cousin Andreas Lazarus (1656–1704) of Sulzbach, who helped procure material from Belgium, was especially important to Imhoff's work.[85] Another relative, Friedrich Wilhelm Imhoff (1671–1717), served as Jakob Wilhelm's man on the ground in

Florence and elsewhere in Italy during his travels in the early 1690s.[86] Other members of the patriciate, often related to the Imhoffs by marriage, also helped out when Jakob Wilhelm needed information. Members of the Volkamer, Peller, and Wurfbain families are all on record as supporting Imhoff by procuring and forwarding important genealogical information.[87] The Imhoff family's long-standing tradition of networking and information-mongering benefited Jakob Wilhelm's projects tremendously.

Aspiring to Noble Status and Connecting to Courtly Culture

While Imhoff and his peers did not face serious threats to their social status from below, they would have been very much aware of the patriciate's own ambitions of upward social mobility. Roughly mirroring developments elsewhere, the patriciate of Nuremberg clearly aspired to noble status. One path to this objective was ennoblement by the Holy Roman emperor. Admission to the imperial nobility conferred titles without feudal territory. Among the Imhoffs, at least one (Catholic) branch acquired the rank of "imperial baron" (Reichsfreiherr).[88]

An alternative way for patricians to obtain noble status was to claim that they had been "old" nobility all along, a strategy that was adopted to integrate individual patrician families into the Franconian imperial knights, a heterogeneous group of old but minor noble families. Nuremberg patricians began to claim noble origins for themselves and their peers by the early sixteenth century.[89] Citing such early self-aggrandizement, one Nuremberg patrician, Philipp Jakob Haller von Hallerstein, even defended his peers' aspirations to nobility in an academic publication printed in 1684.[90] A few years later, in 1696, the Nuremberg patricians finally secured an important decree from the emperor that acknowledged the patriciate as "old" nobility. Taken together with the ownership of noble estates in Franconia, this decree translated into official recognition of the noble status of at least some patrician families, including some Imhoffs, by the Franconian nobility.[91]

Other Nuremberg patricians moved out of the city and attached themselves to one or more of the many nearby princely courts. In the seventeenth century, the former mercantile elite began to transform into courtiers, whose numbers again included several Imhoffs.[92] The forementioned Andreas Lazarus Imhoff became chancellor at the small nearby court of Sulzbach. Baron Johann Joseph von Imhoff zu Untermeitingen (1667–1738) had a career in the service of various Catholic potentates, most important the emperor himself.[93] When the emperor rewarded the English general John Churchill (1650–1722),

Duke of Marlborough, after the battle of Blenheim by naming him an imperial prince and granting him the newly created fiefdom of Mindelheim, Marlborough in turn chose Johann Joseph von Imhoff to be his local lieutenant. For ten years, while the English duke owned Mindelheim, Baron Imhoff served Marlborough as the chief administrator of this small southern German territory. The career of yet another distant cousin, Rudolf Christian von Imhoff (1660–1717), was even more distinguished. He served the dukes of Wolfenbuttel as a diplomat in Barcelona and Vienna, helping to coordinate the duchy's international policy.

The patriciate's focus on courtly and noble honors became even more evident in 1749, when the city fathers decreed that all publications about the patricians had to list their "civil and military charges" and "dignities at the courts of the emperor, kings, electors, and princes."[94] It had become natural for patricians to look to the nobility and their own courtly lives for social prestige and identity. Although Jakob Wilhelm Imhoff always claimed that he was "a bad courtier" and never showed much enthusiasm for his family's noble aspirations, he clearly shared his class's fascination with the aristocracy.[95] Even though he did not cultivate obvious noble aspirations, he retained his clan's habitual feeling of parity with at least the lesser nobility. From his perspective, work on noble genealogies was a genuine extension of his own social status.

Patricians and Genealogical Research

Life as a Nuremberg patrician not only shaped Imhoff's general social and cultural views, but also profoundly influenced his specific interest in history and genealogy. Notwithstanding a few statements to the contrary by prominent contemporaries, Nuremberg was a city teeming with historiographical activity.[96] Like most major European cities, including the imperial free cities of the Holy Roman Empire, Nuremberg was home to a lively tradition of urban historiography, harking back to the late Middle Ages.[97] The patriciate, including several members of the Imhoff family, was actively involved in commissioning and updating works of urban historiography.[98] To them, the history of the city was intertwined in numerous ways with their own family history. Private and public, patrician and civic memory intersected.

In addition to fashioning a public image and identity for the city as a whole, patrician historical culture, including genealogy, also served prominently to create and express familial reputations. In ways reminiscent of the nobility's understanding of auto-genealogy as a tool of legitimating self-definition, Nuremberg patricians also turned to genealogy as a means of touting their social and cultural excellence. A rich tradition of genealogical research and family history in both text and image emerged in Nuremberg and other free imperial cities. Genealogists equipped patrician elites with spectacular, imposing pedigrees. In the process, genealogy also helped the patricians to clarify the contours of their kinship and to establish internal structures and divisions within their families. Not unlike the nobility itself, patrician dynasties slowly but surely transformed into complex lineages. As time progressed, many patrician lineages were proudly remembered as multigenerational conglomerates of individual households and semiautonomous branches, consisting of thousands of individuals, both long dead and still living. Genealogy played a key part in conceptualizing and memorializing the social structure of urban elite families.[99]

Patrician Genealogy

The most important manifestation of patrician genealogical culture was the "Geschlechterbücher" (genealogical books) that most Nuremberg patrician families produced, especially in the early sixteenth century and beyond. These were elaborate books in manuscript form that presented genealogical and archival information about the family in a variety of formats and styles. They were often lavishly illustrated and obviously intended to impress.[100] Other materials, ranging from notes to drafts, letters, and drawings often accompanied and complemented these central works, illustrating the demanding and complex process of creating a coherent familial memory. Commemorative books ("Gedenkbücher," "Memorial") merged personal reflections, genealogy, family history, the history of Nuremberg, and a broad range of other historical information into a single work, thus anchoring both the author and his family in local and general history.[101] Some of these manuscripts were in continuous use within these families and occasionally even passed from one family to another.[102]

As they followed broader trends in early modern historical culture, patricians fabricated historical narratives to support their social and political agenda. They narrated fantastical stories about their families and often rewrote history to serve their purpose. Johann Bartholomäus Imhoff, for instance, a contemporary of Jakob Wilhelm, told an origin story of his family that presented their contemporary claims to nobility as historical fact. The family's medieval translocation from Lauingen to Nuremberg, he claimed, was not the attempt of a merchant family to enlarge its trading network, but now was supposedly a typical example of an old noble dynasty moving from the countryside to a blossoming urban center.[103] By no means did the patriciate's critical approach necessarily intensify with the passage of time. Despite the prominence of humanistic culture in Nuremberg and despite the fact that many patricians openly endorsed, and personally practiced, critical philological methods, patricians nevertheless instrumentalized these new tools to create ideologically attractive versions of their history. Fantastic constructs of urban and family history remained popular well into the eighteenth century.[104]

The Imhoffs, too, had a rich tradition of compiling documentary evidence for their family history.[105] Extensive manuscripts from the early eighteenth century not only included elaborate family trees but also presented lists of family officeholders, highlighting the dynasty's political importance and social prestige.[106] Several individuals engaged in projects of family history, including the most important Imhoff practitioner of family genealogy, the bibliophile Christoph Jakob. In 1700, he became the city's chief archivist and held an advantageous position to study local history.[107] Christoph Jakob actively researched his ancestors and left behind large quantities of working papers. In addition to his own copious library of genealogical literature, he rummaged through various archives and seems to have traveled widely through southern Germany in search of evidence.[108] Family members turned to him when they needed genealogical information. Jakob Wilhelm also benefited from Christoph Jakob's expertise.[109]

Not only family members, however, studied the history and genealogy of the Imhoff family. External authors also produced panegyrical works on occasion. The Nuremberg advocate Leonhard Wurffbain (1581–1654), for instance, wrote a brief work about the family in 1637, on the occasion of the death of Andreas Imhoff III, outlining the origins of the dynasty and highlighting the high-ranking government posts that numerous family members had held.[110] Wurffbain was probably close to the family; in all likelihood, he obtained most of his material directly from them. An equally isolated and even more celebratory pamphlet about the family appeared fourteen years later, in 1651, most likely penned by Johann Georg Ulmer, an otherwise little-known law student in Altdorf.[111] In propagandistic fashion, the author presented the Imhoffs as living incarnations of the civic marriage of arms and letters, military prowess and scholarly responsibility.

From both his own and the everyday practice of his peers' families, Jakob Wilhelm Imhoff thus would have casually learned many ways to explore and celebrate family history. The Nuremberg patriciate combined scholarly approaches, including complex archival research, with hastily produced panegyrics and ideologically driven myths. A similarly multifaceted approach to family history would later shape Imhoff's own genealogical vision. Although he clearly counted himself among the new, scholarly breed of genealogists, he nevertheless countenanced the tradition of identity-building through auto-genealogy, sharing his class's predilection for celebratory historiography. It would not have been too difficult for him to relate to the propagandistic designs that his noble subjects pursued through genealogy.

Imhoff's Father

While Jakob Wilhelm's intellectual trajectory owed much to his class and wider family, the strongest direct influence on his genealogical work came from his father. Many of the cultural and political attitudes of his class converged in Wilhelm Imhoff III (1622–90). He seems to have enjoyed great esteem among his peers, not least because of an unusual dose of personal piety.[112] In a time when rhetorical praise flowed from every mouth, the eulogies he received were exceptionally enthusiastic. Wilhelm was described unanimously as a highly educated man interested in several fields of research, yet so exclusively committed to public service that he never published anything (nor, probably, intended to publish). As a young adult, starting in 1642, he set out on an extensive three-year journey across Europe, visiting the western territories of the Holy Roman Empire (Hanau, Frankfurt), the Netherlands and Belgium (Leiden, Antwerp, Mechelen, Brussels), England (London), and several parts of France (Paris, Lyon) before coming home via Switzerland.

Wilhelm III returned with a burning desire to expand his knowledge of geography, history, and—especially—genealogy.[113] This triad of disciplines often figured prominently when writers highlighted the practical appeal of humanism. Statesmen, or so it was thought, required *useful* forms of humanistic learning in order to steer the ship of state, and these three fields served as key examples of such learning. Contemporaries insisted that, by embracing history and genealogy, Wilhelm Imhoff completed his maturation from a young scholar to a learned politician. His learning had become practical and thus could be called "prudence" (*prudentia*).[114] In fact, Imhoff Sr. had merely put into practice what Nicolaus Rittershausen, the genealogist mentioned in chapter 1, taught at the University of Altdorf: the concept that genealogy could serve as a practical tool for politicians in assessing contemporary affairs.

The politician's pragmatic fascination with and the academic's analytical interest in genealogy eventually merged as Imhoff Sr. and Rittershausen began to collaborate intensively. Although few details are known, there is no doubt about Wilhelm Imhoff's crucial contribution to the production of the third, definitive edition of the Altdorf professor's *Genealogiae*.[115] Imhoff Sr. was also deeply committed to preserving and promoting the legacy of Rittershausen after his death. Although Jakob Wilhelm Imhoff never connected genealogy as closely to political prudence and administrative acumen, he nevertheless professed his admiration for his father's fusion of genealogical learning and political versatility.[116] Even more important, as we will see later, he inherited his father's commitment to continuing Nicolaus Rittershausen's vast projects.

Nuremberg emerges from these reflections as a major place of knowledge production circa 1700. Like other urban centers of its time, Nuremberg boasted exceptional cultural infrastructure, making it a very attractive location to anybody involved in the creation and dissemination of knowledge. Although Imhoff himself hardly ever talks about Nuremberg as a privileged place of knowledge, he nevertheless took full advantage of the city's amenities. He connected with numerous visitors who passed through the city and extracted valuable information from them. Most of his publications appeared with leading Nuremberg printers, honoring the city's prominent publishing business. Nuremberg was, moreover, full of books, home to many rare tomes that were hard to find elsewhere. Finally, the city's academic, political, and economic networks gave Imhoff valuable access to other sites of knowledge in Germany and beyond, helping to integrate most of Europe, from the Baltic to southern Italy, from England to Bohemia, in his genealogical information circuit.

The urban elite, Imhoff's own peer group, contributed heavily to making Nuremberg such an important place of knowledge. Being a patrician in Nuremberg entailed conspicuous engagement with scholarship, art, or literature. Even though a few observers in the mid-eighteenth century occasionally claimed that "our patricians . . . no longer value erudition at all," this was certainly not true of Imhoff's generation.[117] The city teemed with patricians interested in the arts and sciences. Jakob Wilhelm internalized an easy familiarity with scholarly activities from his relatives and peers that extended to writing. Nuremberg patricians were "literary families," comparable in many ways to the dynasties of French writers studied recently by Neil Kenny and Oded Rabinovitch.[118] Imhoff acquired the epistemic habits, work routines, and scholarly practices necessary for his career as a genealogist directly from his peers and relatives.[119] Research, writing, and publishing were familiar activities to a patrician like Imhoff.

Cities emerge in this chapter as fertile ground for genealogical scholarship, in particular for encyclopedic genealogy. There were strong local traditions of genealogy, especially those connected to patrician families. It would have been hard even for the most casual visitor to the proud, republican cities of Europe to miss the enormous importance of history for civic identity. The Imhoffs and their peers were deeply engaged in curating their public image, and they used historiography, including genealogy, very consciously as a means of self-fashioning. The urban context of Nuremberg would have provided any genealogist with a rich supply of role models and paradigms for his work.

Local patrician genealogy shared many features with its much better-known cousin, noble genealogy. Both social groups used family history in roughly comparable ways to express who they were and what they aspired to be. Despite the fact that urban discourses arguably had a soft spot for meritocratic virtue-over-pedigree arguments, patrician families nevertheless developed a dynastic consciousness that resembled that of the nobility.[120] Like nobles, patricians relied on genealogy to secure their social legitimacy. By Imhoff's time, this approach typically combined a willingness to narrate grandiose, albeit empirically dubious origin stories and an outspoken commitment to critical antiquarian scholarship. Patrician families engaged in lineage building just as the nobility did, and they used similar cultural tools to do so, including genealogy.

Nevertheless, genealogy as done in the city had certain idiosyncrasies. Imhoff's urban background gave him an outsider's perspective on the nobility. Despite the patricians' conspicuous self-assimilation to the (lower) nobility, urban leaders remained a very different kind of elite. Cities were not courts, and the noble houses of Europe, especially those of sovereign princes, remained a world apart. An outside perspective on the nobility characterized Imhoff's project. For better or worse, genealogists in urban settings need not work *for* the nobility, but could write *about* the nobility. In Imhoff's project of encyclopedic genealogy, this translated into a potentially open-ended collection of material on *all* noble families, paving the way toward a collective understanding of the nobility.

Thus, in many ways, Imhoff's elite urban background shaped his unique perspective on genealogy. There was, however, a potential downside. The civic and patrician framework could also be stifling. Imhoff sensed—and deplored—a certain amount of parochialism in this mindset. To some contemporaries, autonomous free imperial cities looked increasingly like quaint relics of a bygone era, comfortably limited to local politics at best. Imhoff once complained that Nuremberg's civic republican ideology endorsed a sentiment he attributes to Cicero, namely, that "no one should be too curious about foreign peoples and states." Imhoff begged to disagree and indeed shifted his own focus to neighboring countries.[121]

Although he was profoundly influenced by Nuremberg's intellectual culture and knowledge economy, Imhoff was not a local historian. His grand tour had familiarized him with the European world of courts, a world to which he did not naturally belong but one that he found both vital and fascinating. He concluded that the nobility warranted a transregional approach. Imhoff's projects thus extended far beyond his home and avoided the danger of civic parochialism. His project of encyclopedic genealogy thus combined a patrician's self-confident social outlook and local background with a much broader European perspective. In Imhoff's work, the formative influence of the urban cultural traditions of Nuremberg and intimate familiarity with courtly culture merged.

Genealogy and the Nobility

Despite its republican traditions, the free imperial city of Nuremberg paradoxically also granted Jakob Wilhelm Imhoff access to the nobility of the Holy Roman Empire. Numerous noblemen visited or resided in the city, and patricians like Imhoff would have encountered them on a regular basis. For instance, Nuremberg was home to a group of Protestant nobles from Austria and Bohemia, who had fled their homeland after Emperor Ferdinand II issued the militantly pro-Catholic Edict of Restitution of 1629.[1] By Imhoff's time, these erstwhile refugees were well integrated and now entrenched in the political, economic, and cultural life of Nuremberg and Franconia. Among them were members of the von Stubenberg family, who also socialized with the Blumenorden (although they did not become members).[2] One young scion of the family, Count Georg Augustin (1628–92), an erudite collector of books, prints, and manuscripts, used his connections to other noble families to procure information about several Bavarian and Austrian families for Imhoff.[3] Count Gotthard Helfried von Welz (1654–1724), Imhoff's "long-time patron," also supported his genealogical projects.[4]

In some ways, these two noblemen represented only a minority within their class. By no means did all aristocrats share their enthusiasm for genealogy. Nor should we overestimate the genealogical knowledge of individual noblemen. Imhoff, in fact, encountered numerous nobles who knew little about their own families and cared even less. Notwithstanding such individual blind spots (widespread as they may have been), however, there can be little doubt about the fundamental connection between the early modern aristocracy and genealogy. Narratives of pedigree and descent were fundamental to the nobility's self-definition. Genealogy helped give noble families a social profile and was instrumental in legitimizing their social and political influence. Accordingly, early modern noble houses engaged frequently in sophisticated campaigns to promote, legitimate, and celebrate preferred visions of their family histories. As a major instrument of such aristocratic propaganda and self-fashioning, genealogy was very much present at the many noble courts of early modern Europe. From the minor nobility to princely houses, aristocrats relied heavily on genealogy to articulate their identity.

The paramount importance of genealogy ensured that noble families usually closely monitored relevant information and tried to control the public circulation of genealogical knowledge as much as possible. Generally, noble families remained the prime source of knowledge about themselves. Family-related information was largely kept (and still is) within the family. Although, as we have seen, an increasing amount of material on princely and noble genealogy circulated publicly in different genres of print and manuscript literature, it would have been impossible to write competently about genealogy without tapping noble reservoirs of knowledge. In short: in spite of, or perhaps rather, because of the fact that noble families curated their own bodies of genealogical knowledge so passionately, it was of the utmost importance to cooperate closely with them whenever one launched a new genealogical project. For all intents and purposes, establishing a positive working relationship with the nobility was virtually indispensable to any serious genealogist.

Imhoff was a master at this. He adroitly won over aristocrats of various ranks, drawing them into complex negotiations about information exchanges and publishable genealogies. His ability to engage the nobilities of Europe was one of the most defining features of his work. This chapter analyzes how Imhoff managed to collaborate so efficiently and amicably with many nobles. It highlights the modus operandi by which the genealogist approached noble courts and studies the cultural basis of their intensive collaboration with Imhoff. Although this chapter does not scrutinize individual noble genealogical projects as such, the following sections demonstrate how nobles pursued their own genealogical agenda by cooperating with Imhoff. It becomes clear that many families hoped to harness Imhoff's initiatives for their own purposes. Their cooperation with the Nuremberg genealogist was neither selfless nor dedicated primarily to historical truth per se.

Imhoff's peculiar approach to genealogy, originating outside of the world of courts, made his project especially attractive for many nobles, and for several reasons. On the one hand, encyclopedism lent genealogical narratives a certain air of impartial authority. The fact that their genealogies were included in such reference works, so nobles hoped, would stamp them with an authoritative seal of approval. In the case of the German nobility, this is especially obvious in the case of the "imperial counts," a group of intermediary imperial nobles who frequently felt that their social status was threatened. To them, Imhoff's *Notitia Procerum* resembled something like an independent stamp of approval. Moreover, this opportunity came in a surprisingly cost-efficient form. While some noble houses were certainly willing and able to devise and finance extravagant genealogical projects of their own, they nevertheless appreciated the fact that the new media on the genealogical information market, including universal genealogies, could be powerful and economical surrogates for more expensive auto-genealogical undertakings.[5] Hence, they willingly incorporated projects like Imhoff's in their portfolio of self-promoting resources. Their openness to and occasionally even their enthusiasm for Imhoff's projects clearly indicate how quickly nobles across Europe learned to appreciate the benefits of the emerging genealogical information market. The fact that Imhoff came from a free imperial city may have added to the appeal of his works to many European houses, since many nobles in both Germany and Italy were traditionally outspoken proponents of the concept and structures of the Holy Roman Empire.

Many noble families readily supported Imhoff in light of these considerations. In the first part of this chapter, I explore how this cooperation played out in Italy and Germany, the two best-documented regions. In the next section, I discuss in detail how Imhoff's books promoted the genealogical self-perception of the nobility itself. His works often express key aspects of aristocratic genealogical identity and thus supported the internal management of lineage politics by individual families. For all these reasons, numerous aristocratic families considered Imhoff's books valuable platforms for presenting and promoting their lineages.

Cooperating with an encyclopedic genealogist like Imhoff, however, presented not only opportunities but also unprecedented risks. Insofar as Imhoff wrote from a secure and independent urban—that is, hetero-genealogical—standpoint, and insofar as he primarily followed the genealogical publication culture of the emerging information market, his willingness to condone noble wishes and behavior had its limits. Cases from Germany and Italy show that the relationship between Imhoff and his noble contacts was rife with complex ambivalences. Although most nobles constantly indicated their social superiority in their correspondence with Imhoff, the genealogist sometimes proved difficult to control. Even as most nobles couched their cooperation in terms of patronage, day-to-day events often demonstrate that Imhoff was capable of placing his noble correspondents under significant pressure of his own. His personal situation as a "private person" living in a free imperial city granted him invaluable leeway.[6] As a Nuremberg patrician, he was not *someone's* genealogist in particular but could rather be *anyone's* genealogist. Encyclopedic genealogy, as a genre and media format, dovetailed nicely with his position of independence.

Another potential side effect of Imhoff's encyclopedic approach was that his books introduced and promoted an outside perspective on the nobility. It might now appear to readers as a coherent social class defined by specific customs and practices. Books like Imhoff's contributed in unintended and unanticipated ways to the "creation" of *the* nobility as a unified social body, consisting of numerous individuals who all adopted roughly similar modes of behavior. While the potentially subversive implications of this perspective were far from obvious circa 1700, its inherent possibilities nevertheless deserve a closer look. I discuss the new gaze on the nobility afforded by Imhoff's project at length in the final section of this chapter.

Collaborating with Nobles

Imhoff's impressive knowledge of the aristocracy stemmed to a large extent directly from the nobles themselves. His ability to tap into internal dynastic knowledge, as one-sided and ideologically biased as it may have been, was one of his greatest achievements. Numerous families actively supplied valuable information to Imhoff, opening their caches of otherwise strictly confidential information. Letters and dossiers reached him from small manor houses and large courts across Europe. Many archival records extant today still document how carefully and extensively princely and noble administrations engaged with Imhoff's works.

There were varying degrees of cooperation between the nobles and Imhoff. Sometimes the exchange of information was swift and unproblematic, while on other occasions difficult negotiations were necessary in order to establish profitable collaboration. Some nobles satisfied Imhoff's thirst for information proactively and to his full satisfaction. A few even eagerly anticipated new editions of his works, longing for a new chance to promote their ideas with Imhoff. Several actively tracked Imhoff's upcoming projects.[7] Others, in contrast, were more passive and gave their support only after the genealogist himself had asked them for help. Yet no matter how eager or reluctant these aristocrats may have been in cooperating with Imhoff, cooperate they did. There is no example of a family that explicitly refused to assist him. Even in difficult cases, Imhoff succeeded in eliciting at least some information from the European aristocracy.

Forms of Collaboration: Cases from Germany

Arguably the most basic form of cooperation occurred when families corrected errors of fact in Imhoff's books or updated his information by forwarding news about recent dynastic developments, such as marriages or deaths.[8] The counts of Solms provided detailed information about their sundry offspring, because, in the second edition of the *Notitia Procerum* of 1683, Imhoff had mentioned the family's children only summarily.[9] Friedrich Ernst zu Solms-Laubach (1671–1723), president of the Imperial Chamber Court (Reichskammergericht), remained committed to updating the *Notitia Procerum* well after 1700.[10] The high quality of Imhoff's information and his impressive ability to stay abreast of contemporary data derived in part from the ongoing involvement of many noble families in his research.

Other families engaged much more intensely with the genealogist, seeking to impose their genealogical vision on him. In the best case, this resulted in reciprocal cooperation. In 1693, for instance, Ferdinand Gobert (1645–1708), Count of Aspremont-Lynden, publicly acknowledged Imhoff's positive impact on the self-image of his house. He cited Imhoff's book on French genealogy of 1687 (*Excellentium Familiarum in Gallia Genealogiae*) as an important source for the family's most recent auto-genealogical publication.[11] In recognition of the genealogist's contribution, the count forwarded a deluxe copy of this new book to Nuremberg.[12] The copy seems to have arrived just in time for Imhoff to make use of it in finalizing the 1693 edition of the *Notitia Procerum*.[13] The relevant chapter (IX,8) was significantly longer than in the previous edition, and Imhoff explicitly cited the count's gift as a major source. A brief and fragmentary family tree was lifted directly from the original.[14] Sometime between 1693 and 1699, however, doubts regarding this courtly genealogy seem to have plagued Imhoff. As he reported in the fourth edition of the *Notitia Procerum* (1699), though, he eventually verified the count's published version by reviewing some original charters.[15] Given the fact, however, that the Aspremont genealogy was fictitious, Imhoff, in reality, had fallen prey to a vicious circle of corroborating citations: Aspremont-Lynden had used Imhoff, who in turn cited and eventually authenticated the count's printed, yet erroneous genealogy.

On other occasions, cooperation was a little more complicated. The

Seinsheim-Schwarzenbergs, from nearby Franconia, provide a telling example. In 1696, Count Ferdinand Wilhelm Euseb von Schwarzenberg (1652–1703) approached Imhoff. He had learned about a forthcoming new edition of the *Notitia Procerum* (1699).[16] The count had Imhoff's previous version of the family history checked against internal documents in the family archive and found it satisfying overall.[17] Nevertheless, Ferdinand Wilhelm Euseb thought that "further information" (*pleniorem informationem*) might still be in order. His staff would contact Imhoff and present him with a range of additional details for inclusion in the future edition. The count had a trusted Nuremberg intermediary deliver a summary of relevant archival documents drafted by his counselors.[18] Imhoff took the intervention gracefully and promised to "observe" this additional information in the future.[19] He indeed carefully archived the count's dossier in his files.[20]

Other nobles engaged with Imhoff's work by editing "their" chapters word by word, occasionally sending their revisions back to Nuremberg, in the hope that Imhoff might completely rewrite these sections. In Aurich, eastern Frisia, circa 1684, Chancellor Johann Heinrich Stamler (1632–92), went through a manuscript copy of relevant passages from Imhoff with quill in hand, annotating the text.[21] Likewise in Hanau, a very detailed rewriting of Imhoff's passages on the local counts was produced circa 1690, in which the counts indicated what they expected "their" chapter to contain.[22] As in Aurich, these alterations are documented in a series of interconnected drafts and clear copies, suggesting at least some sort of internal administrative deliberation about the *Notitia Procerum*. Some nobles went so far as to suggest precisely where in his work Imhoff should discuss certain information.[23]

Sometimes these responses to Imhoff's work betray specific ideological or political concerns. Some information packages came with an explicit caveat. The Metternich family, for instance, provided Imhoff with valuable information, but also set specific rules as to how Imhoff could use it.[24] And in 1732, when news of a posthumous fifth edition reached the ducal court at Stuttgart, government officials immediately took note.[25] The counselors found Imhoff's (respectively, Köhler's) account of the house essentially unobjectionable, yet they nonetheless requested a review of the section on Count Leopold Eberhard of Mömpelgard (1670–1723). This infamous member of a junior branch was notorious for his extravagant private life, which culminated in extramarital affairs, divorce, and a juicy mésalliance.[26] He failed to produce a legitimate heir, however, and the Stuttgart branch claimed Leopold Eberhard's territory after his death, but the matter remained unresolved for decades. Imhoff's depiction of the episode was thus a highly sensitive issue. The ducal counselors understandably turned to Johann Jakob Moser (1701–85), professor of law at Tübingen University, to scrutinize Imhoff's account, but he found nothing objectionable.[27]

Moser's positive evaluation of Imhoff's narrative is consistent with the genealogist's typically adroit handling of even difficult situations. Very rarely was Imhoff accused of propagating an openly tendentious narrative. One such case occurred in 1724, concerning Franconia, Imhoff's own *patria.* Several smaller territories in Franconia belonged to junior branches of the house of Hohenzollern, who reigned in Berlin. Neighboring powers in Franconia, including the imperial free city of Nuremberg, had always been anxious about Berlin's local influence. These anxieties took on newfound urgency, when, beginning in 1706, the Brandenburg-Bayreuth branch of the Hohenzollern faced extinction and the mighty king of Prussia, Frederick I (1657–1713), sought to secure the inheritance for the main branch.[28] In this context in 1724, the pro-Prussian government of Brandenburg-Ansbach opposed Imhoff's allegedly pro-Nuremberg, anti-Brandenburg narrative in the *Notitia Procerum.*[29] They complained that Imhoff's account "leaned largely in favor of Nuremberg," which they found utterly unacceptable. Imhoff allegedly represented the Hohenzollern genealogy in a perfunctory and confused manner ("kurz und confus"). The genealogist was also accused of excessively criticizing the reign of Albrecht Achilles, a major fifteenth-century Hohenzollern figure in Franconia. From a pro-Prussian perspective, Imhoff's presentation was fundamentally flawed, and the heated comments in the Hohenzollerns' internal documents demonstrate how carefully they registered and discussed books concerning them that appeared on the genealogical information market.

Collaboration with Southern Italy

Imhoff's project also had significant appeal in (southern) Italy. This may have derived at least in part from his obvious commitment to the Holy Roman Empire, its cultural traditions, and its political framework. Harking back to the complex medieval history of the Italian peninsula, replete as it was with the interference of numerous German emperors, a large number of Italian territories and noble titles were still formally linked to Vienna in Imhoff's time. Several Italian princes, including members of the Cybo and Borromeo families, owed parts of their territories at least nominally to the empire.[30] Toward the end of the seventeenth century, as the Spanish succession crisis was heating up, Italian ties to the imperial court actually intensified. A growing number of Italian nobles gladly accepted imperial offices and titles, which they viewed as additional safeguards in increasingly conflict-ridden times.

In consequence, many of Imhoff's Italian correspondents proudly incorporated imperial references into their public image. Andrea Giuseppe Gittio, for instance, one of Imhoff's most important contacts in Naples, had worked in Vienna, served as imperial counselor, and become an imperial baron.[31] Tommaso, prince of Aquino and Castiglione, paraded his status as a "prince of the Holy Roman Empire," highlighting his connections to Vienna and Germany.[32] Numerous other Italian princes, such as the house of Cybo, featured the imperial eagle in their coat of arms.[33] For these men and their peers, the pro-imperial attitude generally associated with free imperial cities would have made Imhoff, the Nuremberg patrician, an ideal partner.

The Italian nobles' motives to cooperate with Imhoff thus were no less pragmatic than those of their German peers: Italian families and their patriarchs exploited Imhoff's need for information to promote their own public image. His books were "not only like a loud trumpet, but also like a clear and true attestation in praise of the prince."[34] Neapolitan patriarchs gave information to the genealogist "so that you [sc., Imhoff], with your most erudite pen, can give it [sc., the family] greater splendor."[35] Helping Imhoff was an opportunity for self-promotion.[36]

Many Italian nobles expressed their wishes very candidly. On several occasions, for instance, Prince Antonio de Giudice Cellamare (1657–1733)—through Girolamo Maria di Sant'Anna as intermediary—sent Imhoff material about his and other Neapolitan families. In return, he let the genealogist know that he wanted to receive "explicit mention."[37] The family also used Imhoff (always through Sant'Anna's letters) to promote remembrance of individual family members. His letters to Nuremberg, for instance, drew attention to the prince's brother Michele, who was killed in action fighting for the Spanish in Catalonia in 1697. Imhoff duly printed a brief biographical note about him in 1710.[38] Furthermore, one of Antonio's sons let Sant'Anna know about the recent promotion of his uncle Cardinal Francisco de Giudice (1647–1725, Antonio's brother) in the service of the Spanish crown, "so that you [sc., Imhoff] may mention it."[39]

The Giudice prince was but one of a larger group of well-educated and culturally engaged Neapolitan princes around 1700 who cooperated eagerly with Imhoff. Giudice's peer and contemporary, Carmine Niccolò (1671–1726), prince of Santobuono, was equally important for the gestation of Imhoff's works.[40] He likewise came from a very old family, the Caracciolo, and later became famous as the Spanish viceroy in Peru. As a young man, Santobuono was deeply engaged in literature and scholarship. He wrote poetry and forged connections to many Neapolitan academies.[41] He was also dedicated to the study of genealogy. He forwarded relevant material to Imhoff, again relying on Sant'Anna as an intermediary. The prince himself supposedly translated some Spanish manuscripts into Italian for Imhoff and produced a magnificent large-scale genealogical table of his family.[42] Imhoff depended on Santobuono's dossiers not only for his own Caracciolo family tree, but also for his accounts of other southern Italian families. In particular, Santobuono helped Imhoff learn more about the parents of Saint Thomas Aquinas, whose maternal family relations were a matter of considerable debate at the time. Saints were a welcome addition to any family tree; hence, the question of St. Thomas's mother was of considerable genealogical significance.[43] Carmine Niccolò suggested with new evidence (and probably correctly) that the great theologian's mother actually hailed from his own family, the Caracciolo. Imhoff adopted this flattering version and printed it with an explicit reference to the prince, thus directly promoting the Caracciolo's new claim to fame.[44]

The Italian princes approached Imhoff as courtly patrons even more explicitly than their German peers. On the one hand, they framed their contributions and requests in the courtly language of civility. Imhoff would do the Neapolitan nobility an "honor" out of "courtesy," if he agreed to consider their wishes and ideas. On the other hand, they insisted—more forcefully than many German nobles—on his public gratitude. Sant'Anna, Imhoff's man on the ground in Naples, insisted that Imhoff thank his sources, for instance by sending personalized thank-you notes. The genealogist was instructed to mention that "the information" provided by Antonio de Giudice "pleased me very much" and that he was "obliged" for the princely favor.[45] On another occasion, Sant'Anna insisted that Imhoff notify him of the receipt of genealogical information and to do so in "such fashion that I can show your letter to the Prince."[46] In such cases, it was advisable that the genealogist comply. Sant'Anna often spelled out the obvious for Imhoff: "I will confess the truth to you: the prince merits all deference and attention, both on account of his various qualities and talents and on account of his high birth and parentage, insofar as his relatives and kinsmen include the duke of Modena, the duke of Parma, and the duke of Mirandola, whose daughter was his wife"; hence Imhoff should "accommodate the prince's feelings."[47] Such advice (or rather implicit threatening) was not in vain. In most cases, Imhoff accepted the stipulations of ceremonial decorum. He appreciated the fact that willingness to play by the nobility's communicative rules was key to his success. On the basis of this understanding, the princes cooperated willingly.

Imhoff Engages the Nobility

Cordial relations notwithstanding, the balance of power between genealogist and nobility fluctuated. In no way was Imhoff simply on the receiving end of his relationship with the nobility. Although he depended on the nobility's cooperation, they had no coercive power over him and his research. Most nobles understood that they had little hope of coaxing Imhoff to cooperate other than by asking nicely and playing along.[48] In some cases, noble accommodation went very far: some Neapolitan families, for example, even promised to help offset printing costs "if you [sc., Imhoff] were willing to honor them in your book."[49] Although the power relations between themselves and the German author were hardly symmetrical, the nobility could at best resort only to soft power to nudge Imhoff toward compliance.

Some noblemen had to learn the hard way that they could not micromanage Imhoff from afar.[50] In a remarkable letter from July 1700, for instance, Sant'Anna relayed to Nuremberg the disappointment of Tommaso V (1669–1721), prince of Aquino-Castiglione. The aristocrat complained that his family had been omitted from Imhoff's first book on Italian genealogies in 1701 (his *Historia Italiae et Hispaniae Genealogica*), which had presented all families that presumably descended from the Lombard king Desiderius (eighth century).[51] This angered Tommaso d'Aquino, who thought his family belonged to the illustrious circle of descendants of Desiderius. Imhoff should have waited for relevant information about the Aquino to arrive in Nuremberg, he argued.[52] Imhoff, however, simply adhered to his schedule, assuming that the nobles would accommodate his working regimen, not vice versa. His decision to publish without including the Aquino family did not reflect any fundamental opposition to their standpoint. On the contrary, in 1702, Imhoff readily included the family in his next book that concerned Italy (his *Corpus Historiae Genealogicae Italiae et Hispaniae*). In practical terms, however, it was the genealogist, not the nobles, who controlled what appeared in print, and when.

This autonomy distinguished the Nuremberg genealogist from courtly genealogists, who worked much more immediately under the nobility's control. Occasionally, the independent genealogist even had the upper hand. Sometimes, Imhoff could use the prominence of his work to make information flow as he needed it. His status gave him enough confidence to approach individual counts and nobles on his own terms with very specific requests about family-related data. He frequently sent questionnaires to princes and nobles, for instance, soliciting information about ancestors, marriages, and children and their partners (and potential grandchildren). These documents usually consisted of several specific questions, with empty spaces left for the respondents to write down their answers. For example, "Are the princesses' two sisters Johanna Elisabeth and Charlotte still alive and unmarried?" "Did Prince Heinrich of Dillenburg have other children than the following eight?" And on another occasion: "On which day did Prince Wilhelm Moritz of Nassau die?"[53] Imhoff could be quite persistent, occasionally approaching families with such queries several times over the years.[54]

There were also clear limits to Imhoff's willingness to cooperate. In several cases, Imhoff simply ignored the input of nobles, allowing his text to stand as it was. In the case of Count Schwarzenberg mentioned above, for instance, Imhoff did precisely that—nothing at all. He did not even correct small factual errors in the chapter, even though he had ample time to do so.[55] Whether this was the result of negligence or more specific political motives, it is impossible to tell, but either reason would indicate a certain distance from the explicit wishes of a prominent nobleman. To a certain extent, Imhoff was independent enough to do as he pleased, for whatever reason. Count Ferdinand Wilhelm Euseb certainly grumbled, but there was little he—or any other noble—could do to coerce Imhoff.[56] There is no case on record in which the nobility simply withdrew its support. Long gone were the times when it was feasible to refuse to cooperate with noncommissioned genealogists outright.[57] The *public* nature of genealogical knowledge circulation, to which Imhoff's works contributed so much, would have made the omission of a noble house very conspicuous. Absence from a genealogical *encyclopedia*, which strove to be all-inclusive, might raise doubts about a family's status. Boycotting Imhoff was not a realistic option.

Hence, to a surprising degree, it was the genealogist, not the nobility, who set the agenda. Imhoff had undertaken a project that the nobility could influence but not control. This created a unique balance of power between himself and the nobility, one that differed greatly from the situation of a court genealogist. It also differed from the power structures that underpinned administrative and judicial genealogy, whereby officials like d'Hozier were invested with institutionalized discretional authority. In Imhoff's case, the relationship between genealogist and nobility rested on a fluid, yet cooperative alignment of interests in which their reciprocal give-and-take was roughly balanced. The best the nobles could do was to play along and hope to convince the genealogist to agree to their wishes. On the basis of such a subtle arrangement of mutual dependence, the pragmatic logic of *do ut des* functioned well.[58]

Genealogy and Dynastic Identity

Everyday cooperation between noble families and Imhoff was predicated on the assumption that his works might serve as useful platforms for promoting a dynasty's genealogical identity. The nobility felt (correctly, by and large) that they could rely on Imhoff to promote their own vision of their family's history. Occasionally, Imhoff even forwarded drafts of individual chapters and had them approved in advance.[59] Often enough, he acceded to the nobility's wishes and readily accepted courtly revisions more or less entirely. To a significant extent, the content of Imhoff's works reflected the nobility's standpoint.[60] For example, he dropped a reference to French suzerainty over some of the territories owned by the counts of Hanau, when it embarrassed the family in the late 1690s.[61] By and large, Imhoff's version of encyclopedic genealogy followed aristocratic genealogical agendas in many ways, mirroring the variety of functions that early modern genealogy served. The following sections discuss some points on the nobility's agenda in their dealings with Imhoff.

A Mirror of the Holy Roman Empire: Imperial Counts and the Notitia Procerum

Among the German imperial nobility, the so-called imperial counts (*Reichsgrafen*) looked to Imhoff's *Notitia Procerum* with particular fondness. The counts were a distinctive group of sovereign nobles, albeit over small territories for the most part, whom legal criteria distinguished both from the high nobility above them (princes) and low nobility below (especially the free imperial knights). They acknowledged no superior other than their feudal overlord, the Holy Roman emperor himself in Vienna. Perhaps their most distinctive feature, however, was their right to vote with the estates in the Imperial Diet (*Reichstag*), albeit only collectively. This right separated them from the free imperial knights below,[62] and it was this special status in the complex social and constitutional framework of the Holy Roman Empire that gave the counts their distinct elite position. Since they often lacked other sources of power, however, they were quick to feel slighted by the much more powerful territorial princes whose peers they otherwise claimed to be. The social status of numerous families of imperial counts thus felt anything but secure. Many lived with a considerable burden of status anxiety.

Given their somewhat fragile and contested status, the counts welcomed Imhoff's book as an authoritative attestation of their social relevance. In their eyes, his attention amounted to a powerful and highly public endorsement of the Holy Roman Empire's complex hierarchy of noble rank and their exalted place in it. The work's table of contents reflected the institutional and social framework of the empire. After discussing the emperor, the electors, the ecclesiastical and secular princes, Imhoff dedicated books 6 to 9 to the imperial counts.[63] The counts had rarely been treated so extensively before. It was no surprise, then, that they viewed works like the *Notitia Procerum* as a means to propagate the "splendor" of their families. Imperial counts felt that Imhoff's activities were of great "utility" (*Nutzen*).[64] Count Friedrich Eberhard of Hohenlohe (1699–1737) expressed this view most clearly in 1724: "Part of the luster of the status of imperial count derives, inter alia, from the fact that one may find precise accounts of its old houses along with their marital alliances with royal, electoral, and princely houses in such genealogical books. Vice versa, it is deplorable that hitherto we have, time and again, been too reserved in this regard and have, for completely inscrutable reasons, not adequately accommodated those [sc., authors like Imhoff] who have undertaken such work."[65] Imhoff's book—together with other contemporary publications—gave the imperial counts positive publicity, and at a relatively cheap price.[66]

Imhoff's everyday cooperation with the imperial counts reflected the group's embattled and insecure social status. Men from numerous noble houses claimed to be imperial counts but were not universally recognized as such. These families went to great lengths to secure their inclusion in the *Notitia Procerum* with Imhoff. The Burgraves of Kirchberg, for instance, pleaded with him to be mentioned as imperial counts in the fourth edition (1699) and forwarded him a dossier of material to that end.[67] Although Imhoff had previously excluded the family, he indeed added a new chapter on the Burgraves of Kirchberg, after they had convinced him that things had changed for them. By 1698, the reigning Burgrave's mother was the last heiress of the imperial county of Sayn-Hachenburg. Although they were not yet imperial counts, Imhoff recognized that the Kirchbergs "owned a part of the County of Sayn" and thus would become counts in the foreseeable future.[68] Inclusion in the *Notitia Procerum*, therefore, anticipated the Kirchbergs' "second rise to prominence" as the counts of Sayn-Hachenburg.[69]

Others, however, were not so lucky. The von Galen family also actively sought inclusion in the then-forthcoming fourth edition of the *Notitia Procerum*.[70] Imhoff commented on their efforts in a letter to his confidant Heinrich Günther von Thulemeyer and explained why he chose not to grant their request:

> Several years ago, [you] forwarded me a text from the von Galen family in which I saw that said lords asked to be included in the *Notitia Procerum* among the imperial counts of the district of Wetterau. Now, as the new edition goes to press, I considered this request, but I found myself obliged to inquire [in Regensburg] at the Imperial Diet [*in comitiis*] whether the family really has a seat and vote in the said College of Counts. Since the response I received was "hitherto no" and that the estates under their immediate authority may indeed be found in the register of the free imperial knights, but are not in that of the counts, the gentlemen will hopefully not hold it against me if I break with them in that passage and avoid giving offense to said illustrious college [sc., of imperial counts]. I will, however, seize the opportunity to say something, as much as can be said, about this family in connection with imperial counselors, among whom one Herr von Galen is serving.[71]

On these grounds, Imhoff rejected the von Galens' request. He denied them a prominent position in the *Notitia* and instead kept them relegated to the appendixes. He disregarded similar initiatives from other families as well, again presumably after deciding that they were not imperial counts. The Haunsberg family, for instance, tried in vain to achieve inclusion into Imhoff's *Notitia*.[72] In 1708, the counts of Virmond likewise requested inclusion and a "dedicated chapter" about their family in a future reedition of the *Notitia*—unsuccessfully, as it turned out.[73]

Imhoff also had to adjust to the upward mobility of some counts. Several branches of the counts of Nassau received the title of "imperial prince" (*Reichsfürst*) in 1688—in the *Notitia*, this would entail moving them to a different chapter. Rumors about this significant development had reached Imhoff by 1692, as he prepared a new edition of the work. But rumors were not good enough, and they moreover put him in a tight spot. If the rumor proved true, but he failed to act, he would offend an important family; if he acted on baseless gossip, however, he not only would offend the princes, but his entire enterprise would be in danger of ridicule. Imhoff did the sensible thing and wrote to the Imperial Diet at Regensburg, asking for official announcements.[74] He must have eventually received sufficiently reliable evidence that the counts' promotion had indeed happened. The 1693 edition of the *Notitia* (correctly) treated all branches of the house as princes and no longer contained anything about "counts" of Nassau.[75]

The growing authority of *Notitia Procerum* made it desirable for status-conscious aristocrats to be included. Moreover, as the book mirrored the empire's elaborate social hierarchy, it was all-important to them that they were treated precisely in the right chapter. Hence, families took great pains to influence where they appeared in the book. In consequence, Imhoff unwittingly, yet quite effectively, came to police the distinctions between the different groups of imperial nobility. His placement of individual families—among "princes" or "counts"—was an important and authoritative acknowledgment of a family's public standing. This was why the counts lent him such energetic support.

Supporting Identity Narratives

The nobility also supported the genealogist because they appreciated his nuanced handling of the many fantastic legends about the earliest history of their families. Imhoff took a broadly acceptable, moderate approach to the contested tradition of origin stories. On the one hand, he strove to avoid an impression of ridiculous credulity and explicitly distanced himself from some patently untenable origin myths. For instance, he dismissed a story linking the Visconti to Aeneas, Jupiter, Venus, or the imaginary counts of Angera, as "mythical origins."[76] In such cases, he advocated careful scrutiny and methodological discipline.[77] Most families circa 1700 supported this judicious pruning of the wildest fictions.

More important, however, Imhoff also generally declined to embrace the full rigor of philological skepticism. His tact in dealing with origin stories would have commended him to the nobility even more emphatically. Sometimes, Imhoff deliberately overlooked the results of critical scholarship and simply repeated mythical origin stories that other genealogists had disavowed. He ignored, for instance, recent criticism of the traditional genealogy of the Grimaldi family. While Imhoff rejected (albeit in measured words) the most fantastical version of their origin story, which claimed that the ruling dynasty of Monaco went back to the eighth century, he nevertheless repeated the more circumspect, yet still entirely fictitious claim that Grimaldus, the progenitor of the family, had allegedly received the city as a fief from the tenth-century emperor Otto I. He should have known better: D'Hozier and other French scholars had demonstrated years earlier that this claim was based on forged evidence.[78] Despite his excellent connections with France, Imhoff was either out of sync with contemporary scholarship or simply did not care. Instead, he repeated the well-established story.

Even more spectacularly, Imhoff fully endorsed the Viscontis' spurious claim that they descended from Desiderius, the eighth-century king of the Lombards. He supported this story even against mounting scholarly opposition. He knew from Muratori that the authorities in Milan had disavowed this narrative, and he also knew about Muratori's own harsh criticism of the Desiderius story.[79] Nevertheless, Imhoff's sympathy for the myth never waned.[80] He stuck with it even after a stinging critique of his published version of the story—one of the very few public rebukes of his work—appeared in 1708, written by Leibniz's former aide Johann Georg Eckhart (1674–1730).[81]

Why Imhoff endorsed genealogical fictions that he knew, or could have known, were inaccurate is a difficult question to answer. In both cases, the families were probably pleased to see a prominent international author repeat origin stories that they continued to use for aristocratic self-fashioning. Yet there is no evidence that either the Visconti or the Grimaldi openly encouraged Imhoff to promote these problematic tales. Perhaps Imhoff himself felt the allure of such stories and sympathized with the nobility's ongoing fascination with them. They mirrored, at least faintly, similar concerns in his own and in other Nuremberg patrician families, who likewise had a soft spot for tailor-made legitimating origin stories. Or perhaps Imhoff was, in the end, just not a Muratori or d'Hozier, lacking both the scholar's and the bureaucrat's uncompromising commitment to a critical approach.

On other occasions, Imhoff's laissez-faire attitude toward critical scholarship may have even offended the nobles involved. In 1714, for instance, the German counts of Nassau residing at Usingen had Imhoff's version of their family history, as related in the third and fourth editions of the *Notitia Procerum*, checked.[82] They angrily realized that the genealogist had perpetuated an outdated origin story, rehashing the hypothesis that the family first rose to prominence during Julius Caesar's campaigns in France and Germany. Not only was Imhoff not au courant with the most recent scholarship, the counts complained: he also seemed to have no compunction about perpetuating outrageous family origin legends.

Regardless of how the different families felt about Imhoff's treatment of origin stories, it is obvious that he applied a flexible, somewhat arbitrary combination of innovative scholarship and traditional fascination to self-aggrandizing genealogical narratives. Eckhart, his fiercest critic, captured Imhoff's inconsistency with respect to noble identity narratives quite aptly in his rebuttal of the Desiderius fable: He granted that the Nuremberg genealogist had made significant contributions to the field of genealogy, but he nonetheless insisted that Imhoff lacked a coherent "system" that operated "strictly in accordance with the ancient sources and objective truth": Imhoff still propagated "childish myths" and missed the "truth."[83] For all his rhetoric about scholarly criticism, Imhoff's approach to the dilemma of dynastic antiquity was ultimately ambivalent. Under the veneer of antiquarian erudition, there often lay a more relaxed, genteel attitude toward genealogical fact. Many nobles would have been grateful for Imhoff's moderate approach to their cherished, identity-bolstering origin stories.

Structuring Houses, Mapping Lineages

The nobility also welcomed Imhoff's genealogies as public and authoritative representations of lineage structure and internal dynastic divisions. Noble houses could be approached on different scales; one could focus either on the entire dynastic community, both past and present, or on one of its sundry smaller units. Noble houses were imagined as a patchwork of institutionalized subdivisions called "branches" or "lines." Branches had at least some degree of autonomy within a larger family network, but, ideally, they never lost their connection to other branches and the dynastic community as a whole.[84] Genealogies were instrumental in expressing and managing these familial structures, because the distinction between different branches or lines was by no means self-evident. In fact, demarcating boundaries between these subdivisions was often a complex and conflict-ridden affair. The genealogist took great pains to calibrate the relationships between houses and branches correctly, and he usually followed his noble informants' lead as to how internal family structures should be presented. Both his narratives and his genealogical diagrams paid careful attention to intradynastic divisions.

Visual signs were often the most effective means of conveying distinctive branch identities. Imhoff (and his informants) placed graphic signs on genealogical diagrams to mark boundaries between individual branches. When he was producing a section on the Falckenstein family, for instance, Imhoff worked from a manuscript stemma in which the generational sequence between Wilhelm Wyrich von Daun, Count Falckenstein (1613–82), and his sons was interrupted by three boxed headers, indicating that from that point on three lines (*lineae*) had to be distinguished. Imhoff's own printed version followed suit, though in textual form only, noting in italics that henceforth "three branches" (*tres lineae*) existed.[85] Clearly, the different branches wanted to highlight their distinctiveness, without losing sight of their connection to the broader family framework.

Readers urgently needed such branch-identifying indicators. While it was good to have an overview over an entire house, it was even more important in practice to understand who belonged to what branch, so that individuals could be correctly linked to their most important immediate family. Occasionally, readers even visually carved up the grand tapestry of totalizing genealogical tables by circling branches with colored, hand-drawn lines. Several examples in the Imhoff papers attest to this practice. Someone (perhaps Imhoff himself) consistently highlighted familial subunits with green ink, for instance, on the complex family trees of the dukes of Bavaria and the Palatinate or those of the princes of Anhalt-Köthen.[86] Obviously, visual representations of entire noble houses had limited practical appeal when their internal subdivisions were missing. Imhoff consistently highlighted the dynamics of intrafamilial separation and also the (re-)unification of branches in his narratives.[87]

Individual branches were the lead actors in most of the historical drama that Imhoff related. While the idea of an overarching "house" served as a unifying framework for its smaller, semiautonomous units, this framework nevertheless often remained largely abstract. In comparison to the richly textured narratives of individual branch-lines, dynasties as a whole received only a few general remarks about origins, coats of arms, and geographic extent. The predominance of branches over houses was especially conspicuous in the printed genealogical diagrams of Imhoff's mature works on the European nobility. He usually prepared separate tables for individual branches and often required readers to turn a page for each new branch, thus utilizing the materiality of the printed book itself to indicate social divisions.[88] The Florentine Medici, for instance, needed no fewer than nine double-page tables, all dedicated to individual branches.[89] Each branch, furthermore, received its own chapter of narrative text. Complex noble houses thus appeared as a series of single branches. Although the notion of one overarching house—"the Medici"—remained ever-present in Imhoff's discussion, his works did little to make these broader entities palpable. It was probably a reflection of the everyday preferences of many of his noble informants that dynastic branches were the primary focus and guiding organizational principle of Imhoff's narrative and visual presentations.

Junior Branches and Their Hierarchies

While all nine branches of the Medici family in the example above shared a common identity, they were hardly equal in relevance and prestige. The hierarchy of branches—that is, which were "junior" and which were "senior"—was

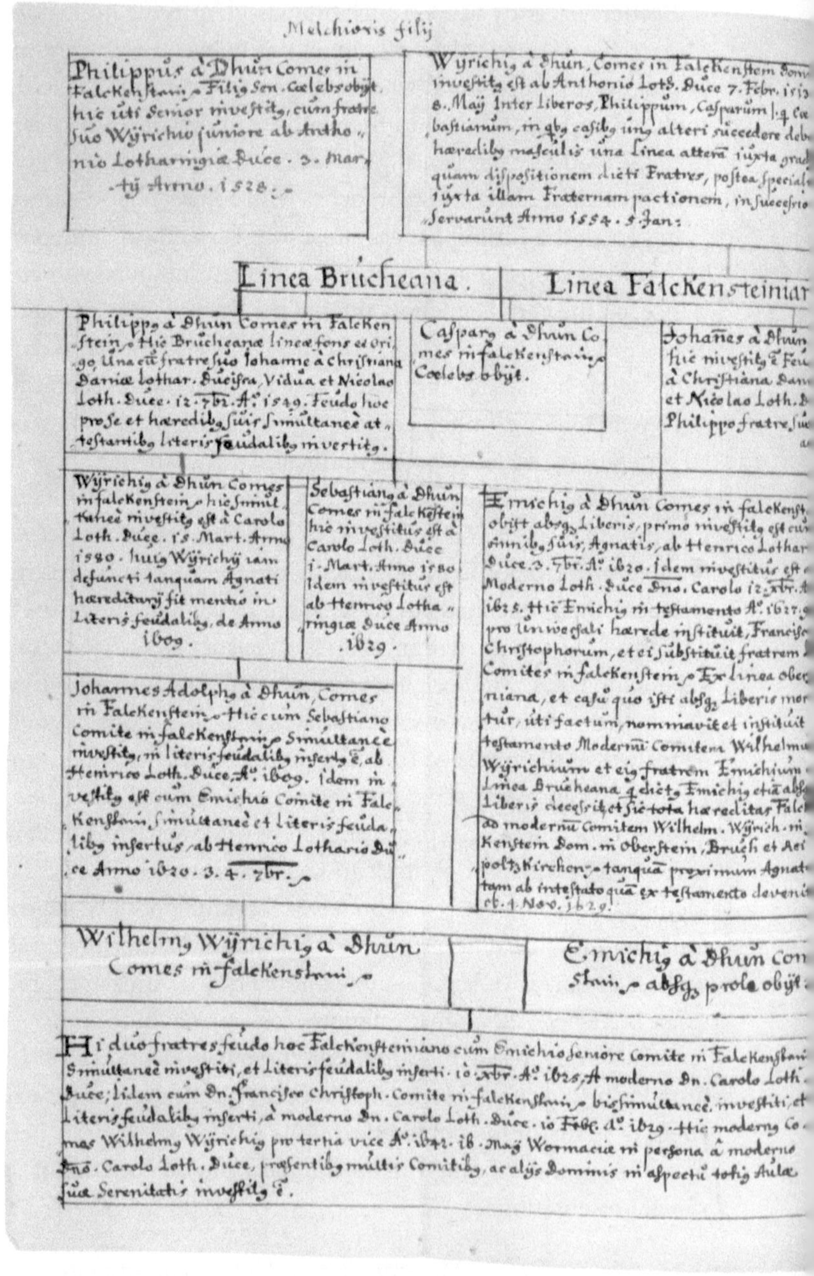

Diagrams often highlighted the separation of different family branches, as here in a manuscript genealogy of the counts of Falckenstein. GNM, Généalogies F (vol. 10), unpaginated (Falckenstein).

...berstein et Bruch
...ternè disposuit. 1567.
... Johannem et Se.,
...t nullis superstitib9
...ativam, hæreditet,
...s confirmarunt, et
...onformarunt et ob.,

Linea Obersteiniana

...m Falckenstein,
...alckensteiniano
... Ducissa, Vidua,
... 7br. A°. 1549. cū
... de linea Bruche

Sebastianus à Dhun Co.,
mes in Falckenstein.

...idonia à Dhun Soror Emichij Comitissa
...n Falckenstein nupsit Dn. Aurelio Comiti
...n Raspurg alias Löwenhaubt, ab hac progeniti
...unt Iohannes Casimirus, et Steino Comites
...n Raspurg. hæc Sidonia non solum, quod bene
...otundum, ante fratrem Emichium obijt, ideoq;
...uccessio feudalis, ad illam non pvenit, nec ad
...lij posteros per saltum exclusis agnatis, trans.,
...erri potuit, viventis siqdem nulla fuit hæredi.,
...as, sed etiam ujata pacta erecta et observantiā
...nter Illustres Comites Germaniæ consuetam,
...onsensū Dn. Mariti Aurelij de Rasp. alias Löwenhaubt
...579. 16 Jun. solenniter corporali juramento
...astricto renunciavit, ita ut quam diu hæredes
...asculi de ista trina linea nempe Falckenst: Oberst.
... Bruckeana, quæ in Comite moderno D. Wilhelmo
...yrichio adhuc illæsa extat, supersint Comites de
...asp. ultra tenorem renunciationis, nihil præ.
...endere vel exigere possint.

Philipps Francis.,
cus à Dhun
Comes in Fal.,
ckenstein.

... Falcken.,
... 1642.

Francisc. Christoph. à
Dhun Comes in Falckenstein e
coelebs obijt Anno. 1636.

Loth. a Dhun Comes in Falc.,
kenstein. Coelebs obijt ante fra.,
trem suum Franc: Christoph e

...ti duo Fratres ex Linea Obersteiniana cum Emichio seniore, moderno Comite Wilhelmo
Wyrichio, ac eiq fratre Emichio, iam pridem defuncto ab atte dicto Loth. Duce. 17. 7br. A°.
1625. Simultaneè investiti, et Literis feudalib9 inserti sunt. Iidem investiti sunt, à mo.,
derno Lotha. Duce Dn°. Carolo. 10. febr. A°. 1629. His mortuis tota hæreditas Comitatq
Falckensteiniani cum appertinentijs ad Lineam Bruckeanā, et in præsentem sæpè dictum
Comitem Wyrichium Wilhelmum devoluta est atq; devenit.

498 LIBER VI.

diam Falckensteinii dominii (quod in confiniis Palatinatus Inferioris ad montem Donnersberg dictum jacet) *partem* in dotem attulit, quam ille, accepta ab Imperatore investitura, ad posteros transmisit: ea enim tempestate *Feudum* id erat *immediatum Imperii*, & demum circa an. 1458. in *subfeudum Lotharingicum* mutatum fuit; postquam Fridericus, & posteà anno 1495. Maximilianus I. Impp. Ducibus Lotharingiæ illud concesserunt, cum illa expressa conditione, ut masculos ex familia Daunio-Falkensteinia provenientes, tanquam legitimos successores feudales, nomine Cæsaris & S. R. I. subinfeudarent. Primus itaque qui feudo Falkensteinio *per subinfeudationem D. Lotharingiæ* investitus fuit, *Wirichius* II. primi trinepos, in *libello memoriali* Ratisbonæ an. 1653. exhibito, memoratur. Hunc excepit filius *Melchior*, qui cum uxore *Margarita, Wilhelmi C. de Virnenburg, Dn in Falkenstein*, & *Francisca Rodenmacherana* filia, *alteram Falkensteinii medietatem* accepit, & de integra Dynastia à *Reinhardo D. Lothar.* an. 1487. investitus est. Huic successere *Philippus*, & *Wirichius* III. quibus *Antonius Dux Lotharing.* an. 1528. feudum contulit: ex eis *Wirichius*, *Philippo* fratre defuncto, hunc inter filios, quos quatuor habuit, succedendi modum an. 1567. constituit: ut deficientibus in una linea maribus altera, secundum prærogativam graduum, hæreditatem cerneret, hancque dispositionem filii ejus ratam habuerunt. Ab iis verò *tres lineæ* ortæ sunt: nempè à primogenito *Philippo* **Broichiana**; à *Johanne* **Falckensteiniana**; à *Sebastiano* **Obersteiniana**; quartus qui *Caspari* nomen gessit, cœlebs obiit.

3. *Johannis* posteritas primum gradum non superavit, sed in filio ejus *Emicone* defecit. Is prole destitutus an. 1627. d. 9. Sept, testamentum condidit, quo *Franciscum Christophorum* agnatum ex *Obersteinia* linea hæredem ex asse nuncupavit, eique *Lotharium* fratrem ejus substituit; quod si verò ambos sine prole decedere contingeret, jussit, ut hæreditas sua *Wilhelmo Wirichio*, & *Emiconi* è *Broichiana* linea traderetur. Erat *Emiconi Falckensteinio* soror *Sidonia*, *Axelio Levenhaupt C. de Rasburg*, Sueco genere, nupta; ea hæreditati fraternæ pro se hæredibusque suis sollemni juramento præstito renuntiavit, atque ante fratrem diem obiit. Sed *Obersteinia* etiam linea intra breve temporis intervallum ad interitum decurrit, atque in *Sebastiani* nepotibus *Francisco Christophoro*, & *Lothario*, an. 1636

extin-

Although Imhoff's printed genealogy of the counts of Falckenstein does not use similar visual signs, the distinct branches are nonetheless highlighted: see the explicit reference to "tres lineae" printed in italics slightly below the middle of the page. From Imhoff, *Notitia Procerum*, 2nd ed., 1687, 498. Munich, Bayerische Staatsbibliothek, 4 Geneal. 40, urn:nbn:de:bvb:12-bsb10358492-2.

hotly contested in many noble houses. Genealogists inevitably became entangled in such debates, since their discussion of individual lineages could hardly avoid questions of precedence and eminence. Imhoff, too, could not prevent his books from being read as endorsements of intrafamilial hierarchies. The aesthetically pleasing principle of "one branch per double-page" created a sequence of branches that was read as reflecting social hierarchies. Imhoff's works distributed social capital, and quite conspicuously.

The Carafa, discussed in Imhoff's book on Italy and Spain of 1702 (*Corpus Historiae Genealogicae Italiae et Hispaniae*), provide an interesting case. Among the numerous branches, there were the princes of Roccella and the dukes of Forlì, both descended from two brothers who lived in the early fifteenth century. However, it was unclear to historians which of the two brothers was older; hence, it was also unclear which branch was senior. In his own narrative, Imhoff duly quoted both sides of the argument and did not explicitly decide the question. His tables, however, made his opinion clear. The princes of Roccella were discussed in "tabula II," while the dukes of Forlì did not appear until "tabula V." With this arrangement, Imhoff adopted the standpoint of Biagio Aldimari and Giacinto Falletti (1661–1722), two advocates of the position of the Roccella-Carafa.[90] Clearly, that family branch had succeeded in persuading Imhoff to give *their* version of family history. Circulating relevant information to the right people was a key strategy for success in the ongoing genealogical rivalry between noble families and branches.

Individual branches did what they could to secure themselves an advantageous position in Imhoff's works. The Portuguese ambassador to Paris, Luis Álvares de Castro (1644–1720), Marquis de Cascais, for instance, forwarded a dossier about his family to Imhoff—which successfully resulted in the inclusion of his branch of the Iberian Norona family, even though Imhoff himself thought the family "lacked material to report."[91] Elsewhere, Imhoff included individual branches because they were important to himself or his political connections. In 1712, he explicitly highlighted the de la Corzana branch of the powerful Spanish house of Mendoza, because the patriarch of the branch, Diego Hurtado de Mendoza a Sandoval (1650–1720), was fighting for Habsburg interests in Spain during the ongoing War of the Spanish Succession.[92]

There were limits, however, to Imhoff's interest in junior branches. If reasons for detailed discussion of a junior branch were lacking, its members had to content themselves with summary treatment. "Now that we have finished our account of the reigning branch of the Medici tree, we shall proceed to the remaining branches, which present our pen with less material," he once de-

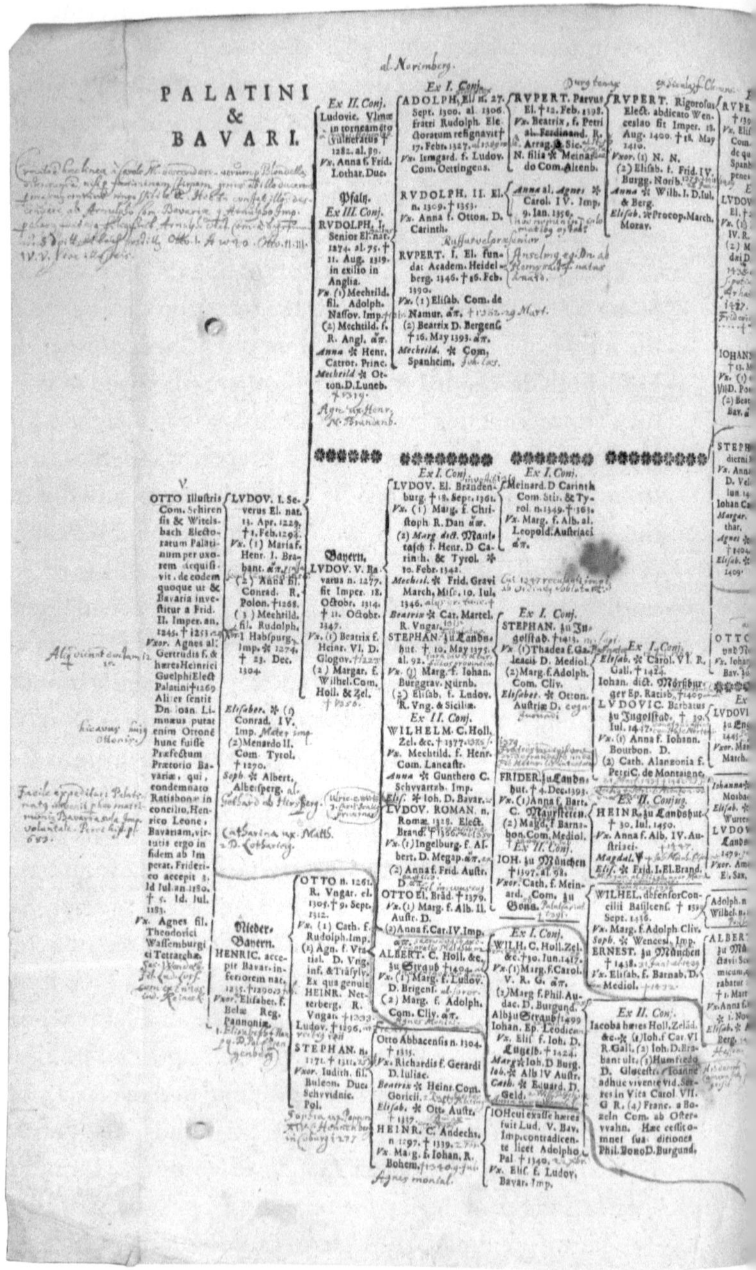

Lines (originally in green ink) drawn on a printed table of the genealogy of the dukes of the Palatinate and Bavaria separate distinct dynastic branches. GNM, Généalogies PAR–PZ (vol. 18), unpaginated (Pfalz).

..1579 cum reversis exstat ex Sacro bello ... capti & tribus ... liberati.

Ex I. Conjug.
Rupertus dict. der Engelländer quia mater fuit ex Anglia † 1426. n. 1406.
Ex II. Conj.
LUDOV. IV. Mitis. El. n. 1424. † 13. Aug. 1449.
Vx. (1) Amal. f. Alb. El. Brand.
(2) Margar. f. Amadæi D. Sabaud. ✱ 1445.

Ex II. Conj.
PHILIPP dict. Posthumus Elect. n. 14. Iul. 1448. † 28. Febr. 1508.
Vx Marg. f. Ludov. Div. D. Bavar. ✱ 11. Mart. 1474.
Mechtild. ✱ (1) Ludov. C. Wurtenb. (2) Alb. VI. Austr. D.

LUDOV. V. El. n. Iulii 1478. † 26. Mart. 1544.
Vx. Sibyll. f. Alb. IV. D. Bav. ✱ 1511. ἄπ.
Philip. n. 7. May 1480. Episc. Freis. 1499. Episc. Zeitz & Naumb. 1517. † 1541.
RUPERT. n. 14. May 1481. † 15. Sept. 1504.
Vxor. Elisab. f. Georgi Divitis D. Bavar.
FRIDER. II. El. n. 9. Dec. 1482. *Primus Evangelicus* † 26. Feb. 1556.
Vx. Doroth. f. Christiern. II. R. Dan. ἄπ.
Georg. n. 10. Febr. 1486. Ep. Spir. 13. Feb 1513. † Sept. 1529.
Heinr. n. 15. Febr. 1487. Episc. Worm. 1523. Succedit fratri Phil. in Freis. 1542. † 1552.
Iohan. nat. 1488. Ep. Ratisb. 1507. † 1538.
Wolfgang. n. 31. Octob. 1494. † 2. Apr. 1558.
Otto Heinr. n. & † 1496.
Æmil. ✱ Georg. I. D Pomer. † 1524.
Helena ✱ Henric. D. Megap.
Elisabet. ✱ (1) Wilhel jun. Landg. Hass. (2) Philipp. III. March. Bad. ✱ 1522.

Rupert. † immat.
OTTO HENR. El. n. 10. April. 1502. ult. Elect ex hac linea. † 12. Febr. 1559.
Vx. Susanna f. Alb. IV. D. Bavar. vid. Cas. March. Brand. ✱ 1529.
Philipp. Bellicos. dictus quod Viennam contra Turcas defendit. n. 12. Nov. 1503. † 4. Iul. 1548.

Frider. Victorios. stirps. Com. a Lewenstein Administrator Electoratus & Tutor Philipp. Posthumi quem adoptavit. nat. 1. Aug. 1425. † 1426.
Vx. Clara von Tettingen.
Rupert. n. 1427. Elect. Colon. 1463. † 1480.

Frider. H. zu Scharffeneck. † 1474.
Ludov. H. zu Scharffeneck. n. 1452. † 1524.
Vx. Soph. Böcklein. Horum filius fuit Frider. primus Comes a Lewenstein.

Chur-Pfaltz.

Ex I. Conj.
FRIDER. III. El. n. 14. Febr. 1551. Reformat. † Oct. 1576. **A.**
Vxor. (1) Mar. Marg. f. Casimiri March. Brand. ✱ 12. Iun. 1537.
(2) Amal. Com. Neumarck. ✱ 25. Apr. 1569.
Elisab. ✱ Com. Erbach.
Georg. n. 20. Feb. 1518. † 17. May. 1569.
Vx. Elisab. f. Wilh. Landgr. Hass. vid. Ludov. Bipont ἄπ.
Richard. nat 25. Iul. 1521. † 13. Ian. 1598.
Vx. (1) Iulia f. Wilh Com. Wied.
(2) Æmil. f. Christop. D. Wurtenb.
(3) Anna Marg. f. Georg. Iohan. Com. Palat. zu Lützelstein.
Wilhel. n. 4. Iul. 1576. † immat.
Sabina ✱ Lamoral. Com. Egmond. Spiræ in conspectu Carol. V. Imp.
Helen. ✱ Com. Leining. & Westerb.
Magdal. ✱ Philip. Com. Hanov.
Brigett. Iohanna. Ottilia. Maria. Cath. omnes moniales.

Ex I. Conj.
CHRISTOPH. R. Suec. Dan. & Norvv. fact 1439. † 6 Ian. 1448.
Vx. Doroth. f. Ioh. March. Brand.
Adolph. Frid. Alb. Iohan. Otto † omnes immat.

FRIDER. zu Simmern und Spanheimb n. 1417. † 28. Nov. 1480.
Vx. Marg. f. Arnold. D. Geldr.
Margar. † 23. Nov. 1426.

IOHAN. nat. ( ) † 1509.
Vx. Susanna f. Ioh. Com. Nass. Sarap.
Rupert. Ep. Ratisb. nat. 1492. † 19. Apr. 1507.
Steph. Præpos. Colon.
Wilhelm Canon. Trev.
Frider. Canon. Colon.

IOHAN. II. zu Simmern/ Graf von Spanheimb Præses Cameræ Spirens. n. 20. Mart. 1486. al. 92. † 13. May 1557.
Vx. (1) Beatrix f. Christpoh. March. Bad. † 1535.
(2) Maria f. Iacob. C. Otting. ἄπ.

LUDOV. Niger zu Zweybruck und Veldentz † 19. Iul. 1489.
Vx. Iohan. f. Anton. D. Croiæ.
Anna ✱ Vincent. Com. Mörsens.
Rupert. fit Ep. Argent. 1439. † 18. Oct. 1478.
Iohan. Ep. Monast. & Magd. 1464. † 1475.
Stephan. Can. Colon † 1481.
Iohan Canon. Argentor.

David. † immat.
Caspar. n. 1458. † 1481.
Vx. Amal. f. Albert. El. Brand. ✱ 1475. ἄπ.
Cathar. Abbat. Trev.
ALEXAND. n. 1462. † Octob. 1514.
Vx. Marg. f. Crat. Wolfgang Com. Hohenloh. ✱ 1499.
Albert. Ep. Argent. † 1523
Philip. Can. Arg † 1469.
Iohan. Canon. Argent.
Samson n. 1474. † 1480. de turre cadens Biponti
Marg. Abbatis. dicitur ✱ Phil Nass.
Elisab. ✱ Iohan. Lud. Com Nass.
Iohanna & *Anna* Vestal.

Otto † 7. Apr. 1499.
Rupert. Ep. Ratisb. 1457. † 1465.
Albert. Ep. Argent 1478. † 1506.
Iohan. Præpos. August.
Marg. ✱ Reinard. Com. Hanov.
Doroth. ✱ Langr. Pfrundentz & Liechtenb.
Barbara, *Anna* } Vestales.

Elisabeth ... Philipp C. Reineck.

LUDOV. zu Zweybruck. n. 1502. † 3. Dec. 1532.
Vx. Elisab. f. Wilhel. Landg. Hass. ✱ 1525.
Georg. Canon. Trev. & Colon.

Zweybruck und Veldentz.

WOLFGANG. zu Newburg. n. 26. Sept. 1526. † 11. Iun. 1569. In Gallia militans pro afflictis Protestantib. **B.**
Vx. Anna f. Philip. Landgr. Hass. ✱ 6. Sept. 1544.
Christina † virgo 1534.

Lützelstein.

RUPERT. zu Lützelstein. † 1544.
Vx. Ursula f. Iohann. VII. Reingrav.
Iohanna Vest. n. 1499. † 1520.
Marg. Vest. † 1542.
Catharina nat. 1510. ✱ Ott. Com. Rietberg.

GEORG. IOHAN. nat. 11. Apr. 1543. † 6. May 1592. **C.**
Vxor. Anna Mar. f. Gustavi R. Suec. ✱ 26. Oct. 1563.
Anna ✱ Carol. March. Badens. & Hochberg. † 12. May 1586.
Ursula ✱ Ulrico Com. Falckest.

GEORGIUS Dives fundat Acad. Ingolstad. 1471 † 1502
Vx. Hedvvig f. Casim. R. Polon.
Marg. ✱ Philippo Elect. Palatino 11. Mart. 1474.
Elisab. †

Ludov. zu Landtshut. † ante patrem.
Rupert. † immat.
Elisab. ✱ Rupert. Com. Pal.
Margar. †

Iohannes regnat una cum fratre Sigism. Bav. Sup. † 28. Nov. 1473.
Sigismund cedit regimen fratri Alberto 1465. † 1501.
ALBERT. IV. Sapiens D. utriusque Bavar. zu Landshut. nat. 15. Dec. 1440. al. 47. † Mart. 1508.
Vx. Kunigund. f. Frid. III. Imp.
Elisab. ✱ Ernest. El. Sax. 1484.
Marg. ✱ Frid. Gonzag. March. Mant.
Christoph. n. 5. Ian. 1450. † 1493.
Barbara Virgo Vestal.
Wolfgang. † 1514.

WILHELM. D. utriusq: Bav. n. 13. Nov. 1493. † 22. Mart. 1500.
Vx. Maria Iacobe f. Philipp. March. Bad.
Sidon. † immat.
Sibylla ✱ Lud. V. Elect. Pal.
Ludovicus n. 1495 † 1545.
Sabina ✱ Ulrico Duc. Wurt. 1511. † 1564.
Ernest. Arch. Salisb. & Ep. Passav. n. 1500. † 1560.
Susanna ✱ (1) Casim. March. Brand. (2) Ott. Heinr. El. Palat. † 1543.

Theodor. † immat.
Sidonia ✱ Philiberto March. Badens.
ALBERT. V. Dux utriusque Bavar. n. 29. Febr. 1528. † 29. Oct. 1579.
Vx. Anna f. Ferdinandi I. Imper. ✱ 1546.

Carol. n. 6. Sept. 1547. † immat.
Maria Maxim. n. 29. Iun. 1552. †
Maria n. 1553. ✱ Car. Arch. Aust. 1571.

Wilhelmische Bayerische Linie.

WILHELM. n. 29. Sept. 1548. † Febr. 1624. **D.**
Vx. Renata f. Franc. I. D. Lotharing. ✱ 22. Feb. 1568.
Frider. n. 1551. † immat.
Ernestus nat. 17. Dec. 1554. Episcop. Freising. 1565. Hildes. 1573. Leodie. 1580. Arch. & Elect. Colon. 1583 † 7. Febr. 1612.
Ferdinand. n. 30. Ian. 1550. † 30. Ian. 1608.
Vx. Maria petembeckia ✱ 27. Sept 1588. ex hoc matrimonio orti sunt Com. de Wartemberg & Schaumburg. Interque alios Francisc. Wilhel. Ep. Osnab. factus Card. 1650.

clared.[93] Elsewhere, he openly said that he was "not interested" (*non nobis curae est*) in further information about branches he considered insignificant and that he therefore mentioned only in passing.[94] As with the Norona family, Imhoff wrote "There is nothing to relate" about one branch of the Spanish Guzmán family.[95]

From the point of view of the marginalized branches, of course, such neglect was highly unfortunate. Sometimes, their leaders approached Imhoff, begging for fuller coverage. The Prussian branch of the Waldburg family, for instance, felt that it had been overlooked by most genealogical publications, including every edition of Imhoff's *Notitia*. In 1703, they asked for a fuller entry. Imhoff, however, did nothing. He probably did not deem the branch worthy of further elaboration.[96] His well-established sensitivity to issues of intradynastic affairs thus did not entail that every lineage was equally important to him. On the contrary, there were clear limits to Imhoff's pursuit of encyclopedic completeness.

Pruning Illegitimate Branches

When a little-known junior branch sent a request for inclusion in an encyclopedic genealogy, the author had to tread very carefully. Early modern genealogy was full of unfounded pretensions to dynastic affiliation. The bewildering plurality of lineages within many noble houses might make it difficult to keep track of their many lines. It often happened that families "discovered" forgotten branches that had dropped out of sight. No one could assume that complete knowledge of the divisions and branches of a given house was always easy to obtain and everywhere accessible.

This opened the door to fraud, and it frequently happened that obscure individuals and their families illegitimately claimed to belong to a prominent house. Very many such fraudulent claims were bolstered by arguments that hinged on family names. Identical or nearly identical names had long counted as a reliable proof in early modern genealogy that a family was related to a noble house. Even in Imhoff's day, many humble families would "change [their names] either on account of the baseness of their origins or to claim an inheritance," thus posing as a branch of a more noble, more prominent family, the Zurich theologian Johann Heinrich Ott indignantly wrote.[97]

Ott could have cited Imhoff's own informant Biagio Aldimari, in Naples, as an example of the practice: Aldimari "ennobled" his family by associating it with the similarly named noble house of Adimari.[98] Similarly, in the Italian town of Cosenza, a range of families by the name of "Aquino" claimed to be related to the prominent house of d'Aquino (of the aforementioned Prince Tommaso), thus aspiring to eminent aristocratic status. Prince Tommaso, however, vehemently disagreed. From his point of view, there was only one acceptable and "genuine branch" (*vero ramo*) in Cosenza (which had produced a prominent commander of the Knights of Malta). All other families living in that city by the name of Aquino were "not of the true Aquino family," according to the prince.[99] Dynastic names were an easy inroad for strangers to self-affiliate with prominent families. Imhoff fully appreciated the dangers of such fraudulent genealogical claims. He probably ordered the reprinting of several pages of one of his books to make sure the d'Aquino's Cosenza-connection was properly discussed in his work.[100]

Imhoff and his noble informants thus had to make difficult decisions about the identity of individuals and families that claimed to be related to the eminent houses of the European nobility. Genealogical knowledge, in that sense, was knowledge of "who's who"—that is, knowledge about which individuals (and branches) were members of a noble house and which were not. By curating such knowledge, genealogists policed the boundaries of the nobility and excluded infiltrators who posed as "branches" of a house without evidence. Genealogical works like Imhoff's encyclopedias shaped and institutionalized social structures.

Who Counted? Women, Infant and Child Mortality, and Other Issues of Completeness

The genealogical "who's who" of the members of noble houses was highly selective in many more ways. Neither the nobility nor Imhoff were necessarily interested in covering *every* individual in a family. Information should, and did, flow only for those persons deemed worthy of commemoration. Genealogies built dynastic identities not just positively, by highlighting certain family members, but also *ex negativo*, by excluding numerous other members. Imhoff's informants were often very explicit about this need for selectivity. In March 1692, Bernardo Benvenuti, one of Imhoff's secondary Italian contacts, agreed to forward material about a number of illustrious houses, including the Strozzi, Salviati, and Medici. Before he sat down to compile the relevant dossiers, however, he wanted to know what the Nuremberg genealogist actually had in mind: "I need to understand if you want the entire trees or only those branches containing the most illustrious persons of the families."[101] In the case of the Strozzi, for instance, their entire genealogical tree consisted of over 1,500 individuals, all of noble rank, but not all engaged in public affairs, since "many have lived privately on their estates."[102] Genealogists needed to think about the right amount of data to include. Selectivity was unavoidable.

Different criteria could be applied to decide who was in and who was out. Sometimes, limitations were necessary for practical reasons. A superabundance of family members might make it impossible to devise a genealogy at all.[103] Benvenuti proposed public impact as a criterion for selection. Far more frequently, however, genealogists trimmed the vast number of members of a noble house to manageable proportions by tacitly applying two other criteria: age and gender. Imhoff and his peers openly acknowledged that genealogies regularly overlooked women and (young) children. Women from problematic backgrounds, for instance, were often unaccounted for, as Bohuslav Balbín once wrote to Imhoff.[104] Sometimes, women were even considered "unimportant," as one informant of Imhoff frankly admitted.[105] Similarly, children who had died at a very young age often faded from genealogical memory after a brief moment of grief.

One of Imhoff's fact sheets brilliantly illustrates this genealogical nonchalance. A list of children of Count Simon VII of Lippe (1587–1627) survives among his papers.[106] The list mentions three daughters, Juliana, Ursula, and Magdalena. A handwritten note, perhaps by Heinrich Georg Thulemeyer, is appended to these names: "These three young ladies have doubtlessly died in their early years, as I cannot remember seeing them mentioned [in an official genealogy on display in the Castle of Detmold]. . . . Should they have reached puberty (which I doubt), it seems very likely that they never married."[107] The point here is that nearly all of this was wrong—evidently nobody made an effort to keep track of the count's children. In reality, there were only two daughters, not three: Juliana Ursula (1617–30) and Magdalena (1620–46). The two girls had lived for thirteen and twenty-six years, respectively, yet their lives had been all but completely forgotten.

Tellingly, the annotator casually inferred a young death from this genealogical misinformation. In reality, however, the two young women had probably been forgotten because they indeed remained unmarried, a condition that facilitated their descent into oblivion in a culture where genealogical thinking about family was ineluctably tied to procreation. This was not merely a question of gender (or age), however: the counts of Lippe also forgot some of their male scions just as easily. Count Simon, in fact, also had had a son, Johann Ludwig (1618–28). He, too, had completely vanished from his house's memory by Imhoff's time. Ten years of life were not enough to leave behind a lasting memory of him. "No one remembers Count Simon ever having had a child called Johann Ludwig," the commentator noted. Two other boys, who died at four and ten years of age, respectively, also went unremembered.[108] Nothing but faint and implausible rumors or empty names survived of such youngsters. Genealogical memory was far from complete, and even the more rigorously investigative and inquisitive approach advocated by Thulemeyer and Imhoff could not easily penetrate this silence.

Whatever such examples may contribute to current discussions of the feelings of early modern parents toward their offspring, they make abundantly clear that children of either sex who died prematurely were especially likely to disappear from genealogical memory.[109] Genealogists, moreover, tended to overlook the family members who either failed to make a public impact or did not produce children. The Imhoffs took this same approach, biased toward male and childbearing-female members, in their own family historiography. Leonhard Wurffbain's 1637 genealogy of the family, for instance, explicitly omits unmarried and childless family members.[110] He cited practical reasons: the work would have become too "expansive" otherwise.

The genealogical marginalization of entire categories of relatives, however, came increasingly under attack. Imhoff's friends warned him against the practice and tried to convince him to enumerate all the "sisters and daughters of kings."[111] Sometimes, these suggestions served an obvious political rationale. One friend pointed out that, in England, women and their offspring could ascend the throne, in marked contrast to France, where Salian law formally forbade female succession.[112] Elsewhere, women were prominent because genealogists had relied on proofs of nobility as sources—proofs of nobility that, by design, included both parents of each individual. This was the case with Spener, who promised to pay even greater attention to women in his narrative genealogies—at least to those who lived up to a certain age and married into other families.[113] Heinrich Meibom also argued for the extensive coverage of women.[114]

Imhoff himself was relatively more attentive to women and young children than some of his sources. He corrected, at least in part, the false account of the children of Simon VII quoted above.[115] In his *Notitia Procerum*, Imhoff mentioned that at least three of the count's sons had died very young (including, presumably, Johann Ludwig), although he did not name or date them. He also provided more detail on the marginalized women. Although Imhoff himself also overlooked Juliana Ursula, he at least rescued Magdalena from oblivion and very briefly told her story. She started life as an abbess, but then left the convent for dynastic reasons to marry a count of Rietberg. The young woman died, however, before the marriage could take place.

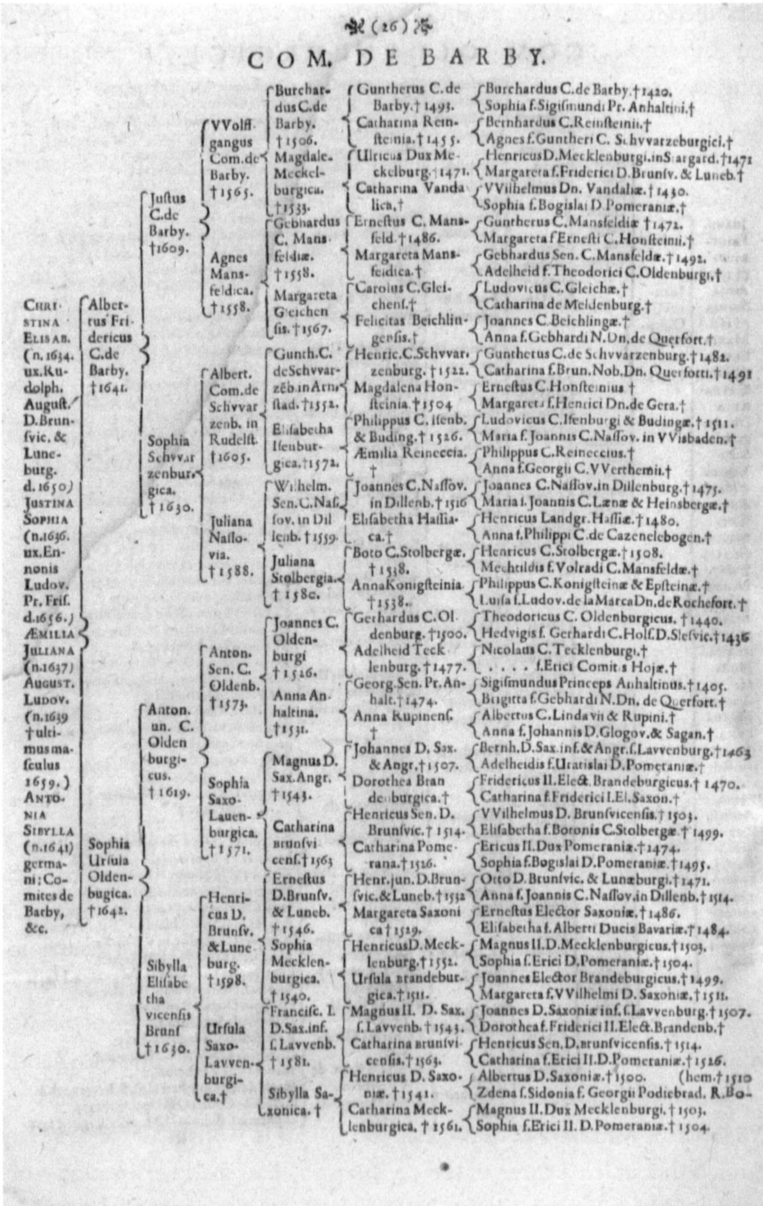

A sample page from Spener, showing a printed "proof of nobility" genealogy that includes both male and female ancestors, from Spener, *Theatrum*, 1:26. Munich, Bayerische Staatsbibliothek, 2 Geneal. 163 a-1/4, urn:nbn:de:bvb:12-bsb10328321-5.

As these examples illustrate, early modern "dynasties," as determined and presented by genealogists like Imhoff, were highly malleable entities, both with respect to their external limits and internal structures. Imhoff and his peers constantly made choices about who to remember and in what capacity. Although he often adopted nobles' own ideas about their dynastic structure and history, Imhoff nevertheless also applied criteria of his own, as, for instance, when he distinguished between important and unimportant branches. As seen in the case of the imperial counts, Imhoff was generally well-aware of his influence. The genealogical decisions he made in his scholarly study may not have had direct institutional or legal power, yet everyone understood that those decisions had significant influence on how his readers would view the noble houses under discussion.

Genealogy and the "Invention of the Ancien Régime"

Despite—or, perhaps, precisely because of—Imhoff's largely sympathetic portrayal of individual nobles and noble houses, his work may even have had unintended and, in the long term, potentially subversive consequences. Works like his made the early modern nobility visible to readers in new ways. They helped define a class, a collective that was shaped by shared values and practices. In a way, genealogies functioned like biographical or bio-bibliographical dictionaries, defining a literary canon, a scholarly field, a social profession.

Biographical encyclopedias were one of the most important new genres of eighteenth-century European culture.[116] Specialized lexicons dedicated to specific social groups—by region or profession, for instance—became crucial scholarly resources during the Enlightenment. They had a utilitarian purpose: to provide readily available information to people interested in biographical research. More important, however, these reference works must be understood as representations of the social, scholarly, and political world. They described and constituted social groups. By bringing together many individuals who belonged to one faith, profession, or region, for example, biographical dictionaries created collectives. Individuals, often presented in alphabetical order, appeared as instantiations of broader social categories, whether these were religious (a dictionary of Jansenists), artistic (a dictionary of famous authors), or social categories (a dictionary of nobles). In every case, biographical dictionaries helped define large social groups by enumerating their members. If theoretical treatises defined or analyzed social groups through abstract consideration, biographical handbooks and dictionaries did so by providing large numbers of pertinent examples.

Imhoff's books fulfilled a roughly similar function for the nobility. While they were not dictionaries in the strictest sense, his works nevertheless contained a wealth of easily accessible information. The narrative sections gave brief biographical sketches of noteworthy nobles, much as a dictionary would have done. The clear organization of his books made it relatively easy for readers to look up individual nobles. Standing in for a subgenre that was de facto unavailable at the time—dictionaries of nobles—works like Imhoff's played an important part in instructing a wider public about the nobility.[117] In doing so, the impact of encyclopedic genealogy mirrored the effect of other dictionaries and encyclopedias: it demarcated social groups. Imhoff's publications contributed to the concept of "the nobility" as a collective, transcending individual families. They did so, however, from an empirical perspective. Genealogies were relatively unconcerned with abstract discussions about the essence of nobility; instead, they fleshed out abstract ideas by providing abundant biographical detail. They provided thick descriptions, not definitions of nobility, and contained many stories that exemplified, for better or worse, what the nobility was. Yet Imhoff's standardized layout and parallel discussion of countless families also normalized the abundance of everyday noble experiences. One could potentially distill the idea of "typical" noble life from his works.

What image of the nobility and their everyday life might have emerged from reading Imhoff's books? By and large, Imhoff stayed focused on traditional indicators of rank. He portrayed individual nobles on the basis of their titles, offices, and military or diplomatic achievements. Office-holding and reigning heads of noble houses were included almost by default. Imhoff's concept of the nobility also remained dominated by males. To a reader of his works, the aristocracy might easily appear as a large group of men whose attention was focused primarily on questions of prestige and possessions, piety and power.

Imhoff did little to highlight other aspects of the nobility's cultural impact. He seems to have had no interest in literary and intellectual endeavors, for which many nobles might have been mentioned. Giovanni Pico della Mirandola (1463–94) and his nephew Gianfrancesco Pico della Mirandola (1469–1553), famous philosophers and writers of the Tuscan Renaissance, are mentioned only in passing and without reference to their literary accomplishments.[118] The humanist author Enea Silvio Piccolomini (1405–64) received extensive treatment not as a scholar or literary figure but rather in his prominent capacity as Pope Pius II.[119] Vittoria Colonna (1492–1547), the renowned and admired Renaissance poetess, does not appear in Imhoff's narrative at all, even though Antonio Bulifon, Imhoff's important Neapolitan contact, played a significant part in Colonna's editorial afterlife.[120] Imhoff could have known about her, but that was not congenial to his understanding of the nobility. His encyclopedia was not a survey of writers or scholars, but of powerful dynasts, ruling princes, and military leaders.

The genealogist topped off this otherwise rather monotonous picture of the nobility as an office-seeking, title-hungry, and war-mongering social group with an occasional sprinkling of more colorful stories of aristocratic extravagance. Imhoff had an eye for the nobility's antics, and readers might well be forgiven for associating the nobility with the endemic abuse of power after they read some of the juicier passages Imhoff included—for instance, the cynical nobleman who killed his allegedly adulterous wife (Niccolò III d'Este [1383/4–1441]).[121] Then there were the confirmed adulterers. Imhoff's works abound in discreet but nonetheless explicit information about illegitimate children. He also recorded bizarre stories of aristocratic feuds, including grotesque episodes in which nobles had mutilated servants to avenge their master's misdeeds. Imhoff dedicated several lines, for instance, to a late-seventeenth-century case from southern Italy that involved the maltreatment of a duke's servant at the hands of another duke; the offending prince was then mutilated in turn after he was roused from his marital bed in the middle of the night—all because of an alleged minor infringement of hunting privileges. It is typical of Imhoff's narrative choices that he also spoke positively about the duel that the two parties later staged to settle the feud.[122]

By characterizing the nobility and its way of life as an essential feature of early modern social, political, and religious history, genealogical literature contributed significantly to casting the aristocracy, for better or worse, as a symbol of the status quo. It paved the way toward the conception of the "ancien régime," to which later generations could turn with a wide range of ideological goals. Scholars agree today that its emergence as a specific social construct of prerevolutionary Europe first occurred at the end of the eighteenth century.[123] After 1789, contemporaries needed a concept that could capture the embattled status quo from which the progressive political approaches could be sharply distinguished. While many observers circa 1800 looked back to the ancien régime with fond nostalgia, it is nevertheless clear that the notion itself originated in the context of the French Revolution. The catchphrase evoked the bygone world of aristocratic domination and decadence and aptly designated the object of revolutionary ire.

Imhoff's portrait of the nobility contains not a hint of subversion or sarcasm. Quite the opposite: his accounts of noble biographies and lifestyles were not intended to ridicule or criticize their subjects. Yet, we may well wonder whether such books might not eventually be read in such a subversive way. If genealogical literature began as a key format for celebrating the nobility, it could also serve as a rich source of material for confrontation, criticism, and contempt. Unsurprisingly, several Enlightenment *philosophes* savagely attacked and ridiculed genealogy.[124] Certain progressive milieus also distanced themselves from lavish genealogical representations, which they regarded as outdated symbols of a world of undisputed aristocratic privilege and at odds with more recent meritocratic notions of social distinction.[125] Despite the fact that Imhoff would have recoiled from any such broad condemnation, by presenting the nobility as a large, coherent body of individuals, dedicated to a narrow range of "noble" activities and their own self-aggrandizement, his own and the works of other encyclopedic genealogists may unintentionally have given rise to such views.

The European nobility cooperated with Imhoff proactively and, occasionally, even with great enthusiasm. German and Italian houses circa 1700 accepted the existence of new genealogical media and the emergence of an unprecedented genealogical information market. They obviously attempted to ensure that their interests were served in this new environment. There is little evidence of any form of fundamental opposition to Imhoff's endeavors. Many nobles discovered instead that the new literary forms, including the genealogical encyclopedia, might benefit them greatly by propagating their dynastic identity. Accordingly, in almost every instance, the nobles who cooperated with Imhoff's projects attempted to instrumentalize the genealogist more or less directly for their own purposes. Encyclopedic genealogy could function admirably as a tool of dynastic propaganda. Auto- and hetero-genealogy were not in absolute opposition.

This chapter, however, also highlights the difficulties that nobles faced when trying to influence Imhoff. The Nuremberg genealogist worked from a comfortable position of relative autonomy. Not even the most eminent princes could easily coax Imhoff into compliance. As a few surviving examples demonstrate, Imhoff could be curt and impolite as well. If he did not feel constrained by practical needs and did not need a family's cooperation to obtain additional information, Imhoff's willingness to accede to the nobility's wishes could quickly reach its limit.

Although the relationship between genealogist and nobility was not exactly one of equals, their asymmetrical power relations were prone to fluctuate. Sometimes Imhoff had the advantage and set the agenda for the nobles. His genealogical projects could exert a certain force of attraction, eliciting a flow of information from the noble courts to Nuremberg, as individual aristocrats felt that cooperation was inescapable. Imhoff also sometimes openly prodded nobles to cooperate, startling them into unexpected actions. Omission from his encyclopedia was not a realistic option in most cases. At least to some extent, the nobility had lost control over when and how it engaged in genealogical ventures. The changing status of genealogy in European knowledge cultures circa 1700 may not have diminished its potential as a tool for shaping noble houses and articulating dynastic identities, but the shifting media landscape nonetheless recontextualized noble genealogy in major ways.

The "Genealogical Brotherhood"

Less than twelve months after he published *Regum Pariumque Magnae Britanniae Historia Genealogica* (Genealogical history of the kings and peers of Great Britain) in 1690, Imhoff felt compelled to publish a ten-page appendix to the work.[1] At long last, he had received important additions to his English collection from Sir Henry St George, the younger (1625–1715), Clarenceux King of Arms. This new material necessitated extensive changes to his previously printed genealogical tables. The *Appendix ad Historiam Genealogicam Regum Pariumque Magnae Britanniae* published in 1691, is no more than a list of addenda and errata.

St George's extraordinary willingness to help seems to have made a profound impression on Imhoff. In the short preface with which the brief pamphlet begins, Imhoff cites his example in an exhortation to all his readers to imitate the English scholar and forward material to Nuremberg. The genealogist even includes a relatively long list of noble houses for which he needed information. As he prepared a new edition of the *Notitia Procerum*, Imhoff invited all his readers to collaborate and share their knowledge with him. This public call for genealogical information was broadcast even more widely when the German language journal *Monatliche Unterredungen einiger guten Freunde von allerhand Büchern und andern annehmlichen Geschichten* (Monthly discussions between several good friends about all manners of books and other pleasant subjects) printed a brief summary of the Latin preface to the *Appendix* a few months later.[2]

The target of Imhoff's public appeal for help in 1691 was not primarily the nobility itself. Important as the flow of information from aristocratic families was, it was insufficient on its own. There was too much that noble dynasties either did not know or knew only imperfectly, even about themselves. Imhoff realized that he needed additional sources of information. But to whom could he turn? In the preface to his *Appendix*, alongside bureaucrats and government officials, Imhoff explicitly addresses a diffuse group of genealogically informed readers: "Adeste . . . Viri Populares!" (To aid, Men of the People!). As this chapter demonstrates, the genealogist's intuition was correct. There were indeed many people from various walks of life who could and did contribute to his projects. In fact, the amount of material that emanated from Germany and Europe is staggering. Genealogical knowledge in all kinds of forms—texts, fact sheets, tables, excerpts, copies of printed genealogies—arrived at Imhoff's desk on a regular basis for many years.

As Imhoff's correspondence shows, early modern genealogists shared sources and helped each other access sites of knowledge that were otherwise closed; they engaged in heated debates about specific empirical questions and discussed methodological issues; they shared concerns for their field and assured one another about the utility of their beloved pastime. Their exchange was both practical and affirmational. The participants in this flow of information constituted, in the words of Imhoff's French informant Charles René d'Hozier, a pan-European "genealogical confraternity"; or, in contemporary scholarly language, a "community of records."[3] The following reconstruction of this republic of genealogists reveals an understudied world of genealogical exchange. Although it broadly followed established patterns of scholarly correspondence, genealogical communication nevertheless had its own peculiar structures, difficulties, and conventions.

The first thing to discuss is the geographical extent of this communication. Most of this chapter is organized geographically so that the full range of this community clearly emerges. Imhoff's more than one thousand surviving letters reached most of western and central Europe, directly or indirectly. When studied in a comparative perspective, however, it also becomes clear that the intensity of this transnational flow of genealogical information varied dramatically in different parts of Europe. For some regions, Imhoff relied on multiple correspondents. In Naples and parts of Germany, for instance, he often asked different informants to research the same families. In many other cases, however, Imhoff had only a single crucial contact to whom he could turn. For France, Iberia, and Bohemia, for instance, Imhoff was limited to a single particularly valuable local intermediary. Some regions, moreover, remained effectively off-limits, Poland being a case in point. Venice likewise received very little coverage.

Second, the geographical arrangement of the following sections reflects the fact that genealogical knowledge was predominantly regional or even local in nature. Imhoff's own letters certainly demonstrate that early modern genealogists could cooperate on a transnational, even pan-European scale, but they also show that such a wide-ranging and integrative approach was relatively rare. Many of Imhoff's contacts, in contrast, focused only on a single region or a single family. Genealogy was a highly local affair to them. When it came to genealogical knowledge, Europe—as reflected in Imhoff's correspondence—was parceled across a number of more or less independent regions. Imhoff's information base, thus, might best be described not as a single, integrated network, but rather as a set of parallel, largely independent clusters of individual regional informants.

This was not a consequence of any broad parochialism among genealogical scholars. Rather, it resulted from the peculiar nature of the object and research needs of genealogy. On the one hand, the European nobility, especially at the highest echelons, was inherently pan-European and transnational. Leading dynasties could usually be found in numerous realms, and such houses easily transcended national boundaries, for instance, through marriages or estates. On the other hand, however, the everyday life of the vast majority of nobles and noble families was firmly grounded in local realities. A focus on the activities of local branches was the rule, as illustrated in the preceding chapter. This is evidently true of the lower nobility and of most of middling noble families; yet even the numerous branches of high noble houses were often very localized.

The local embeddedness of noble families meant that original genealogical research also had to be local. Relevant sources usually remained with the families, on site. This was not only true of written documentation stored in castles and manor houses, but it applied especially to immoveable sources of family history, such as architecture and artwork. Detailed source-work on noble families, thus, had to be conducted in person, on site. This is precisely why court genealogists were so valuable to Imhoff: they stayed in one place for a long time as they worked through large bodies of sources. It was natural and necessary that Imhoff's informants have a local or regional outlook.

Imhoff's own broader perspective was largely a consequence the literary genre of his work. Encyclopedic genealogy required a transregional, potentially even pan-European, overview. Not least for practical reasons, then, his genealogical approach necessitated a position removed from the myriad local contexts of genealogical research. Imhoff appreciated the local embeddedness of original scholarship on noble houses precisely because he personally was unable to pursue this line of inquiry. It was impossible for him to travel to all the places he wrote about and inspect their local archives and monuments. In contrast, he needed to compile, via correspondence, the numerous local findings of genealogical inquiry and then assemble and aggregate them into a broad survey.

Third, the diversity of the inhabitants of the *respublica genealogica* was as impressive as its geographical extent. Imhoff contacted many kinds of informants: scholars and academics, administrators and pastors, Catholics and Protestants. Genealogical information obviously could be found in very different settings, albeit in different amounts. The degree to which Imhoff's informants were involved in his projects, however, varied greatly. Many of his correspondents contributed marginally, supporting Imhoff only once or twice and providing nothing but stray pieces of information.[4] Several others, in contrast, had a sustained influence on his work, and a select few friends were of truly vital importance.

Imhoff's correspondence makes it clear that the republic of genealogists was a relatively informal subset of the larger republic of letters. His letters frequently betray his difficulties in bringing his circle of informants together. Few models of pan-European genealogical correspondence were available to him in the 1680s and 1690s. Routinized transnational cooperation on genealogy seems to have been a relatively recent, perhaps also relatively rare phenomenon, in contrast to far more institutionalized correspondence in other highly specialized fields.[5] It was never easy for Imhoff to make first contact with his many informants. Although some of his networking clearly adhered to the established rituals of the republic of letters, and although he made use of a number of prominent intermediaries, on many occasions Imhoff had to improvise to establish successful genealogical communication. Very frequently, he openly relied on contingencies and chance encounters, as we will subsequently see. Imhoff's correspondence should alert historians to the profound influence of random meetings on scholarly connectivity.

Fourth, a multitude of different perspectives on genealogical knowledge emerges from the variety of people involved in the field. Different correspondents had different kinds of genealogical material available, which was produced for different purposes and with different audiences in mind. They forwarded Imhoff everything from sporadic personal information (dates of birth, names, dates of marriages, etc.) to sophisticated historical narratives about entire noble houses. The material they passed along came in turn from a wide variety of sources, such as archival documents, inscriptions, parish registers, or personal knowledge. An examination of Imhoff's correspondence thus uncovers some of the most common media in which genealogical information was stored.

The members of this genealogical brotherhood cooperated for different reasons, from personal to scholarly to propagandistic. Given this variety of perspectives, it is hardly surprising that the letters occasionally hint at the divisive potential of genealogy. Though relatively uncommon, there were nevertheless moments when contemporary political friction between European states directly affected the circulation of information in the republic of genealogists. Despite the obvious capacity of genealogy to unite informants from various contexts for years in a community of dense communication, Imhoff's letters also indicate how ideologically sensitive genealogical issues continued to be. Fierce controversies over dynastic identity and political expedience simmered just beneath the surface at all times.

The Holy Roman Empire and Northern Europe

The German contributors to Imhoff's books fall in two large categories: men whom Imhoff addressed primarily as academics or scholars, and those he approached primarily on account of their positions in the state or church administration. This division of Imhoff's informants into bureaucrats and scholars is certainly not a neat one. Many of his scholarly supporters also held government positions, while many bureaucrats obviously also cultivated scholarly interests. Nevertheless, the two groups provided slightly different, albeit, at times, overlapping, types of information, according to the social and epistemic environments they came from.

Administrators often possessed firsthand information about recent dynastic affairs on account of the fact that they lived and worked in close proximity to nobles. They learned about many details of family history en passant as they witnessed events of dynastic significance or managed oncoming crises. To them, genealogy was a resource for handling family matters and external relations. Knowledge of aristocratic genealogy was extremely important to their daily work, but they usually did not cultivate it per se.

Scholarly informants, in contrast, often considered the study of family relations part of a broader commitment to studying the past. Genealogy, to them, was not (only) practical knowledge, indispensable for managing the privileges and politics of a noble house; it was, in addition to any practical uses, also of interest for its own sake. By Imhoff's lifetime, the investigation of the past had become a major means of creating a "scholarly" persona. Antiquarianism was a widely shared language of refined learning, and genealogy was naturally part of it. Even scholars who did not count as genealogical experts were thus often seriously interested in intricate questions of dynastic history and in a position to help Imhoff, who could reliably expect knowledge of a specific kind and form from his scholarly informants. Parcels of information they sent regularly included commentary on original sources and discussion of complex historical problems. Whereas information provided by administrators was often especially useful for adapting to recent events, scholars were Imhoff's preferred interlocutors for the study of the more distant past.

Politicians and Officials

In his above quoted preface from 1691, Imhoff foregrounded the important role that administrators and other people concerned with "accounts and letters and archives" might play in his project.[6] One of Imhoff's most impressive achievements was indeed his ability to harness not only courtly, dynastic forms of genealogy (as seen in the previous chapter), but also its legal, administrative, and institutional manifestations. Many *deductiones* and similar items, along with copies of legal documents of genealogical import, such as contracts or wills, can be found today among his genealogical dossiers.

One political context that provided Imhoff with a deep reservoir of administrative genealogical material was the Imperial Diet (*Reichstag*), which met in the free imperial city of Regensburg, about 150 miles east of Nuremberg. Originally, the Diet met only sporadically and in different southern German cities, including Nuremberg. In the aftermath of the Thirty Years' War, however, as many issues remained unresolved after the Peace of Westphalia, the session of the Diet at Regensburg in 1663 had failed to come to a proper close. Instead, the meeting continued indefinitely and eventually the Diet became a de facto standing institution. Permanence quickly transformed the Diet into a buzzing center of constant political debate, gossip, lobbying, and news-mongering.[7] Large numbers of printed and manuscript documents of all kinds circulated in Regensburg to influence the Diet's decisions and to promote individual causes. These writings often contained personal, dynastic, or genealogical information, as, for instance, when family branches brought their internal conflicts before the Diet. A large amount of such literature reached Imhoff's desk.

He had several contacts in Regensburg, including Count Stubenberg, who had transferred his private residence in 1671 from Nuremberg to Regensburg, as well as several of the city's envoys.[8]

Imhoff's quest for genealogical knowledge also occasionally extended to other institutions of the Holy Roman Empire that also served as (temporary) hubs of information. A piece on the Hohenzollerns, for instance, reached Nuremberg "by communication of the envoy of Hohenlohe to the regional diet in Ulm, by the intercession of Dr. Multzius, envoy of Öttingen at the regional diet."[9] On other occasions, Imhoff successfully tapped into governmental channels of interdynastic communication to acquire information.[10]

In a few such cases, we know how Imhoff's contacts were first established. Family ties helped. On one occasion, for instance, Heinrich Günther von Thulemeyer (1642–1714) approached his father to obtain further information about Duke Ferdinand Albrecht I of Bevern from the duke's counselor, whom his father obviously knew.[11] On other occasions, such cooperation resulted from chance meetings. In late 1690, for instance, Philipp Albrecht Ortt, a high-ranking official in the county of Hohenlohe, visited Nuremberg. There he seems to have met Imhoff, possibly in an administrative context. This (chance) encounter resulted in Ortt's contributing several details to Imhoff's collection.[12] Ortt revived an earlier tradition of the transmission of genealogical information to Nuremberg that a previous official, one Georg Sigismund Sömer (1679, 1682) had initiated. Ortt's role, in turn, was later assumed by yet another bureaucrat from Hohenlohe, one Secretary Koch (1706). The case of Hohenlohe thus illustrates how Imhoff's genealogical networking integrated chance events and long-standing, ongoing administrative contacts.[13]

Significantly, the cast of Hohenlohe officials who informed Imhoff about the family of the counts also included a parish priest, Tobias Francke (dates unknown) of Michelbach.[14] Although they seldom appear among Imhoff's correspondents, local (Protestant) pastors deserve particular mention, because they had privileged—indeed almost exclusive—access to one particular type of source that had enormous potential for genealogists: church registers. In theory, every parish was supposed to register the baptism, marriage, and death of every parishioner, including nobles. The concept of parish registers can be traced back to antiquity, yet its implementation remained spotty virtually everywhere for centuries. Not until the sixteenth century did parish registers become an everyday reality in most places.

In principle, parish registers are every genealogist's dream come true: they are full of personal data. In practice, however, these books were largely inaccessible to early modern genealogists; parish priests normally stored them in their parish "archives."[15] Accessing church registers would have entailed identifying and approaching local priests, a process not widely practiced by genealogists at the time. In the few cases when Imhoff actually received excerpts from church registers, for instance, regarding the counts of Limburg, they probably reached him with the help of governmental officials. Imhoff would not have known personally about pastors Frieß and Müller in the little towns of Markt Einersheim and Sommershausen, nor would he necessarily have known that they were the sites of important baptismal churches to the local nobility. In all likelihood, these local priests had searched their church registers for information about the house of Limburg at the behest of the counts' powerful administrators.[16] Only locals would have known where the counts were baptized and who the current pastors were, and only high-ranking politicians would have had the power to order the latter to divulge information.

What Imhoff received from administrators and politicians was predominantly information about current dynastic affairs, such as fact sheets that listed the offspring or recent marriages of reigning princes. Such contacts were less likely to contribute comprehensive surveys of family history. This was predominantly the contribution of a different group of correspondents.

Scholars, Historians, and Genealogists

Imhoff also received help from a large number of historians and genealogists in the Holy Roman Empire. Gottfried Wilhelm Leibniz, who was then working as a court historian for the Guelphs in Hanover, was easily Imhoff's most famous early modern German contact.[17] Leibniz was among the most sophisticated German genealogists and historians of Imhoff's lifetime in terms of methodology and critical acumen. His employment by the princes of Hanover notwithstanding, Leibniz exemplified the stereotypical critical and scholarly researcher, dedicating decades of his life to in-depth research on one family. He made frequent research trips to archives and libraries and spent a great deal of energy on the discussion, dating, and interpretation of individual documents. His thorough approach to genealogical research made the Hanoverian scholar the most prominent exponent of "modern" genealogy to Imhoff.

Yet although Imhoff admired Leibniz in a general way, he could not serve as an immediate inspiration for Imhoff's everyday work. Their projects were too different, and the notoriously secretive Leibniz was not particularly eager to share his innovative research with Imhoff, a compiler and much less rigorous historian. Although the two men had met by chance when Leibniz passed through Nuremberg in 1688, Imhoff did not resume their conversation until four years later. And it was a third party, Antonio Magliabechi, who alerted Imhoff to Leibniz's important ongoing work on the Este family, since the project had not yet been highly publicized around 1690.[18] Imhoff's correspondence with the great scholar was always polite but remained superficial and inconsequential. Leibniz, in turn, always cherished a well-informed source and kept up to date about Imhoff's projects, as their respective circles of correspondents overlapped to a significant extent.[19] In the end, though, Leibniz had only limited impact on Imhoff's work, most of which derived from Leibniz's few published works rather than from personal communication with Imhoff.[20]

Philipp Jakob Spener (1635–1705), arguably the most prominent encyclopedic genealogist at the time, was far closer to Imhoff's genealogical vision. Imhoff had met Spener, the future head of German pietism but at the time still a budding theologian, in Strasbourg early on his grand tour in 1670. Their meeting may have deepened the Nuremberg patrician's genealogical interests, and Imhoff subsequently drew significant inspiration and information from Spener. Imhoff's publications frequently cite Spener's works, and the name "Spener" appears even more frequently in Imhoff's papers.[21] Imhoff even adopted Spener's reading list, in part.[22] The two men maintained a sporadic correspondence over the decades after they met. Their letters are cordial and discuss genealogy, yet Imhoff probably drew more information from Spener's printed works than from his letters. This was, perhaps, because their correspondence began to intensify just as Spener had begun to withdraw from scholarly pursuits and take on more demanding religious and political responsibilities as a leading church official in Dresden.[23] Spener's profound impact on Imhoff could hardly be inferred from his letters to the genealogist.

It thus was not Imhoff's most famous acquaintances who made the most important personal contributions to his projects, but rather a broad group of less illustrious genealogists and historians. There were a large number of regionally oriented historians in early modern Germany who were authorities on the many noble families in their geographical area of expertise—a group of scholars incidentally whom Leibniz also valued for their wealth of knowledge.[24] These men dedicated their lives to exhaustive research in local repositories, publishing specialized treatises and editions of relevant sources. Imhoff relied on their works and personal communication to cover both the remote past and the contemporary genealogical status quo of many German noble houses. What he could not find in their publications, they often eagerly supplied in manuscript form and letters.

Imhoff himself did not belong to this academic scholarly milieu, although he had spent a significant amount of time at Altdorf University. Instead, he gained access to this milieu through a series of intermediaries. One early influence-broker was the Zittau schoolman and playwright Christian Weise (1642–1708), as attested by half a dozen letters he exchanged with Imhoff in 1682. Imhoff may have learned about Weise from his Latin treatise on newspapers and journalism, the third edition of which had appeared in 1676. It contained explicit corrections of one of Rittershausen's tables, and Weise's critical assessment of contemporary genealogical literature in Germany read almost like a blueprint for what Imhoff eventually achieved.[25] Weise himself provided little specific genealogical information,[26] but he discussed the conceptual framework of Imhoff's research at an early stage and helped put his projects on the right track by situating Imhoff's fledgling initiatives in the context of current developments in the field.

Weise also introduced Imhoff to a close-knit group of genealogically minded scholars in Thuringia and neighboring territories. This region's political fragmentation and dynastic complexity—a number of families continued to partition their territories among (male) heirs while their branches continuously intermarried—as well as the prominence of historical and political studies at Jena University after 1650, fostered a buzzing historical culture in the region.[27] Caspar Sagittarius (1643–94) was arguably the leading member of this Thuringian circle of local historians and genealogists. He corresponded not only with Weise, but eventually also with Imhoff[28] and contacted additional local historians in Coburg and other smaller cultural hubs.[29]

In addition to Weise's connections, Imhoff could also tap into Thuringia's historical culture through patron-client networks based in Nuremberg. It remains unclear precisely what kind of relationship connected the Imhoff family to Christoph Wegleiter, for example, but the young man wrote deferential letters back to Imhoff (and his father) from Jena where he studied in the early 1680s before moving to England. In Jena, Wegleiter obliged his patron by acting as a local information broker, not only contacting some of the same regional historians that Weise had mentioned, but also adding names to Imhoff's already impressive list of informants.[30]

While the historians of Jena preoccupied themselves with Thuringia and the territories under the Ernestine branch of the house of Wettin, a parallel group of scholars in Leipzig and environs looked to the other half of Saxony, to the electoral seat of Dresden and the territories governed by the Albertine branches of the family. Imhoff probably gained access to this cultural milieu via his pietist network, thus seamlessly fusing religious and intellectual pursuits. In all likelihood, it was Imhoff's pietist acquaintance Johann Christoph Frauendorf (1664–1740) who connected the genealogist to a young man in Leipzig, Christian Wagner (1663–93).[31] Wagner was the son-in-law of the famous jurist Jakob Thomasius (1622–84); he published on oriental languages, translated works by the French expert on genealogy and heraldry François Ménestrier, SJ,[32] completed Caspar von Lohenstein's novel *Arminius* (a seminal work of the German baroque), wrote libretti for the opera, and made substantial contributions to *Acta Eruditorum*, the widely read scholarly journal. He also was a knowledgeable historian and provided Imhoff with high-quality genealogical information about the Saxon electors and their families, including the minor branches of the house. An intimate knowledge of regional history enabled scholars like Wagner to access the sometimes obscure, sometimes still unprinted local historical sources that all genealogists depended on.[33]

It was perhaps Wagner who introduced Imhoff to Christian Juncker (1668–1714), a former student in Leipzig and later teacher and rector in Schleusingen, Eisenach, Altenburg, and Dresden, as well as a correspondent of Leibniz.[34] Juncker also provided Imhoff with excerpts from local sources, forwarding, for instance, information culled from manuscripts he had acquired on the death of genealogically minded local scholars.[35] It seems, though, that genealogy was not really Juncker's priority. He may have had his own, hidden agenda when he cooperated with Imhoff. Without ever losing sight of Imhoff's predilection for genealogy, Juncker eventually exploited the patrician's prominent position in Nuremberg to promote his own projects, which included an important treatise about the study of medieval history, a history of the county of Henneberg, and a numismatic biography of Martin Luther. The process of researching past and present dynastic histories often resembled a bartering of information and exchange of favors.

Whereas Thuringia and Saxony were particularly well covered, informants from other parts of Germany were somewhat scarce. Heinrich Günther Thulemeyer (1642–1714), a scholar and politician initially working from Heidelberg, covered the important regions of Hesse and the Palatinate for Imhoff all by himself.[36] Further to the south, in Württemberg, Johann Ulrich Pregitzer (1647–1708) was Imhoff's only contact.[37] The two men even met several times. The Pregitzer clan seems to have had a personal connection to a Tübingen branch of the Imhoff family.[38] At one point, Pregitzer's homonymous son (1673–1730) worked for Imhoff as a proofreader (*corrector*).[39] Despite these connections, and despite their shared interest in encyclopedic genealogy, few items in Imhoff's genealogical collection can be traced back directly to the Pregitzers.[40]

Finally, a Bohemian Jesuit, the scholar and historian Bohuslav Balbín (1621–88) in Prague, warrants special mention.[41] In him, Imhoff encountered another archival worker par excellence. For decades, Balbín took advantage of his very close ties to many noble families to visit the archives of the Bohemian nobility.[42] Balbín's cooperation with Bohemian aristocrats was so smooth that some contemporaries pondered whether Balbín's genealogies were potentially flawed: "Balbín got himself so many friends from the fact that he serves the interests of many," one critic remarked.[43] Balbín attempted to integrate both the courtly and scholarly approach to genealogy, incurring both the advantages and drawbacks of such a combination. Still, it is beyond doubt that the Jesuit was an expert on Bohemia, and Imhoff depended on him. Without Balbín, it seems, Imhoff would have been unable to learn about the region, since few alternative sources of information are on record. The two men exchanged about a dozen often lengthy and detailed letters, and Imhoff publicly thanked the Jesuit for his contributions.[44]

As generous as Balbín may have been with his information, he had his limits. It had little to do with confessional differences between the Catholic Jesuit and the Lutheran pietist. Throughout Imhoff's letters, there are very few indications that the Protestant-Catholic divide ever impeded international cooperation in the name of genealogical scholarship.[45] Balbín's bias was sociopolitical in nature. His scholarship embraced a robust conception of Bohemian identity. Twice in his letters to Imhoff, he explicitly stated that he was discussing the old Bohemian aristocracy exclusively. In other words, "our nobility" to Balbín consisted only of families who had lived in Bohemia before 1619, the year of the fateful Battle of White Mountain at the beginning of the Thirty Years' War, when the Catholic forces of Emperor Ferdinand II crushed the resistance of the mostly Protestant Bohemian nobility to Vienna's Counterreformation measures.[46] Many Austrian families subsequently settled in re-Catholicized Bohemia. Balbín declared—perhaps reacting to broader questions from Imhoff—that he had no interest in these new houses.[47] Although Balbín claimed that this decision was largely for pragmatic reasons, since every genealogist carefully had to manage his workload, it had obvious political implications. Balbín himself descended from a family that belonged to the "old" Bohemian lower nobility and was generally critical of Habsburg policy in Bohemia after 1619, which he considered an invasion of local Czech culture by Germano-Austrian influence.[48] It was not for nothing that the Jesuit repeatedly clashed with the emperor's censors in Vienna.

To conclude, whereas genealogical knowledge normally flowed freely between Imhoff's contacts in the Holy Roman Empire, there nevertheless were limits. Genealogy always existed in a local context, and Imhoff occasionally seems to have misjudged the lay of the land when he asked for the wrong kind of information. Historians and scholars generally found it unproblematic to forward genealogies, but things changed if they feared that their information might reappear in inappropriate contexts. Balbín's acerbic reaction reveals the sensitive nature of genealogy, even though Imhoff had not at all intended to provoke the Jesuit. The potential of genealogy to stoke political and ideological controversy was always lurking beneath the surface. It was indispensable to every encyclopedic genealogist to ensure that the exchange of genealogical knowledge avoided the danger of partisanship. By and large, Imhoff succeeded in this endeavor.

Southern and Western Europe

Charles-René d'Hozier and the French Connection

France was the first country Imhoff turned to after the first edition of his work on Germany had appeared in 1684. He claimed that his interest in France originated in the close social and cultural relationship between the two countries.[49] In reality, however, Imhoff may have been pushed toward France by his desire to anticipate a parallel project by Spener.[50] He also may have relied on memories of his trip to France ten years earlier, when he potentially met some of the men who now helped him carry out his new project.

In the preface to *Excellentium Familiarum in Gallia Genealogiae*, his book on French genealogy published in 1687, Imhoff names a few important contributors. Besides Christoph Nicolai von Greiffencrantz (1649–1715), a well-connected diplomat and important correspondent of Leibniz, who appears primarily to have sent books, Imhoff mentions a few letters (now seemingly lost) he exchanged with Pierre-Scévole de Saint-Marthe (1618–90), the son of a family of genealogists and himself a widely acclaimed historian.[51] At an early date in his research project, Imhoff also contacted Guy Allard (1635–1716) in Grenoble, another prominent specialist on French genealogy.[52] None of these men, however, became Imhoff's chief French source. This role fell to Charles-René d'Hozier in Paris, the owner of no fewer than 230 volumes and ledgers of genealogical materials. He also is duly mentioned in Imhoff's preface from 1687.[53] As we have seen previously, d'Hozier was a typical institutional genealogist, who pursued genealogical information on behalf of the monarchy. Serving as royal *juge d'armes* under Louis XIV, he embodied the king's desire to "know his nobility."[54]

Although Imhoff never held any such institutional authority, the two men nevertheless approached the nobility and genealogy from roughly parallel perspectives. Insofar as d'Hozier was dedicated to critical impartiality by virtue of his office, his genealogical investigations were aloof from (not to say, opposed to) the myriad dynastic interests of individual noble houses. Like Imhoff, he represented genealogical research *about* the nobility, not *by* it. It thus was perhaps unsurprising that the Nuremberg genealogist found the Parisian administrator a very congenial correspondent; indeed, d'Hozier was one of the very few with whom Imhoff occasionally shared personal information.[55] In any case, judging from the available sources, Imhoff never seriously tried to open up a secondary channel of information in France, as he would do in Italy.

D'Hozier's institutional position and scholarly qualifications made him a paramount genealogical authority in France; he thus could open all kinds of doors for Imhoff. On the one hand, the *juge d'armes* was on good terms with many elite French noblemen, including members of the very distinguished houses of Courtenay, d'Estrée, Bouillon, and Salviati.[56] On the other, d'Hozier discussed Imhoff's projects with the intellectual elite of Paris and introduced Imhoff to several scholars and collectors, such as François-Roger de Gaignières (1642–1715).[57] Imhoff appears to have written to Gaignières directly on occasion, but most of the time d'Hozier acted as go-between.[58] Gaignières was one of the most prolific collectors of manuscripts, prints, and drawings circa 1700, for which he relied on a vast network of informants from all over France.[59] Genealogy figured prominently among his interests, and he forwarded important material to Imhoff.[60]

Other scholars in Paris and the provinces contributed to Imhoff's research more sporadically, since d'Hozier approached them only on a case-by-case basis. Although they were not always on good terms, the *juge d'armes* wrote to the eminent scholar Étienne Baluze (1630–1718) in late 1698 or early 1699.[61] Later, the royal genealogist Pierre Clairambault (1651–1740) and the Jesuit historian, philosopher, and journalist Claude Buffier (1661–1737) briefly appear in d'Hozier's letters as informants.[62] For one highly specific task, namely, the reconstruction of the genealogy of the Franco-Italian house of Strozzi, d'Hozier approached Laurent Planelli (or: Pianello) de La Valette (1644–1718) in Lyon.[63] Because the Strozzi had strong ties to Lyon, it was only natural to turn to this royal official and expert manuscript collector for help. Planelli's own life trajectory, moreover, broadly mirrored that of the more illustrious Strozzi.[64] D'Hozier's bet paid off: Planelli de La Valette indeed sent valuable details.[65] On another occasion, in 1699, d'Hozier enlisted the help of Honoré Caille du Fourny (1630–1713), another renowned French historian and genealogist. What's more, du Fourny was the owner of the papers of Père Anselm (1626–94), arguably the most famous French genealogist in Imhoff's lifetime.[66] Imhoff thus made (limited) contact with the legacy of one of the most important genealogists of the seventeenth century—at least posthumously. Insofar as Imhoff held Père Anselm's scholarship in the highest esteem, he would have been especially pleased about this connection.

Remarkably, most of Imhoff's correspondence with d'Hozier and other French genealogists took place *after* the publication of his (only) book on France.[67] Part of the lasting appeal of French genealogy lay in its relevance to Italian affairs, since many families, such as the Strozzi and Salviati, had members and branches living in both countries. Nevertheless, Imhoff's interest in French families generally did not serve any specific research objective. At no point did he intend to update his early work or produce another book on France. Yet information on French nobles continued to arrive at a steady rate. Imhoff's correspondence with d'Hozier and many of his other friends thus had a broader appeal to the Nuremberg patrician: It helped Imhoff to transform into a kind of "living archive" of contemporary genealogy or, to adapt a contemporary term, into a "bureau d'adresse" (a kind of information hub) for dynastic news.[68]

For his part, d'Hozier also remained invested in his correspondence with Imhoff. He took an active interest in Imhoff's ongoing projects.[69] To some extent, this interest may have resulted from d'Hozier's close affiliation with French administrators and politicians. In 1701, as the Spanish Succession loomed large on every politician's mind, the French chancellor, Count Pontchartrain, asked d'Hozier to provide him "the age of every prince in Europe," presumably so that he might have an overview of the state of the international nobility. D'Hozier asked Imhoff for help in procuring relevant information about Germany.[70] The Nuremberg genealogist, seemingly undeterred by the escalating War of the Spanish Succession, which pitted the Holy Roman Empire against France, sent the list of living princes that d'Hozier had requested by late 1702.[71] In 1706, Imhoff furthermore published a French-language primer on the elite Spanish nobility. Although a direct connection to d'Hozier cannot be established, Imhoff's work nonetheless reflected growing (French) interest in Spanish genealogy. Recent events compelled French officials working on

Spanish affairs to navigate a complex and unfamiliar social terrain. They would have found Imhoff's brief survey a helpful addition to their reading list.

Magliabechi, Muratori, and Northern Italian Genealogy

Around 1690, a few years after his work on the nobility of France had appeared, Imhoff turned south, to Italy. His growing expertise as a master of encyclopedic genealogy is obvious in the versatility with which he eventually gained access to relevant material on the peninsula. The beginning of the project, however, was very frustrating. There were few well-trodden paths for genealogical knowledge exchange across the Alps. Initially, Imhoff relied on local contacts in Nuremberg to learn about the great houses of northern Italy. Often, however, these amounted to no more than very tenuous ties, derived from chance encounters. For example, Imhoff intermittently relied on Johann Christoph Wartis, an obscure acquaintance and perhaps a citizen of Nuremberg or family client, to obtain fragmentary information about some Roman families.[72] At another point, Imhoff seized the moment when a friend of his, Johann Peter Holthusen (dates unknown), an otherwise unidentified counselor at the Palatinate court, traveled to Rome with the prince. Imhoff asked Holthusen to provide him with missing information.[73]

On yet another occasion in 1690, a chance encounter with the Swiss legal scholar Johann Jakob (or Jean-Jacques) Battier (1664–1720) in Nuremberg helped Imhoff gain access to Milanese material.[74] For reasons unknown, Battier was well-connected with Andrea Pusterla (1657–1742), head of the Ambrosiana library in Milan from 1683 until 1695.[75] At first indirectly via Battier, and then directly beginning in 1693, Imhoff received genealogical information and manuscripts from Pusterla.[76] All of these early connections, however, important as they were, proved too incidental and tenuous to sustain Imhoff's project. Two years after Imhoff made personal contact with Pusterla in Milan, for instance, the latter left his post. Imhoff's homegrown ties to Italy easily snapped. It was a very frustrating situation for the ambitious patrician. As he noted in late 1690, "This project becomes all the more difficult, insofar as I cannot find authors who have diligently treated this material."[77]

All this soon changed, however, when Imhoff reconnected with the great Italian scholar Antonio Magliabechi in Florence. The two men had met in the early 1670s when Imhoff had traveled through Italy, but they had not kept in touch.[78] Now, perhaps facilitated by some of his Nuremberg contacts, Imhoff approached Magliabechi once again, asking him "to deign to support my project and to inform me about books and men by means of which I might succeed."[79] Turning to Magliabechi to relay genealogical information proved to be a wise decision. Magliabechi is today rightfully considered the most influential "cultural intermediary" between transalpine Europe and Italy of the late seventeenth century.[80] In response to Imhoff's plea for help, the Florentine gradually assembled a reliable circle of informants for the Nuremberg genealogist.

Thus, when Pusterla left his post at the Ambrosiana in 1695, Imhoff made no attempt to remedy the situation himself but simply appealed once again to Magliabechi for assistance. Imhoff's hopes were not disappointed. Magliabechi—who had helped Pusterla to leave Milan for the lucrative position of Abbot of Abbiategrasso—already knew that a promising young Italian scholar, Ludovico Antonio Muratori (1672–1750), had just replaced Pusterla in Milan. It was Pusterla himself who had introduced Muratori and Magliabechi.[81] This was the kind of local network that Imhoff had in Germany but lacked elsewhere. Obtaining insider access to regional scholarly communities in faraway places was one of Imhoff's principal accomplishments. By securing Magliabechi's ongoing cooperation and entrusting the management of genealogical knowledge acquisition in northern Italy primarily to him, Imhoff guaranteed that he could successfully cover the region. As a result, Imhoff quickly overcame the impasse after Pusterla's departure from Milan, and Muratori duly began to send material on the Borromeo and other regional families.[82]

From then on, Imhoff approached other areas of Italy, including Modena and, to a lesser extent, Rome, more or less exclusively through Magliabechi. Prior to 1700, when Muratori himself moved there from Milan, the most important scholar in Modena was Benedetto Bacchini, OSB (1651–1723), a man whom Magliabechi had recommended to Imhoff.[83] Bacchini had served as head of the ducal library since November 1691 and was closely associated with the ruling Este family. Until Bacchini started to send information about this important house in 1693, Imhoff had little material on the Este at his disposal.[84]

Magliabechi also connected Imhoff to Rome, in particular to a man named Prospero Mandosio (1643–1724).[85] Hailing from an old noble family of Umbria, yet trained as a lawyer, Mandosio had a solid reputation as a historian of the city of Rome.[86] Mandosio was very well connected to local scholars in Rome. He conducted wide-ranging and often difficult searches for relevant information for his work, relying not least on a large number of informants that included such luminaries as Carlo Cartari (1614–97), head archivist of the papal archive in the Castello Sant'Angelo.[87] From 1696 to 1702, Mandosio sent Imhoff material on the Sforza,[88] the Barberini and Colonna,[89] and the Uceda,[90] a Spanish family from which the current Spanish ambassador in Rome hailed. It was not without reason that Imhoff praised Mandosio in print.[91]

Imhoff thus benefited not only from Magliabechi's own vast network of contacts but also relied, through the Florentine's intercession, on many more networks further afield, thus integrating a variety of local information conduits in his genealogical projects. Much of the information on Italy that Imhoff used came from Magliabechi's "various friends at these courts."[92] In contrast to d'Hozier in France, whose role for Imhoff was otherwise similar, Magliabechi did not personally procure much information for Imhoff. Rather, he pointed the genealogist toward the most knowledgeable men, and then let them work out the terms of their genealogical cooperation themselves. Whereas Imhoff could cover most of centralized France through d'Hozier in Paris, he needed to cultivate a range of different local contacts in politically fragmented northern and central Italy.

Even concerning his own city, Florence, and its ruling house, the Medici, Magliabechi suggested consulting a local expert, Bernardo Benvenuti (1634–1700), the widely respected librarian of Crown Prince Ferdinando (1663–1713).[93] In light of these credentials, and despite the fact that Magliabechi regarded Benvenuti as an "arrogant" person, he still considered Benvenuti the only scholar capable of furnishing a "perfect" Medici genealogy.[94] After Imhoff approached Benvenuti, the latter ended up corresponding with him directly and sent material to Nuremberg.[95] Hence, with significant input from Magliabechi and Benvenuti, Imhoff successfully covered the important house of Medici.

Imhoff's Florentine contacts tellingly illustrate how informants often viewed genealogical projects from their own local perspective. Imhoff's interests fit into local contexts over which he had no influence and of which he may even have been entirely unaware. Magliabechi's and Benvenuti's willingness to help with the genealogy of the Medici certainly did not spring purely from selfless charity. History in Florence—including genealogy—was a particularly vexed subject for several reasons. First of all, until relatively recently, as late as the fifteenth century, the dukes had been "normal" citizens of a republic—they were anything but ancient nobility. Second, although Florence was effectively a monarchy after 1530, the city still officially claimed to be a republic. Third, the reigning (since 1537) branch of the Medici was in effect a junior lineage that had come to power after the extinction of the preeminent "Magnifico" line. For all of these reasons, historiography was problematic for the Medici, and they were not eager to promote it. Seventeenth-century Florence has thus aptly been called a "state without history."[96] This description also applied to genealogy. Almost nothing about the genealogy of the Medici appeared in Florence in the seventeenth century. At one point, in 1700, Imhoff complained that the family's history had "not yet been sufficiently investigated."[97]

By the end of the seventeenth century, however, Florentine genealogy gradually revived. The house of Medici now faced extinction. None of the sons of the reigning duke, Cosimo III, was likely to have (male) offspring. Moreover, the family was torn apart by intergenerational conflict. Crown Prince Ferdinando, in particular, was utterly estranged from his father Cosimo III. In these circumstances, the aging grand duke took an interest in establishing a narrative about his family's magnificent past. Prince Ferdinando, meanwhile, vigorously supported the growing eagerness of the Florentine nobility to define (and defend) their status more strictly by genealogy. By rallying the nobility, not least by reinforcing social boundaries by genealogical means, Ferdinando bolstered his political influence. He not only sponsored a new institution for the education of young nobles, but, in 1685, the prince also charged Benvenuti to take up an older project and compile a definitive genealogical survey of the Florentine nobility, in effect, closing their ranks to outsiders.[98]

While Imhoff himself almost certainly was not privy to any of these considerations, his requests may nevertheless have thus come at just the right time. His interest in Medici genealogy coincided felicitously with various Florentine factions and interests. The leading historical authorities of both the grand duke and the crown prince could relate to this foreigner's interest in Tuscan genealogy. Local scholars, thus, appropriated Imhoff's initiatives to their own ends as much as he appropriated theirs. From many local perspectives, Imhoff's requests for information about Italian genealogy often served, or were made to serve, local agendas.

Naples: A Key Hub of Genealogical Information

Naples was the crown jewel in Imhoff's collection of Italian genealogical correspondents. "I have very good authors there," the genealogist once noted.[99] At first, Imhoff again relied on his own devices to acquire information. He had visited Naples on his grand tour in the early 1670s, but only for a few days. Perhaps memories of earlier encounters spurred him to turn his attention to the Neapolitan genealogist Carlo de Lellis (?–1687/8) in the mid-1680s.[100] In 1685, Imhoff acquired de Lellis's most important genealogical publications.[101]

In de Lellis, Imhoff encountered a different brand of genealogical scholarship. De Lellis was not, like Muratori, Pusterla, Magliabechi, or Benvenuti, the librarian or historian of a particular prince, but rather worked as an independent scholar and jurist in the orbit of the central government of the kingdom of Naples. Although he hailed from the nobility himself, de Lellis shared many cultural features of the "robed," or *togati*, classes, families that had risen in the administrative and bureaucratic service for the Spanish viceroy since the late sixteenth century.[102] Accordingly, de Lellis drew much of what he knew about Neapolitan history and genealogy directly from the central archival repositories of the government itself, rather than from individual dynastic archives.[103] His extensive archival work made him an especially appealing informant. De Lellis and Imhoff's works, moreover, both adopted the approach of surveying many noble families without attaching themselves exclusively to any one dynasty.[104] Imhoff wrote a first letter to the Neapolitan in late 1687, about three years before he commenced his correspondence with northern Italy.

His timing, it turned out, was terrible: de Lellis had just passed away. But Imhoff was lucky, and a remarkable story of unanticipated cooperation unfolded. Genealogical inquiries took on a life of their own when they were made in the right contexts. The republic of genealogists, albeit little institutionalized, showed remarkable resilience on the local level, even in the event of the death of a prominent member. Imhoff's orphaned letter ended up in the hands of a group of genealogical enthusiasts, many of whom shared de Lellis's background, and these men simply passed his letter on among themselves. Initially, it came to rest with an unnamed "bookseller" (*libraio*).[105] In all likelihood, this was Antonio Bulifon (1649–1714), one of the most dynamic editors and booksellers in Naples at the time.[106] Bulifon was instrumental in connecting Naples to book markets elsewhere. He was French by birth and maintained good contacts with his native country. Since Bulifon moonlighted as a newsmonger and journalistic observer of his times, his bookshop was a widely known place to meet people and hear the latest gossip. Just like librarians, archivists, apothecaries, and barbers, booksellers were important "mediators of the republic of letters." Making contact with Bulifon, if only by accident, was thus a lucky stroke for Imhoff, because Bulifon knew exactly what to do with Imhoff's letter.[107]

Bulifon handed the letter to Andrea Giuseppe Gittio (?–1700), another man of the robe with genealogical interests and extensive archival experience. Gittio took up the task and answered Imhoff instead of the late de Lellis.[108] This was the beginning of a slow but continuous correspondence lasting from early 1688 until Gittio's death in 1700. Gittio remains an obscure figure today.[109] Hailing from a provincial family, his father, Lelio, moved to the capital to pursue a legal career (as also did his brother Michelangelo, Andrea Giuseppe's uncle), thus rising in the ranks of the robed. Andrea Giuseppe followed in his father's and uncle's footsteps. In addition to his widely acknowledged legal expertise, he conducted extensive archival research on genealogy, producing a vast amount of papers. Without a doubt, Imhoff could expect high-quality expertise from Gittio.

It was from Gittio, in turn, that Imhoff learned about yet another Neapolitan genealogist, Biagio Aldimari (1630/39–1713).[110] As with Gittio, Aldimari's aspirations of social advancement led to a lifelong dedication to legal studies and to genealogy, before he eventually entered the services of the house of Carafa, whose history he published in a luxurious book in 1691 (*Historia genealogica della famiglia Carafa*). Whereas Aldimari eventually embraced the role of a dynastic historian, his education and social background also closely aligned him with the *togati*.

Although Magliabechi knew most of these men and corresponded with some of them, Imhoff's connections to this group were not the Florentine's doing. Magliabechi, however, did provide Jakob Wilhelm with one more Neapolitan contact, a highly learned and genealogically inclined, yet otherwise obscure Discalced Carmelite named Girolamo Maria di Sant'Anna.[111] This Catholic monk, who eventually became Imhoff's most important contact in Naples, had much to offer the German Protestant. He was in charge of the library and archive of his order's prominent Neapolitan monastery, and this collection was well stocked with genealogical and historical material.[112] Even more important, Sant'Anna gave Imhoff access to a very different type of source. The Carmelite stood apart from the scholarly circle of *togati* to which most of Imhoff's other Neapolitan informants belonged. He cultivated a wide range of high-profile contacts with the nobility of southern Italy.[113] Although Gittio and Aldimari also had important relationships with local nobles, it was largely Sant'Anna who secured Imhoff access to the internal genealogical knowledge of sundry noble families.[114] Many of the great families of southern Italy, including the houses of Aquino, Cellamare, Gaetano, and Giudice, opened their archives to Imhoff and forwarded him important genealogical material by way of Sant'Anna.

Although Imhoff never explicitly commented on the remarkable diversity of his Neapolitan informants, it is nevertheless obvious that the different groups were motivated by very different reasons to cooperate. In Gittio's case, his desire to publicly display *togati* culture coincided with an excessive concern about his family's rise to the nobility. He was "so infatuated with his own nobility, that he deemed no other activity more worthy than finding sources to establish his own genealogy and that of others," as one eighteenth-century observer noted with polemical condescension.[115] Genealogy was a big part in Gittio's attempt to improve his private fortunes, and this concern, in conjunction with antiquarian learning, led to a lifelong dedication to archival research. Forwarding relevant knowledge to Imhoff further publicized his public persona.

Aldimari's intellectual career followed a somewhat different trajectory. His extensive genealogical research had originally begun in the legal-historical circles of Neapolitan administrative officials. By the time Imhoff arrived on the scene, however, he had long since moved to a courtly context. Aldimari's shift to the Carafa court might explain his rather "secretive" nature. Since he had committed to producing a celebratory work on the Carafa, he probably feared competition from Imhoff, whether it was realistic or not. In the end, Aldimari communicated little knowledge to Nuremberg. Imhoff cited him as a published authority but not as a contributor.[116]

Bulifon, the bookseller who orchestrated part of the genealogical information transfer, would have been interested in helping Imhoff in turn not least because it fit his well-established profile as a facilitator. Sant'Anna, finally, seems to have been motivated by his role as a confidant of the regional nobility; he thus supported their efforts to have their internal self-definition reproduced by a foreign author in prominent publications. Imhoff clearly profited from the numerous different reasons people had for helping him, and it is perhaps his greatest achievement that he effectively cultivated all these different groups at the same time.

The Problems of Iberia

Whereas Imhoff dealt directly with a host of French and Italian genealogists and could, at least indirectly, address specific questions to local noble houses, matters were different in Iberia. Imhoff ultimately published several works on the Iberian nobility, but finding knowledgeable informants for Spain and Portugal was especially difficult. Many of Imhoff's friends shared his frustration. Reliable connections to Spanish scholars and intellectuals were uncommon among Imhoff's comrades. Magliabechi, for instance, had no correspondents there.[117] When Spener published his *Theatrum Nobilitatis* in 1668, he openly admitted that he had good coverage only until 1622, the year when a standard work by Alfonso Lopez de Haro had appeared—de facto Spener's sole source for Iberia.[118] Although Imhoff eventually fared far better, he likewise did not have easy access to Spanish genealogies in the 1680s or 1690s. He began acquiring literature in preparation for future projects around 1690, trying to purchase relevant books all over Europe.[119] It was a frustrating project, however, as Spanish books were rare in most places. His access to relevant literature was more or less random and depended on the fickle goodwill of collectors, owners, and booksellers.

Imhoff tried byways and detours. Since Spanish influence was strong in several territories across Europe, he might have found several potential entry points into Spanish culture. Yet Imhoff hardly turned to contacts in Milan or Vienna to learn about Spain, perhaps because he was not particularly well connected to either city.[120] Instead, he endeavored to take a detour via Naples—also attached to the Spanish Empire—to obtain information about Spain. Imhoff stressed that Spanish and Neapolitan genealogies "needed to be treated together because of the conjunction of blood" that resulted from centuries of Spanish rule over southern Italy.[121] What this meant became evident in 1696, when a new Spanish viceroy, Luis Francisco de la Cerda y Aragón (1660–1711), Duke of Medinaceli, arrived in Naples. Imhoff immediately seized the opportunity and wrote to his friends in Naples, asking them to obtain genealogical information about that important Spanish family from the viceroy's household. This initiative was not without success. At one point, for instance, Imhoff received the important bit of news that the viceroy was without surviving offspring. He had so far failed to father a male heir (a daughter, Catalina [1678–81], had not survived childhood), but his wife had traveled to the baths of Ischia to remedy the situation—unsuccessfully, as it turned out.[122] Such sensitive information could occasionally be obtained by observing the Spaniards living in Naples; yet waiting for individual Spanish nobles to arrive in Naples to inquire about their families was ultimately not a reliable way forward.

It gradually dawned on Imhoff that he needed to find "a person experienced in the genealogies and dynasties of Spain." Since he still had no contacts in Spain, he tried to find someone suitable via Naples.[123] On several occasions, his local friends mentioned Diego Vicenzio de Vidania (1623–1732), a high-ranking Spanish official in the Neapolitan church and administration, as a potential candidate. He would have been an excellent choice: Vidania was a great scholar and an avid collector of antiquities. His genealogical expertise was beyond doubt. Yet, Vidania declined Imhoff's invitation to join his genealogical brotherhood, excusing himself on account of his administrative duties.[124] Whether Vidania may have had other, more substantial concerns—ideological or political opposition to Imhoff's project?—is impossible to tell.

Whatever Vidania's real reasons, Imhoff still needed to find someone knowledgeable about Spain. But who to turn to? Around 1695, d'Hozier in Paris likewise explained that he had little knowledge about Spanish genealogies or access to them.[125]

Perhaps out of frustration, Imhoff asked his other Neapolitan contacts more directly about Spanish genealogies. Sant'Anna seems to have contributed little to Imhoff's knowledge about Iberia. Perhaps his contacts were confined to Italian families, although at least one of Sant'Anna's patrons, the prince of Santo Buono, later made a career in Spanish Central America. Only Gittio responded positively to Imhoff's insistent queries about the Spanish aristocracy. He provided bits and pieces of information, which Imhoff combined with other material. As a result, Imhoff successfully published his first exposition of Spanish genealogies in 1701.[126]

Since Gittio had died in 1700, however, Imhoff's Spanish projects had seemingly reached an impasse. Suddenly, however, an entirely new route to Spain opened up: "When I was at a loss with respect to the families of Spain, having been deprived of a certain person [sc., Gittio] who lived in the south, in whom I had placed all my hopes, there appeared to me from the north, like a *deus ex machina*, an eminent and energetic aide in Gerhard Ernst von Franckenau."[127] Franckenau (1676–1749) was an ambivalent figure: he combined pietist religiosity with heterodox and radical interests, scholarly rhetoric about intellectual vigor with scandalous plagiarism.[128] A young man from the Palatinate, he had just embarked on a diplomatic career serving the Danish crown. From 1698 to 1701, he worked as secretary to the Danish ambassador to Spain, the erudite Baron Friedrich Adolf Hansen von Ehrencron (1652–1711). After they returned from Madrid in 1701, both Ehrencron and Franckenau gained immediate renown as authorities on Spanish erudition, not least on account of the many books they brought back with them.[129] The return of the Danish embassy from Spain, thus, was of great significance to the German *respublica litteraria*. It seemed to open up a scholarly terra incognita in southern Europe.[130]

It remains unclear how Imhoff made contact with the diplomat,[131] but Franckenau almost instantaneously put scholarship on Spanish genealogy on a new footing. Franckenau's very first letter, from early 1702, made Imhoff realize how little he knew. In it, Franckenau criticized the family tree of the house of de la Cerda that Imhoff had just published. A much more detailed pedigree arrived from Copenhagen that must have made Imhoff regret his overhasty publication. In an astonishing feat of last-second editing, he inserted a new version of the de la Certa stemma into his forthcoming *Corpus*

Historiae Genealogicae Italiae et Hispaniae, which appeared in 1702, shortly after Franckenau's first letter.[132] Until 1704, many more genealogies reached Nuremberg, usually in narrative form, since Franckenau rarely drew tables. A good portion of Imhoff's third and final book on Spain, *Genealogiae Viginti Illustrium in Hispania Familiarum* (Genealogies of twenty illustrious families in Spain), published in 1712, derived directly from these letters.[133]

In the case of Spain, Imhoff thus once again relied on a single intermediary who monopolized almost all of Imhoff's information. In contrast to d'Hozier, however, Franckenau was not a native of the country he was reporting about and, furthermore, was no longer even living there. It is difficult to discern how Franckenau acquired his knowledge. Some of his dossiers for Imhoff contain detailed remarks about manuscript sources in Spain, suggesting firsthand familiarity with the original documentation.[134] Franckenau also occasionally alluded to personal observation and conversations he had while living in Spain.[135] Much of Franckenau's information, however, was culled from existing Spanish literature, some of which he owned himself, some of which he found in Ehrencron's much richer collection.[136]

There were clear limitations on the expertise of an outsider. In contrast to d'Hozier and Magliabechi, Franckenau lacked the casual familiarity with their peers that they derived from their lifelong personal interactions. This difference is illustrated, for instance, by the example of Luis de Salazar y Castro (1658–1734), the most important Spanish genealogist at the time.[137] Imhoff, who had tried unsuccessfully to communicate with Salazar y Castro for years, may have hoped that Franckenau numbered him among his personal acquaintances.[138] But Franckenau had likewise failed to make personal contact with Salazar y Castro while he lived in Madrid: he had missed him when he tried to pay the scholar a visit at his home.

The diplomat subsequently attempted to make good this disappointment. Perhaps motivated by Imhoff's keen interest in and appreciation of Salazar y Castro's contributions, Franckenau wrote a formal letter to the famous Spaniard in early 1702. In it, he asked Salazar y Castro for a list of genealogical works about Spain.[139] The Spanish scholar immediately penned a "Biblioteca genealógica española" (Spanish genealogical library) containing 240 titles, which he sent to Copenhagen, and expressed a general willingness to teach the Germans about Spanish genealogy.[140] Epistolary correspondence thus had to stand in for Franckenau's relatively shallow personal connections with Spanish genealogists. After initial difficulties, Salazar y Castro proved to be a valuable resource for Imhoff via Franckenau.

Information about Spain was not only hard to come by for Imhoff, it was also a highly sensitive topic circa 1700. Political conflicts directly affected the flow of genealogical knowledge. Whereas Spain had been effectively off-limits to d'Hozier in 1695, the country appeared more accessible to him after 1702 or so. The Bourbon succession to the Spanish throne immediately improved French communication with Iberia, and Imhoff would profit from it. Suddenly, since some of his "friends" had moved to Madrid "with the king [sc., Philippe V]," d'Hozier could successfully acquire genealogical information.[141] The fact that Imhoff published on Spanish grandees in French in 1707 (*Recherches historiques et genealogiques des grands d'Espagne*, i.e., Historical and genealogical research on the grandees of Spain) also seems to reflect growing French interest in Spanish genealogy.

But as the French gradually became dedicated to understanding the Spanish aristocracy, others of Imhoff's friends grew wary of being associated with the Catholic monarchy. Although he eventually provided significant information about Spain from Naples, Gittio initially took offense at Imhoff's equivocation of Italian and Spanish families. "Moreover, if [you] should request information from me about the kingdom of Naples, that should be fine, but you constantly ask me about Spain as if I was a Spaniard or living in Madrid," he once brusquely replied to the genealogist.[142] It seems that Imhoff had been too cavalier in handling the complexities of southern Italian identities. Spanish families were a sensitive topic to Gittio. This did not preclude cooperation, as he went on to state in his letter, but Imhoff needed to tread carefully. The political dimension of genealogy always had to be taken into account. Inadvertently hitting a nerve might endanger an otherwise well-established genealogical connection.

Trusting Local Knowledge

The preceding survey of Imhoff's regional contacts illustrates an aspect of genealogical knowledge that was of paramount importance: it was highly local. As such, it was legible primarily for insiders and hard to decipher elsewhere. The sources were often intelligible only to those who possessed specific knowledge about local customs and contexts, including languages. Imhoff often depended, quite literally, on interpreters. Personal names were a case in point. For one, there was the well-known difference between vernacular and Latinate forms of names.[143] Then, there were the many different vernacular forms of the same names. Franckenau once had to explain to Imhoff that "Rui" and "Rodrigo" were in fact the same.[144] Gittio similarly alerted Imhoff to the equivalence of different forms of personal names: "Jaime" and "Jacopo," for instance, were interchangeable.[145] He also corrected minor errors in Imhoff's spelling of Sicilian names.[146]

Genealogical research, therefore, could often be conducted only on site. It was people from the region under consideration who could cultivate genealogical knowledge most effectively. The localism of genealogical information was so pronounced that some of Imhoff's friends doubted whether a non-local or "foreign" perspective was feasible at all. Gittio, in Naples, once challenged Imhoff: "It is not easy for foreigners unfamiliar with the land [*non prattichi del Paese*] to write [genealogies], since they have to rely on the reports [*relationi*] that are sent to them and cannot detect mistakes."[147] Genealogists needed to know about local customs and contexts. They also needed to know about local archives and sources, and the local legal framework, not to mention local political sensibilities.

Early modern scholars across Europe grew increasingly obsessed with describing and analyzing "indigenous" natural and cultural phenomena they found at their doorsteps.[148] One of these locally rooted phenomena, scholars like Gittio insisted, was the European nobility. Aristocratic identities were anchored in specific regions—branches were often named for their main territories.[149] Thus came about, for instance, Saxon branch dynasties of "Saxe-Gotha-Coburg," "Saxe-Weimar," and "Saxe-Zeitz"—or, for that matter, the "Imhoff of Untermeitingen" and the "Imhoff of Eastern Frisia" lineages.[150] As the Italian d'Aquino family put it in 1702: "It is a sufficiently certain fact, verified daily by experience, that many families, although they may have the same origin and derive from the same stock, yet have taken up residence not only in different cities of the same kingdom, but also in faraway places and in provinces and kingdoms that are very distant."[151] Or, in the words of Nicolaus Rittershausen: "Concerning the sixteen forebears of the counts of Fürstenberg, it honestly is very difficult to round them all up, since the different realms, where one has to search for them, are quite distant from each other, one branch being in Spain, the offspring of others in Bohemia."[152] How could the genealogist access this local knowledge from afar? Gittio pointed out that there was little Imhoff could do other than trust his informants, for better or worse. Since he was unable to explore local sources himself, the genealogical compiler depended on locally produced "reports" or "memorials."

As everyone acknowledged, this dependency created numerous occasions for error and misinformation. "I was too trusting of friends," Rittershausen once ruefully admitted after he discovered he had uncritically relied on local informants.[153] Imhoff fell into the same trap; he explained one such error to a friend as follows: "I will have initially copied someone else in good faith without further investigation, and since no objection to it was made, it must have been carried over from one edition [of the *Notitia Procerum*] into the next."[154] "Trust," "good faith," and "belief"—these were not necessarily the epistemic virtues that scholars and historians touted as hallmarks of the "new," critical historiography. But there was little Imhoff, based in Nuremberg, could do when he was compiling a genealogical encyclopedia. Like naturalists and astronomers, who habitually accepted secondhand reports as hard empirical evidence when they were sent by the right people, historians and genealogists based their narratives on incoming dossiers produced by locals with direct access to sources and archives.[155]

It is difficult to reconstruct how Imhoff reached the conclusion that an informant was trustworthy. He was certainly not skeptical in principle about secondhand knowledge that had been privately communicated. Imhoff largely accepted the authority of his informants and very frequently trusted his personal contacts more than the printed genealogies. Although he and his friends talked a great deal about the relative merits and faults of their many published sources, few passages in his papers demonstrate concerns about his correspondents' reliability. There is also little indication that Imhoff had an explicit preference for one informer over others. Only rarely did he rank the reliability of his correspondents.[156]

In some cases, Imhoff's confidence in his informants probably stemmed from previous personal encounters, as with Magliabechi and perhaps d'Hozier, whom he may have met during his stay in Paris. D'Hozier's prominent institutional position would also have corroborated his authority, while Magliabechi's authority was broadly established in the Republic of Letters and would not need to be ascertained in detail by Imhoff. For lesser-known individuals, though, such external credentials were often lacking; still, Imhoff worked with whoever was willing to help and eagerly incorporated whatever bits of information he could acquire.

There were ways of mitigating the possible fallout from untrustworthy and unsubstantiated informants. One powerful tool to do so was Imhoff's rhetoric of preliminarity. Throughout his publications, he emphasized that genealogy was a constant work in progress and in constant need of updating and improvement. Similarly, his correspondence abounds with formulations that relativize his genealogical knowledge. He and his peers continually noted gaps in their knowledge, pointed out what was *not* (yet) known, and highlighted the challenging nature of genealogical knowledge in general. As d'Hozier stated, "it is difficult to have everything and know everything" in genealogical matters.[157] Genealogical knowledge was full of "dubia" and "incerta"—doubtful and uncertain information.[158] Correcting errors was a sort of modus vivendi for genealogists, as Spener once said, "I can readily yield to those persons who have pointed out my errors to you."[159] It was almost natural that informants would be wrong on some points: Genealogical data "are matters of fact [*res facti*] that one has to accept just as one would from trustworthy men, and the old consolatory maxim remains true: *errare humanum est*."[160] This focus on the problems and preliminary nature of all genealogies was certainly appropriate in light of the practical and conceptual difficulties of actually conducting genealogical research.[161] Yet this rhetoric of epistemic limitations also relieved the compiler and exculpated him if he had overhastily accepted someone else's erroneous information.

Nevertheless, on several occasions Imhoff's surviving materials attest to a slightly more ambitious practice, showing his dedication to check and expand on incoming local knowledge. Imhoff was by no means simply a passive recipient of what his friends forwarded to him. On the contrary, he peppered them with highly specific and searching genealogical questions.[162] For example, Imhoff had several questions after he received a valuable packet of information about the Este family from Modena in early 1694. He wrote back to the source, Bacchini (via Magliabechi), twice to ask for further details concerning one "Hippolita Estensis" (1620–?), wife of Borso d'Este (1605–57).[163] Bohuslav Balbín's letters to Imhoff often give the impression that he worked through lengthy to-do-lists that the latter had sent him.[164] The Bohemian scholar may have shaped their correspondence by selecting the information he found valuable and important, but the German recipient, though he never fundamentally doubted his informant's methods, still clearly tried to check and amend the incoming data, if only by subjecting it to additional questions and seeking clarification.

The genealogist was very specific on occasion as to what his informants needed to do. Imhoff adopted an extraordinary procedure to micromanage his correspondents and control local information. In 1696, he drew up a genealogical table of the Sforza family and highlighted with colored ink the individuals about whom he required additional information. He sent the document to Mandosio in Rome with the following instructions: "I would like to know whether any of the living Signori Sforzi, marked in red, are married and in what status they live."[165] Mandosio complied and filled in the missing in-

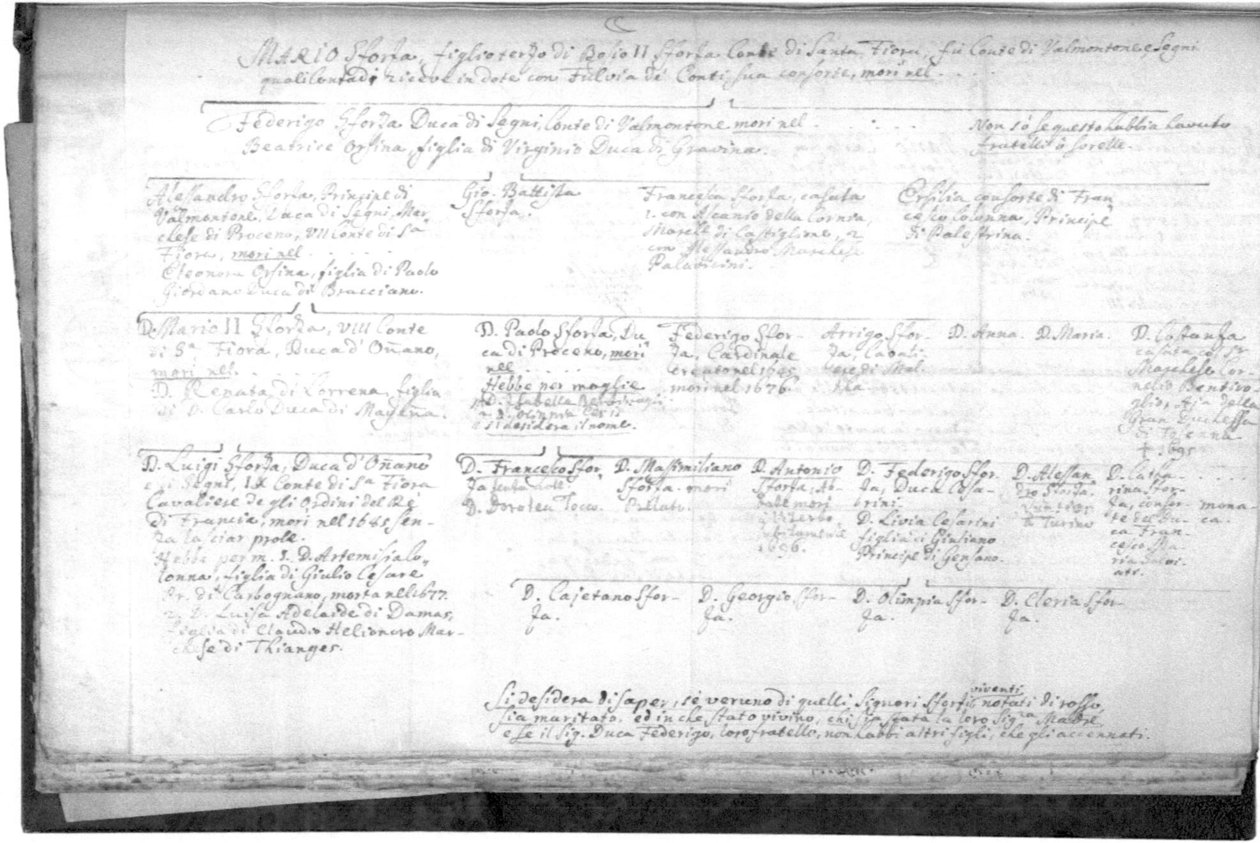

Imhoff's genealogical table with underlining (originally in red ink) indicating individuals about whom he needed clarification and Mandosio's inserted information (lighter ink, different hand). The writing at the bottom of the document, containing instructions for Mandosio, reads: "I want to know whether any of the living Signori Sforzi, marked in red, are married and in what status they live, who their mother was, and whether Signor Duke Federigo, their brother, had other children than those mentioned." GNM, Généalogies S–T (vol. 21), unpaginated (Sforza).

formation in his own hand before sending the document back to Nuremberg.[166] There are several further parallel cases, in all of which Imhoff turns stemmata into a matrix for obtaining additional genealogical data in a controlled procedure.[167]

On other occasions, Imhoff used similar tools to corroborate and improve the information he had. More than once, he replied to letters with detailed lists of "scruples" or "uncertainties" (*dubia*) engendered by incoming information. Franckenau, who received several such requests for clarification, duly replied and substantiated individual facts.[168] The Nuremberg scholar also sent drafts of family trees he constructed to friends for review. Pusterla in Milan, for instance, seems to have received Imhoff's genealogies of the Visconti "so that I can put them in a clearer light."[169] Imhoff may have been a compiler, but he tried as hard as possible to guarantee the quality of his data. Information acquisition for Imhoff, thus, was never exclusively a one-way transfer of data but rather a reciprocal process of ongoing conversation.

Taken together, Imhoff's practices for assessing and checking incoming information were tailored to fit his position as a genealogical compiler working from afar. He was generally inclined to trust his informants, who had either been recommended to him by intermediaries, or whose writings had already convinced Imhoff of their general credibility. This did not mean, however, that the genealogist did not have his own opinions about new material. Compiling information did not mean that Imhoff naively trusted all of it. He was not easily satisfied by what his correspondents forwarded. The fact that he applied a noteworthy degree of scrutiny to many of the dossiers he received is significant. He certainly reviewed almost everything he acquired in the light of other sources available to him. Although that never led him to reach a broader conclusion about the quality of his informants, it nevertheless required detailed criticism. Taken together with his point that imperfect knowledge was inevitable, this mixture of general trust and detailed scrutiny defined Imhoff's approach to the dossiers he received.

The art of harvesting local genealogical knowledge from afar was of the utmost importance for projects of encyclopedic genealogy. Imhoff's mastery of this practice is indeed striking. Few of his peers could boast of similarly wide-ranging contacts to so many local communities and sites of genealogical knowledge across Europe. Many other encyclopedic genealogists compiled information merely by excerpting it from existing encyclopedic literature, such as the works of Rittershausen or Spener, often adding little material of their own. And compilers of genealogical information who actually asked around and tried to contact local specialists were often frustrated. "Very few of the people from whom I wanted some knowledge and news took the trouble to oblige me and send some information," one genealogist complained a few decades after Imhoff's death.[170] "Lack of information" forced many compilers to abandon their projects entirely or scale them back dramatically.[171] Not Imhoff, however. He stood out for his ability to enrich the state of encyclopedic genealogical research by tapping into local conduits of information, thus giving universal genealogy some of the localized and specialized flavor of archive-based genealogical research. His works stemmed from his skill at tying numerous local strains of information together into a pan-European republic of genealogists, no matter how transient and fragile this genealogical brotherhood may have been in the end.

Viewed in their entirety, the variety of informants, collaborators, supporters, and friends whom Imhoff marshaled for his projects is truly remarkable. In many parts of Europe, Imhoff successfully engaged the help of very different men in his quest for genealogical information. Like pieces of a jigsaw puzzle, he fit his contacts together and combined their local and regional expertise to produce a genealogical image of most of western, southern, and central Europe. There were obstacles to overcome, such as the difficulty of finding someone knowledgeable about Spain and Portugal. Rome was also relatively understudied, and Imhoff never made any serious attempt to contact Venetian scholars directly.[172] For Bohemia and England, his contacts were comparatively scarce. Nevertheless, the sheer amount of information that Imhoff successfully solicited is impressive. Judging from the surviving sources, neither Rittershausen nor Spener, Imhoff's two most important role models, ever came close to his range of informants.

Importantly, Imhoff relied on very different types of informants. There were scholars and scholarly communicators like Magliabechi or Thulemeyer. There were officials and administrators, most obviously d'Hozier. Imhoff's circle of friends, moreover, included many "knowledge technicians," such as booksellers and librarians. Diplomats were also coveted contributors who often proved crucial when all other attempts had failed. Imhoff's informants included churchmen and laymen. Several of his correspondents might be called genealogical experts or even "professionals": Aldimari, d'Hozier, and St George Ashe spent much, even most of their time researching and deliberating genealogical questions. Others, however, acquired information relevant to Imhoff in a more accidental manner or as a by-product of other activities, as was the case with many bureaucrats. Nevertheless, administrative circles were often particularly important for Imhoff. To an astonishing degree, these officials could open up hidden caches of genealogical information that noble and princely families had hoarded.

In dealing with this complex and diverse set of men, not to mention the considerable egos of some of them, Imhoff nearly always seems to have struck the right tone. Signs of conflict in his correspondence are rare. This is all the more remarkable insofar as genealogical research was frequently tainted by personal and political loyalties. Yet underlying tensions surface very rarely in Imhoff's letters.[173] Imhoff seems to have been adept at the give-and-take of early modern channels of knowledge. He duly acknowledged his dependence on his collaborators and appreciated their proactive efforts.

Imhoff's amicable and tranquil character served him very well. His urbane style made genealogical information flow across some of the most notorious divisions of early modern Europe. Confessional strife, for instance, was hardly ever an issue. Imhoff counted not only many Catholics among his correspondents, but also several men of the Catholic Church, including monks and members of other religious communities. No anti-Catholic or anti-monastic feelings, however, emerge from his letters to the Discalced Carmelite Girolamo Maria di Sant'Anna or to the Jesuit Bohuslav Balbín.[174]

Imhoff's correspondence connected universities, courts, administrative offices, Jesuit colleges, urban patrician homes, and the institutions of the Holy Roman Empire. Informants from these places of knowledge may have had very different motives for cooperating, but all of them nevertheless contributed valuable information. They all regularly interacted with early modern aristocrats, and they all knew at least something about them. Yet informants of different stripes interacted with the nobility in many different ways. Hence courtly contacts provided genealogical information that often differed from that provided by academics or government bureaucrats. Some contributors only had a few isolated details to offer, while others sent comprehensive genealogical exposés; some information was based on firsthand observation, other pieces of information came from archival research; some concerned the remote past, while other material focused on recent events. The relevant point here is that, for Imhoff, only an inclusive approach to this bewildering variety of information and informants could succeed. Imhoff was interested in *all* forms of genealogically relevant knowledge, whatever the reason why they had been forwarded to him. Only by combining heterogeneous sources could Imhoff complete his genealogical projects. He was truly an encyclopedic genealogist.

The Genealogist at Work

No description of Imhoff's study has been discovered, but we may expect that it was full of papers of various kinds, formats, and quality piled high on his desk and shelves. There were presumably interspersed books and other printed material, such as journals and newspapers. Perhaps Imhoff, like Magliabechi, lived in a "chaos of . . . folios."[1] More likely, however, the genealogist kept his materials more or less neatly organized into little dossiers, one for each family. This, at least, is how they have come down to us.[2]

Imhoff's dossiers included a bewildering array of different written artifacts, as his informants forwarded genealogical information in a wide variety of forms. They sent him both brief fact sheets and large-scale family trees, reading notes and verbatim excerpts, accounts of medieval charters, and the latest gossip. Imhoff collected both original sources and thirdhand information in fourth-hand manuscript copies. None of these items fit seamlessly together. Their information overlapped in many places, but also showed glaring gaps. In some cases, information piled up over the years, as earlier dispatches were superseded by more recent ones, while other bits of information arrived literally as the printing presses were already being turned. Out of this heterogeneous material, Imhoff created his works through a complex process of bricolage, literally—physically—piecing together the many different kinds of information he had acquired.

Encyclopedic bricolage was an intensely material and physical activity, as scholars increasingly realize.[3] The evidence of Imhoff's papers lets us imagine early modern genealogists working with scissors, glue, needles, and slips of paper to bring information together physically. Sophisticated methods of writing information down on paper, especially the use of hand-drawn diagrammatic sketches, were vital to the process of sorting genealogical data into well-ordered structures. Although no coherent description of Imhoff's own modus operandi survives, we can reconstruct how he must have had numerous preparatory papers in front of him, perhaps overlaying partial family trees so that their spatial juxtaposition might enable him to envision a more complete, integrated stemma. For sure, the genealogist would have moved sheets around on his desk in search of details and connections merely waiting to be made. Quite literally, as we will see, he arranged his documents like building blocks to create his genealogies.

The following chapter reconstructs Imhoff's working methods in detail. First, I discuss the tools and strategies he employed to find and organize contemporary genealogical information. In Imhoff's day, this was hardly standard procedure. Whereas scholars increasingly theorized about and even taught the methods and "little tools of knowledge" for studying antiquity at schools and universities, there was little established know-how for studying the recent past.[4] As Imhoff strove to stay abreast of the latest dynastic events, he had to improvise new procedures and develop new ways of storing and digesting genealogical news.

The second part of this chapter foregrounds another physical dimension of encyclopedic bricolage. One of the most important practices genealogists used to assemble genealogical tables from heterogeneous splinters of knowledge was drawing, often in the form of preliminary genealogical sketches. The important role of visualization for *displaying* information, including tables and diagrams, is widely acknowledged today. In the field of genealogy, arboreal images—trees of knowledge, including family trees—were especially popular.[5] Imhoff's papers allow us to add yet another dimension. They demonstrate that the physical act of drawing was also a crucial tool for *creating*, sorting, and curating data during the process of genealogical research. Sketching diagrams helped the genealogist digest information.

The final section brings these discussions of Imhoff's various routines together and illustrates genealogical bricolage with a case study. The pages Imhoff dedicated to the Italian house of Este in his *Corpus Historiae Genealogicae Italiae et Hispaniae* of 1702 can be connected to a range of preliminary manuscript documents. These papers allow us to observe in detail Imhoff's material working habits and the choices he made about what and what not to include. The case of the house of Este reveals genealogical work in action. We can see how many of the conceptual considerations discussed above—including

Imhoff's attention to the hierarchy of dynastic lineages and his constant search for the most recent information—translated into specific material practices out of which, eventually, a finished genealogical narrative and a set of tables emerged.

Keeping Up to Date

There was a growing presentist bias in encyclopedic genealogy circa 1700. Readers expected that Imhoff would "put together *all the living members* of each house" under discussion, as the genre became part of the emerging political information economy.[6] In the words of one of Imhoff's peers, "it is unnecessary to know the arms of families that have long since died out."[7] Rittershausen, it is worth recalling, had completely abstained from discussing genealogical history before 1400. Contemporary information was a unique selling point. Some readers even wanted recent dynastic events to be highlighted with special fonts or graphical signs, so that they might find them more quickly.[8]

As Imhoff would have insisted, contemporaneity was not only very important; it was also one of the greatest challenges of the genre. How might he learn about recent noble births, deaths, and marriages as quickly and reliably as possible? And how could he manage the incoming data? Imhoff proved very enterprising and innovative in his effort to satisfy his readers' presentist expectations. Through trial and error, he eventually found numerous innovative ways to procure, verify, and store up-to-date information.

"Public" News and Print Media

The most obvious starting point for Imhoff was "public" information about the nobility, as he and his friends put it.[9] A significant amount of dynastic genealogical intelligence circulated in a wide variety of printed news media. It was "public" because it was theoretically available to every interested reader. As Imhoff once noted, journalists were increasingly turning their attention to relevant events.[10] Even though journals and newspapers were not dedicated to genealogy per se, they nevertheless contained a significant amount of relevant data, albeit often mentioned en passant, in the course of reporting on political or ceremonial affairs.[11] Imhoff needed to disentangle biographical information from a plethora of news items in order to refashion journals and newspapers as genealogically valuable sources.

The Nuremberg genealogist proved to be a highly proficient user of what was at the time still a new and controversial source of information. He eagerly read journals and newspapers from many places, including the *Gazette de France*,[12] the *Mercure Historique et Politique* from The Hague,[13] the *Historische Remarques der Neuesten Sachen in Europa* from Hamburg,[14] various newspapers from Rome, and the *Gazette Françoise* from Leiden.[15] Dozens of pages of excerpts show that Imhoff systematically perused these newspapers. Although he rarely quoted them explicitly in his published works, it will become apparent that it was often through news media that Imhoff learned of the latest events, often at the last minute.

Yet Imhoff read newspapers not only for current dynastic events but also as a historical source for the recent past.[16] When he researched a specific family, he systematically scoured back issues of one or more journals (or his excerpts of such journals) that seemed relevant and collated all the relevant news items he found in chronological order. The typical results were numerous fact sheets, folios filled with information about a family or house, all reflecting the standard format of news reports: a more or less concise summary of events, for instance, a marriage, a death, or a birth, introduced by a specific date and a location (such as "from Rome").[17]

Imhoff ran into typical difficulties when working with journalistic news media. One problem in particular is obvious from his notes as he tried to exploit the genre's informational riches: the papers reported either too little or too much about individual nobles. No journalist was beholden to genealogical standards of biographical exhaustivity. In 1710, for instance, Conrad Barthold Behrens told Leibniz that he needed to inform Imhoff about two recent births in the royal family of Berlin, because "the names of the two princes born in Berlin last week were not reported in the newspapers."[18] Vice versa, newspaper reporting on individual nobles often was far too extensive to be useful in its entirety. It took hard work to condense all of the excerpted information into concise summaries. Imhoff compressed several pages of excerpts about Lorenzo Onofrio Colonna and his wife, Maria Mancini, for instance, into about twenty lines of printed text. He decided to summarize his equally extensive notes on Filippo and Egidio Colonna, who both appeared prominently in contemporary news, even more dramatically.[19] Imhoff eventually deemed many other events reported and excerpted from news media to be too marginal to merit inclusion in his printed works at all.

There was another reason why journalism alone was an inadequate source of information for contemporary genealogy: it was simply too unreliable, as many contemporaries criticized.[20] Prospero Mandosio, for instance, told Imhoff that he had found "little truth and many lies" in the newspapers of Rome.[21]

Imhoff indeed had encountered such problems in his own experience. He and his friends spent considerable energy verifying and correcting information that was circulated by the news media. The newspapers usually reported the thrust of dynastic events accurately, but they often got minor details wrong, such as dates or place names.[22] On other occasions, their genealogical information was simply false. In 1694, Imhoff informed Thulemeyer that a French journal had retracted an erroneous report about the death of a Hanoverian prince.[23] In 1709, the question arose as to whether the queen of England was really pregnant, as the newspapers reported. No, Fabricius informed Imhoff, he should not trust the news.[24] While such canards may not have troubled most readers, the margin of error for genealogists was simply too great. Misdating a noble's death by even ten days was a serious error from Imhoff's perspective.[25]

The careful study and utilization of news media were among Imhoff's most potent tools in pursuing his fascination with contemporary genealogy. Without the new opportunities for genealogical research that the incipient news circuit (unwittingly) provided, Imhoff and his peers could not have realized their firmly presentist vision of genealogy. Rittershausen had set a precedent for a new, contemporary (political) perspective on genealogy, but he still largely lacked the means to put it into practice, because so few printed journals and newspapers were available to him. Imhoff, by in contrast, typified the ability and willingness of a younger generation to engage the highly ambivalent and complex news media that, in many cases, enabled him to follow the familial developments of the nobility almost to the day of publication.

Tapping into Courtly Communication

If Imhoff relied heavily on news media for information about recent dynastic developments, he showed his true mastery of the field by harnessing additional resources for staying informed. In particular, the genealogist found ways to access so-called letters of notification (*Notificationsschreiben*) and turned them into a genealogical data pool. "Letters of notification" were a contemporary genealogist's ideal source.[26] These were standardized, handwritten notes sent from one court to others on the occasion of significant dynastic events—births, marriages, deaths.[27] With these letters, the European society of princes ensured that basic knowledge about key family developments was widely available to one another in reliable form. Such notifications guaranteed that neighboring noble houses officially took note of the changed dynastic circumstances and adjusted their actions accordingly. The sharing of genealogical knowledge also bound the social elite together, insofar as each newborn member of the nobility, each new marriage, and each death had a ripple effect on existing relations. Letters of notification, although Imhoff rarely acknowledges them explicitly as sources in his printed works, very likely supplied many of the bare facts that Imhoff relayed about the recent dynastic events.

As a non-noble, private individual, Imhoff could never partake in this valuable correspondence personally. Two of his acquaintances, however, gave him access, albeit indirectly. One was Jakob Wilhelm's close relative Andreas Lazarus Imhoff, who served as chancellor at the small court of Sulzbach-Rosenberg. Andreas Lazarus opened up his court's collection of *notificationes* to the genealogist. One of Andreas Lazarus's letters gives a nuanced comparative assessment of these documents:

> Knowing that much of the information that one takes out of the public journals is often false concerning the dates of births and deaths of noble persons, I have come to think it might be of service to you if I sent you the attached extract of our letters of notification, as they are commonly called. In these, you will find exact information about the dates of birth, marriage, or death of most of the princes who are allies of my Master [the Count Palatinate of Sulzbach]. It is true that these letters do not contain the names of newborns, as it is the custom to only announce the birth of a "Prince" or a "Princess," but I expect you will be able to supply what is missing from your own papers. Trust me, it will not be to your disadvantage if you tried to acquire similar information from all other courts. This would make your work more perfect and illustrious.[28]

Andreas Lazarus suggested *notificationes* as an alternative to journalistic sources. As the quotation implies, the court in Sulzbach systematically collected these letters.[29] Imhoff's cousin presumably went through an archival collection of them. On later occasions, he continued to forward new information as subsequent *notificationes* arrived in Sulzbach.[30]

Imhoff's second access point to this courtly communication was Duke Moritz Wilhelm (1664–1718), the reigning prince of the small Saxon duchy of Zeitz and "a lover of genealogy."[31] Moritz Wilhelm was a highly educated, curious, and culturally ambitious prince. History and, of course, genealogy were among his favorite fields.[32] After he had exchanged a few letters with the duke, Imhoff dedicated his English genealogies (1690) to him; personal correspondence between the Nuremberg patrician and the Saxon duke flowed back and forth thereafter until Moritz Wilhelm's death almost thirty years later. Imhoff and Moritz Wilhelm even met in person at least twice to discuss their shared love for genealogy.[33] Imhoff's relationship with Moritz Wilhelm differed from his relationship with most other nobles insofar as their correspondence was not focused primarily on the duke's own family, which came up only sporadically. Instead, despite their very different social stations, the two men shared a lively interest in news about the imperial nobility. Notwithstanding occasional historical references, their correspondence exhibits a strong presentist focus. Many of their letters consist almost exclusively of rather dull reports of "genealogical news" (*genealogische novitäten*). Although we occasionally find gossip here, general observations there, neither side seems to have been motivated by any underlying (political) agenda.

As a reigning prince of the Holy Roman Empire and a member of the house of Wettin, one of the most powerful families in Germany, the duke was included in the circulation of such courtly notifications. He seems to have registered the arrival of each new notification, not least because he was himself highly interested in genealogy. The prince depended on these letters to stay abreast of contemporary dynastic developments.[34] He also quickly forwarded new information to Nuremberg, often explicitly mentioning the latest notifications.[35] Thus, even though the original letters were not sent to Nuremberg, Imhoff managed to keep his genealogical information unusually up to date through his indirect participation in this courtly exchange of dynastic announcements.

These notifications poured forth a continuous stream of news about individual nobles and important dynastic events. If one simply filed the notifications away chronologically, their sheer volume would overwhelm anyone searching for specific details.[36] In order to make such piles of similar letters a viable research tool, Imhoff, his cousin Andreas Lazarus in Sulzbach, and Duke Moritz Wilhelm needed to make them searchable. There is unfortunately little information available as to how this was accomplished. Andreas Lazarus, for instance, did not describe what the previously mentioned "extract" he sent from Sulzbach looked like or how it was produced. Both Jakob Wilhelm Imhoff and Duke Moritz Wilhelm occasionally seem to have gone through piles of recent notifications, compiling lists of events by year or by every four months.[37] Fragmentary as these indications are, they nevertheless show that once genealogists secured access to such communications, they were very capable of transforming notifications into an effective genealogical resource.

Talking to People

There was something more basic than reading newspapers and courtly letters that one needed to do in order to learn the most recent news about nobles' lives: talking to people. Oral communication was of paramount importance for genealogy. True, critical historians who took greater interest in the ancient strata of family history discredited oral testimony, deeming it to be an unreliable and unverifiable transmitter of historical knowledge.[38] To Imhoff, however, insofar as he was interested in the most recent dynastic events, personal conversation was a crucial research tool. In order to learn about living aristocrats and to keep his contemporary genealogical knowledge up to date, Imhoff constantly had to engage in personal exchanges with nobles and their entourage and even eavesdrop on others as they casually spoke about peers and friends. No one expressed this reality more clearly than the Bohemian Jesuit Bohuslav Balbín in a private letter to Imhoff:

> It is impossible to describe how hard it is to find out these minor [genealogical] details, especially those that cannot be gleaned from books, but rather must be sought from the memory of very few living people, by inquiring about them. An infinite number of us knows Kolowrats and Schlicks etc., but scarcely anyone knows how many brothers there were, and who they are, who were the sisters, in what order they were born, who was their father's mother, who were their relations by marriage, their in-laws, their sons-in-laws, and the names of each one. These things have to be gathered from their house itself.[39]

Conversation was so important because the nobles themselves were often the only source of information about recent family developments. Many genealogically important facts had not (yet) been made public, and some never were. Nobles and courtiers knew a great deal about their peers, and they often had their information with authority and confidence, since they may have recently met the people in question, were related to them, or knew someone who was. Such experiential familiarity with the contemporary aristocracy was nearly impossible to find in writing. Imhoff—and his intermediaries—deliberately cultivated conversations wherever they could, engaging diplomats, nobles, and court officials in private exchanges in the hope of extracting precious information.

There is, unfortunately, very little evidence that sheds light on Imhoff's conversational strategies. Since he almost never talked about his daily routines in his letters, it is next to impossible to ascertain whom he met in Nuremberg and what they spoke about. But Imhoff's many middlemen across Europe spoke much more freely about the conversations they had with knowledgeable people, often on Imhoff's behalf, and we may assume that their strategies and experiences roughly corresponded to his own approach. Imhoff's friends often explicitly acknowledged that pieces of new information had originated from "oral" exchanges.[40] Such conversations might yield surprising insights, as the informants casually recounted entire episodes of recent family history in considerable, if not exhaustive, detail.[41] Not infrequently, it was noble women who willingly divulged information about their relatives in private conversation. Chatting with Princess Nicandro (d. 1703), for instance, was Girolamo Maria de Sant'Anna's key to information about the Naples-based Sermoneta-branch of the house of Caetani.[42]

Oral information had another, additional advantage. It was regularly much more direct than published or even handwritten information. If they were caught in the right mood and by the right person, many nobles and officials loved to tell stories about themselves and their families. By encouraging them to talk, one might hear sensitive news that no one would have dared to put into writing. Religious questions, for instance, might be clarified casually and spontaneously—off the record, so to speak. It was "over dinner" that Count Helferich von Welz told Imhoff how he had learned that another count would not convert to Catholicism.[43] There was also the chance that oral exchange might reveal embarrassing and sensational information. One Countess of Wied, for instance, met an (unidentified) informant of Imhoff in the summer of 1680. As they made casual conversation, the baroness started gossiping about her peers. She reported, "In his youth, Prince Ludwig of Baden was enamored with one Amalia von Löwenstein, and he made a promise to her concerning the future [sc., to marry her]. Because he forgot her, though, once he came of age and the Princess of Neuburg came to his attention, she [sc., Löwenstein] entered a convent. His cousins, the two sons of Markgraf Leopold [of Baden], are badly handicapped: the older one is mute, and the other one is not much better off."[44] Whether or not these (largely accurate) details proved relevant to Imhoff's research, the episode clearly illustrates the potential of oral communication. Socializing was a key strategy to learn about the nobility.

Imhoff and his friends were undeterred by gossip. Perhaps such juicy information added to the pleasure of pursuing genealogy in general, and contemporary genealogy in particular. They assumed that most of the information acquired through personal conversation was "absolutely certain."[45] There is no evidence that any of Imhoff's friends were ever fundamentally skeptical of oral information, as they were of printed news media. In fact, they hardly ever talked about oral testimony abstractly as a medium of knowledge exchange.[46]

To them, relying on personal conversation was an inevitable and acceptable means of acquiring information. Genealogists immersed themselves in the widespread flow of oral information in the early modern period.

Nevertheless, Imhoff's friends did not overlook the perils of oral transmission. They were well aware of the cursory nature of private conversation and understood that its spontaneity often resulted in fragmentary, one-sided, or incomplete information. Christoph Wagner, Imhoff's trusted informant from central Germany, described a revealing episode in 1689: "Yesterday, when I had lunch with a most important gentleman in foreign affairs, I heard that Heinrich was now the husband of the only daughter of the prince of Dessau. On coming home, I opened my [genealogical] tables, but I found there were multiple princes of Dessau."[47] Despite the fact that it is difficult to divine the identity of the persons Wagner mentions, the incident neatly illustrates how interlocutors not infrequently entered casual conversation unprepared and may not have had all the relevant details at their fingertips.[48] Oral information was not necessarily wrong, but it was often incomplete or vague. It required clarification, as Wagner discovered when he turned to the books and manuscripts of his well-stocked library. Oral and written information should theoretically corroborate and check each other in a mutually beneficial dynamic.

There were other challenges oral information entailed, as well. Harvesting genealogical information from conversations could be a frustrating process, as the Jesuit Bohuslav Balbín confessed to Imhoff. It was not easy to get ahold of competent people for interviews: "But if they are away from Prague and living in Moravia or Silesia or in Vienna or somewhere else, from where" can the information be obtained?[49] The itinerant lifestyle of the nobility, who traveled from central courts to private estates and back, not to mention diplomatic or military missions, contributed significantly to the flow of information but also might threaten to cut it off. Imhoff and his friends might have to wait patiently for months before important consultations could take place. From late 1707 and to early 1708, for instance, d'Hozier in Paris—and, by extension, Imhoff in Nuremberg—waited at least half a year before he could meet with the Abbé Salviati to inquire about his family.[50]

The oral acquisition of genealogical information was thus certainly not without its difficulties. Yet, for Imhoff and his friends, there really was no alternative. Contemporary genealogy could not forgo oral testimony. On the contrary: personal conversations were a crucial supplement to the portfolio of printed and manuscript media that informed genealogists about recent events.

Contemporary genealogy depended on sources that kept pace with daily family life, and conversations with people close to the nobility were one of the best ways to keep up.

To establish the latest facts about noble families, Imhoff and his friends relied on specific research routines, routines that differed dramatically from the antiquarian methods that were used to reconstruct the more remote genealogical past. A methodology for researching contemporary history, however, had not yet been developed. Whereas many important sources for the study of the past, such as charters, seals, and chronicles, were subjected to ever more rigorous methodological considerations, few sources utilized by critical, scholarly observers of the present day had received systematic, theoretical treatment.[51] Staying genealogically up to date thus remained a process of improvisation, however sophisticated and confident the work of Imhoff and his peers already was.

Pens and Scissors: The Materiality of Genealogy

Compiling genealogies was a profoundly material process. Imhoff literally needed to assemble pieces of information. Compilation entailed the physical handling and arranging of sheets, dossiers, and files. Papers had to be spread out on the scholar's desk so they might be read in tandem. In the case of visual media, we may even speculate whether Imhoff physically superimposed diagrams and family trees, so as to connect them visually. In other cases, his working papers show clear signs that either he or his informants had physically reworked individual papers to suit their purpose. Scissors and glue played an important part, for example, in the creation of a horizontal foldout diagram that detailed the genealogy of the counts of Salm. On many other, less spectacular occasions, Imhoff used glue to attach small, information-bearing notes to larger sheets of paper, perhaps for safekeeping. In short: genealogists worked physically with manuscripts and prints in a process of encyclopedic bricolage.

In order to join fragmentary pieces of genealogical information together, Imhoff and most of his peers relied on one scholarly practice in particular: the production of genealogical knowledge by drawing. Sketching and resketching family trees connected information culled from the incoming letters and dossiers. Drawing and redrawing genealogical diagrams was an important step in the process of consolidating information and creating order out of the many packets of preliminary information with which Imhoff started.

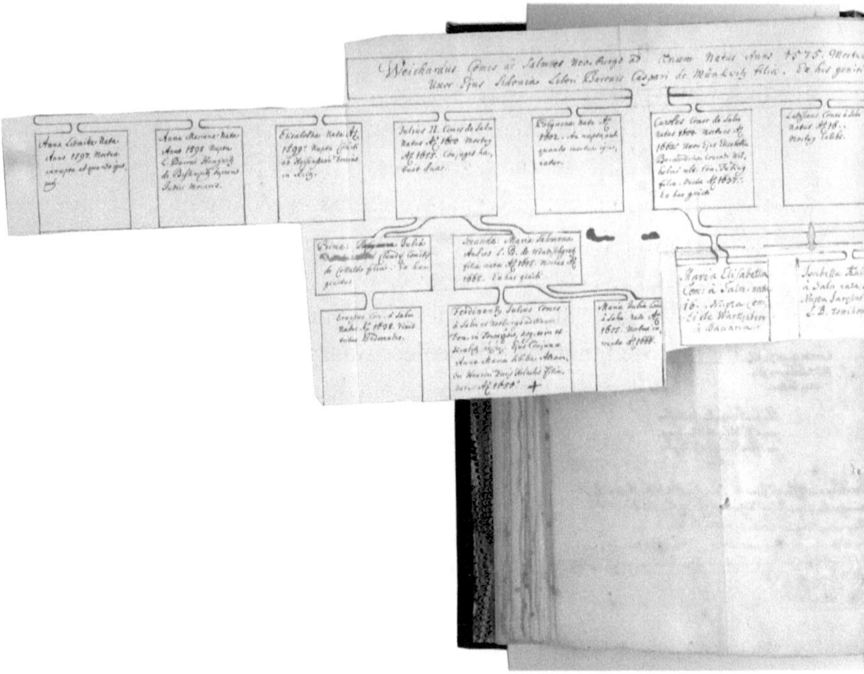

An irregularly shaped genealogical table for the counts of Salm, illustrating the use of scissors and glue by early modern genealogists. GNM, Généalogies S (vol. 20), unpaginated (Salm).

Visualization was extremely important to Imhoff in the process of connecting a vast number of individual nobles to larger social units such as families, branches, and dynasties.

Scholars generally recognize the importance of images for the history of knowledge. They highlight the epistemic power of spatial arrangements on a flat surface in transmitting facts and especially relationships between facts.[52] Visuals make information not only more accessible but also facilitate the hermeneutic process of digesting it. As early modern scholars worked through difficult texts, for instance, they often organized what they considered key content in diagrams, thus taking note of their own progress while reading and expediting future use of the same text by establishing reference points and giving it a clear structure.[53]

Genealogical information was also illustrated with visual media from a very early date. Scholars have studied at length the various types of genealogical diagrams that abound in the sources from the Middle Ages and subsequently.

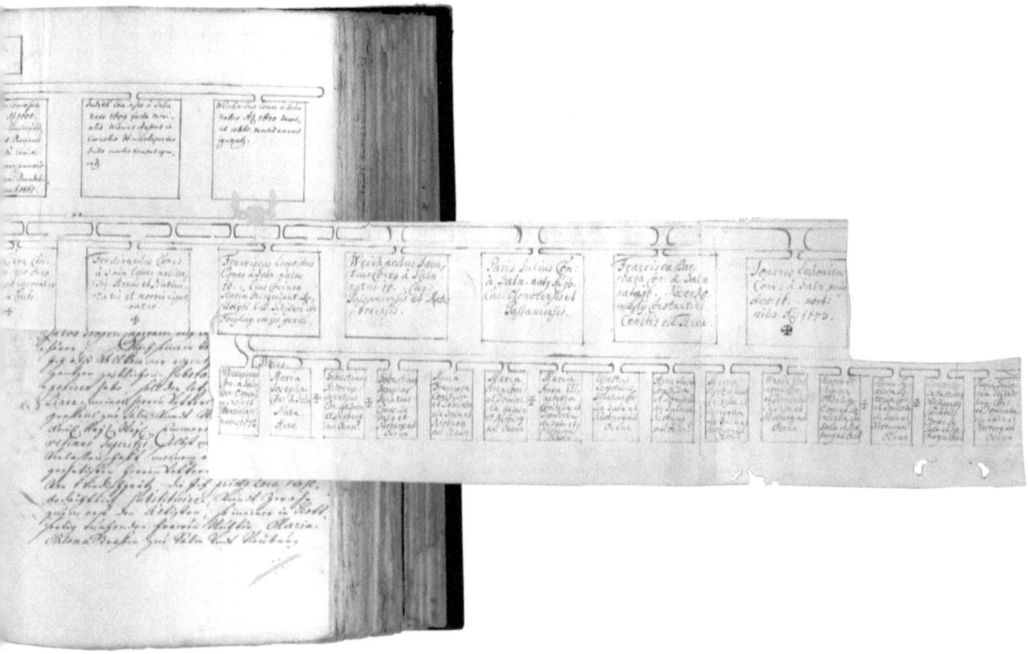

By far the most prominent visual representation of genealogical information is the proverbial family tree.[54] Genealogies often embedded information in real-life imagery: many genealogies were presented as realistically depicted trees. Numerous prints of this kind were produced in seventeenth-century Europe, and Imhoff owned several spectacular examples of the genre. His own works, however, adopted a different type of genealogical visualization, namely, unadorned and sober diagrams. Without any pictorial additions, these genealogical diagrams illustrated noble houses simply by connecting individuals to one another with straight lines and stylized brackets. As will become clear later, Imhoff was a master at designing such genealogical diagrams.

It was this diagrammatic type of visualization that recommended itself to genealogists as a tool for processing information. Consisting only of personal names and simple lines, drawing rudimentary genealogical diagrams was a basic physical operation of genealogical research. They could be done quickly: a genealogist could sketch them en passant, as he was working through his material. Diagrams could also fit in small spaces like the margins of books, and Imhoff occasionally sketched fragmentary diagrams on the pages of historical works while reading, producing mini-stemmata of one or two generations.

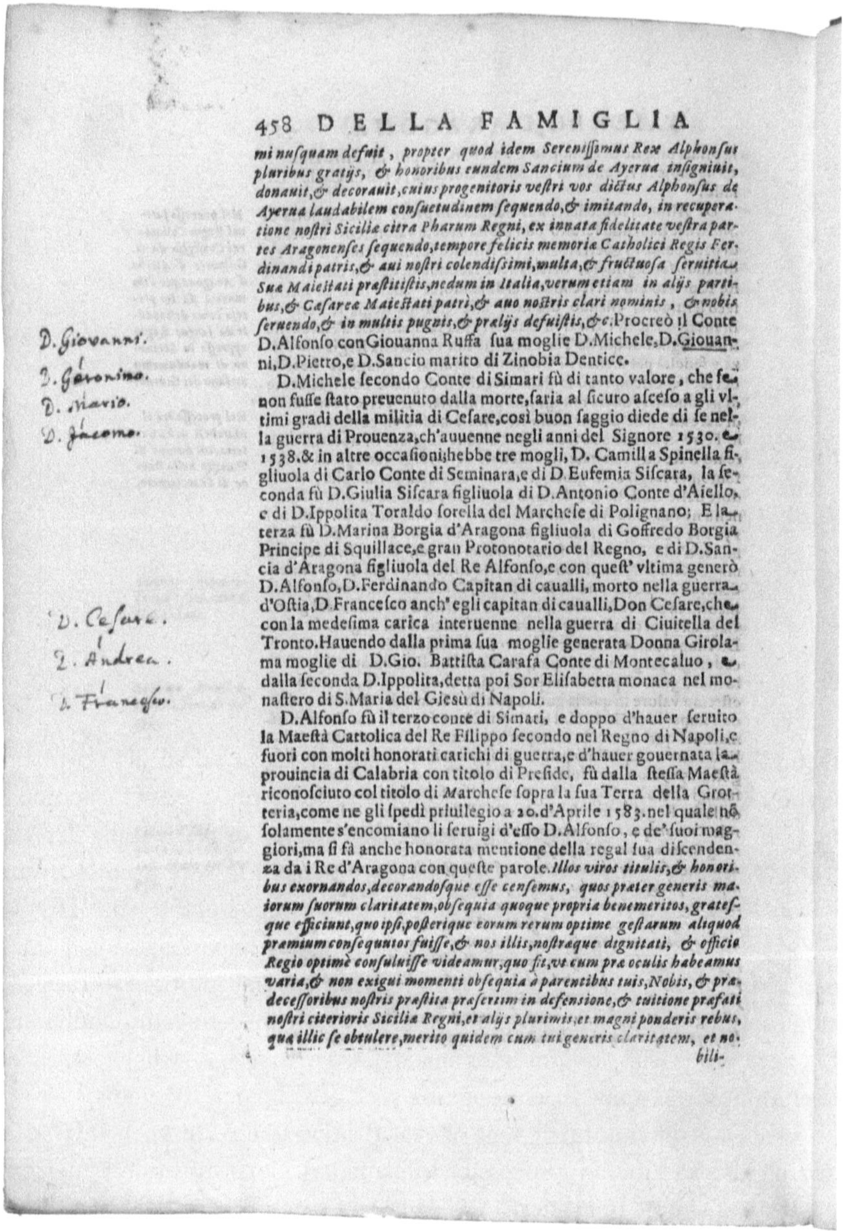

458 DELLA FAMIGLIA

mi nuſquam defuit, propter quod idem Sereniſſimus Rex Alphonſus pluribus gratijs, & honoribus eundem Sancium de Ayerua inſigniuit, donauit, & decorauit, cuius progenitoris veſtri vos dictus Alphonſus de Ayerua laudabilem conſuetudinem ſequendo, & imitando, in recuperatione noſtri Siciliæ citra Pharum Regni, ex innata fidelitate veſtra partes Aragonenſes ſequendo, tempore felicis memoriæ Catholici Regis Ferdinandi patris, & aui noſtri colendiſſimi, multa, & fructuoſa ſeruitia Suæ Maieſtati præſtitiſtis, nedum in Italia, verum etiam in alijs partibus, & Cæſareæ Maieſtati patri, & auo noſtris clari nominis, & nobis ſeruendo, & in multis pugnis, & prælijs defuiſtis, &c. Procreò il Conte D.Alfonſo con Giouanna Ruffa ſua moglie D.Michele, D.Giouanni, D.Pietro, e D.Sancio marito di Zinobia Dentice.

D.Michele ſecondo Conte di Simari fù di tanto valore, che ſe non fuſſe ſtato preuenuto dalla morte, ſaria al ſicuro aſceſo a gli vltimi gradi della militia di Ceſare, così buon ſaggio diede di ſe nella guerra di Prouenza, ch'auuenne negli anni del Signore 1530. e 1538. & in altre occaſioni; hebbe tre mogli, D. Camilla Spinella figliuola di Carlo Conte di Seminara, e di D.Eufemia Siſcara, la ſeconda fù D.Giulia Siſcara figliuola di D.Antonio Conte d'Aiello, e di D.Ippolita Toraldo ſorella del Marcheſe di Polignano; E la terza fù D.Marina Borgia d'Aragona figliuola di Goffredo Borgia Principe di Squillace, e gran Protonotario del Regno, e di D.Sancia d'Aragona figliuola del Re Alfonſo, e con queſt' vltima generò D.Alfonſo, D.Ferdinando Capitan di caualli, morto nella guerra d'Oſtia, D.Franceſco anch'egli capitan di caualli, Don Ceſare, che con la medeſima carica interuenne nella guerra di Ciuitella del Tronto. Hauendo dalla prima ſua moglie generata Donna Girolama moglie di D.Gio. Battiſta Carafa Conte di Montecaluo, e dalla ſeconda D.Ippolita, detta poi Sor Eliſabetta monaca nel monaſtero di S.Maria del Gieſù di Napoli.

D.Alfonſo fù il terzo conte di Simari, e doppo d'hauer ſeruito la Maeſtà Cattolica del Re Filippo ſecondo nel Regno di Napoli, e fuori con molti honorati carichi di guerra, e d'hauer gouernata la prouincia di Calabria con titolo di Preſide, fù dalla ſteſſa Maeſtà riconoſciuto col titolo di Marcheſe ſopra la ſua Terra della Grotteria, come ne gli ſpedì priuilegio a 20. d'Aprile 1583. nel quale nõ ſolamente s'encomiano li ſeruigi d'eſſo D.Alfonſo, e de' ſuoi maggiori, ma ſi fà anche honorata mentione della regal ſua diſcendenza da i Re d'Aragona con queſte parole. *Illos viros titulis, & honoribus exornandos, decorandoſque eſſe cenſemus, quos præter generis maiorum ſuorum claritatem, obſequia quoque propria benemeritos, grateſque efficiunt, quo ipſi, poſterique eorum rerum optime geſtarum aliquod præmium conſequutos fuiſſe, & nos illis, noſtræque dignitati, & officio Regio optimè conſuluiſſe videamur, quo fit, vt cum præ oculis habeamus varia, & non exigui momenti obſequia à parentibus tuis, Nobis, & prædeceſſoribus noſtris præſtita præſertim in defenſione, & tuitione præfati noſtri citerioris Siciliæ Regni, et alijs plurimis, et magni ponderis rebus, quæ illic ſe obtulere, merito quidem cum tui generis claritatem, et nobili-*

D. Giovanni.
D. Geronimo.
D. Mario.
D. Jacomo.
D. Cesare.
D. Andrea.
D. Francesco.

Marginal sketch of a primitive genealogical tree from Imhoff's personal copy of Carlo de Lellis, *Discorsi delle famiglie*, 1:458. Nuremberg, Stadtbibliothek im Bildungscampus, Hist 311.2°.

Small, preliminary stemmata from the dossier on the house of Lobkowitz. GNM, Généalogies L (vol. 14), unpaginated (Lobkowitz).

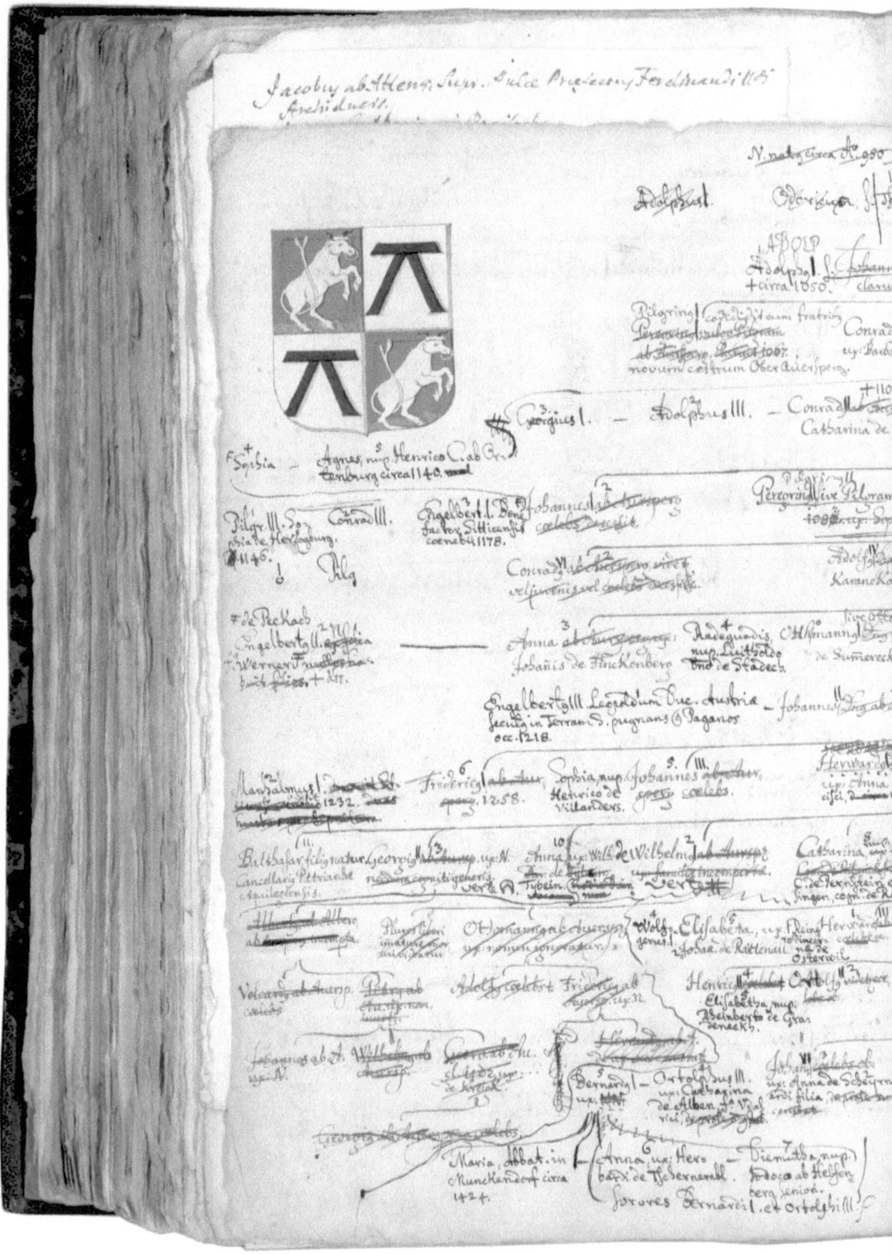

A draft of the family tree of the Auersperg family, full of addenda and corrections. At the right edge in the lower half of the page, a sign points to the continuation on the back (see next figure). GNM, Généalogies A (vol. 6), unpaginated (Auersperg).

Johannis (V) ab Aursperg et Kunegunde Gobernador religi
liberi sunt

...ondardg [5] — Kunegundis ... Ladislai — Anna [6] nup. Francisco ... Dno de Stubenberg — Barbara [7] nup. Friderico Dno. de Gallenberg

... DOMINI AB AUERSPERG.

...a ab Aursperg ... 1027. — ~~Odoricus s. Theodorius~~, qui ...

... +1081 ... Dom. ... 1067. — ~~Adolphus ab Aursperg~~, ~~superstes 1070.~~

... 1083. — Diepoldus, seu Theobaldus ~~...~~ literas a. 1100.

... A. 1050 ... — Johanna [4] ~~ab Aursperg~~ ... virgo — Catharina [3] nupt. Reimperto ab Eppenstein ... +1106.

... — Sophia [3] +1185 nupta Gebhardo VI Com. de Sulzbach. — N. [4] filia nupta Wolfrado de Treven, circa 1186.

... al. Falckenberg al. Finckenberg.

... obijt A. 1215 [13 Dec] ... Elisab. ... 1190. — ~~Elisabetha ab Aursperg~~

... 1246, ... Clara de Fischenthal, ... A. 1228. — Ottomanus II. [3] — ~~Hitzela~~ s. Caecilia [4], nupta N. Dno de Seldenhoven

Tcontra Archiep. Salisburgensem

... obijt 1382 ... Clara, ... Fran. ... virgo. — Petrus [3] ~~ab Aursp.~~ ... obijt. — militarunt Comiti Goritiensi, et ... — Wolfgangus ~~coelebs~~. — Hainzlinus Senior

... Aursperg, suis ex ... nepotibus circa 1270.

Mainhalmus ... ux. ... Hertwig ... Valckenberg

...BARDUS ... d. S. Luce ... Senior, obijt 1304 ... Elisab. de Winden ... 1276. — Joban [4] ~~ab Aursperg~~ ... Vide ⊙ — Elisabetha [9] ... monialis in Michelstetten ... — Volcardus [5] ~~ab Aursperg~~ ... Vide ... — Ottomanus s. Otto III. v. ⊕ verte

... Liebhard [3] ~~ab Aursperg~~ ... obijt, ... — ~~Laurentius ab Aursp.~~ ...

... Cunegunda de ... S. Nicolai ... 1312.

... III. ~~ab Aursp.~~ +1388 ... Anna de Rahen ... 1440. — Fridericus [3] ... + in Italia circa 1330. — Barbara [6] nup. Vlrico de Graben. — ~~Pilgrinus ab Aursp.~~ ...

... [1] ~~ab Aursp.~~ ... de Liechten... — Herrandus [2] ~~coelebs~~ — Volcardus ~~coelebs~~ — ~~Herrandus ab Aursp.~~ — Volcardus ab Aursp. ...

~~Wilhelmus ab Aursp.~~

~~Herrandus ab Aursp. coelebs~~

Such miniature sketches allowed genealogists to remember the most important information of lengthy passages more easily; they worked like other types of marginal annotation, reducing complex narratives to essential facts.

Simple diagrams condensed complicated textual information into easily comprehensible genealogical facts elsewhere, too. Such visual memory aids abound in Imhoff's papers, often recording no more than basic relationships (for example, A begat B). A sheet dedicated to Bohemia's house of Lobkowitz, for instance, features several sketches of two to three generational sequences of descendants.

Writing names and drawing lines on paper did more than merely represent information in an aesthetically pleasing way. Inscribing individuals—names and dates—on a blank writing surface in spatially significant ways vis-à-vis others (such as placing parents above their offspring) could express biological relationships. These in turn helped genealogists establish families and ultimately lineages. Drawing created order, insofar as it placed individuals within a visualized matrix of kinship.

From such small, fragmentary sketches, far more complete genealogical diagrams gradually emerged. It was a process of perpetual trial and error. Diagrams required revision as one's knowledge of familial relationships evolved. The physical acts of drawing and redrawing mirrored and helped spur progress in understanding individual dynasties. Additional people, who had initially been overlooked, had to be squeezed in; brothers turned out to be cousins or vice versa, and so lines needed to be redrawn. Some areas of a page became quickly crowded, while others remained unexpectedly empty. The visual products Imhoff created as he marched toward a comprehensive family tree are accordingly full of corrections, redrawings, marginal additions, and other indications of retrospective graphical rearrangements. Many drafts of genealogical tables resemble visual battlefields. Some sketches in Imhoff's papers are barely legible, let alone intelligible, to us now. The visual messiness of a draft of the family tree for the counts of Auersperg, for instance, illustrates perfectly how genealogical knowledge was gained in bits and pieces. The fact that preliminary diagrams quickly began to buckle under the weight of incoming information is material testimony to the complexity of the process of establishing genealogical knowledge.

The retrospective sorting of siblings into their proper birth order may serve as a telling example of this complexity. Imhoff initially had written down the children of a particular count of Auersperg in no particular order, only

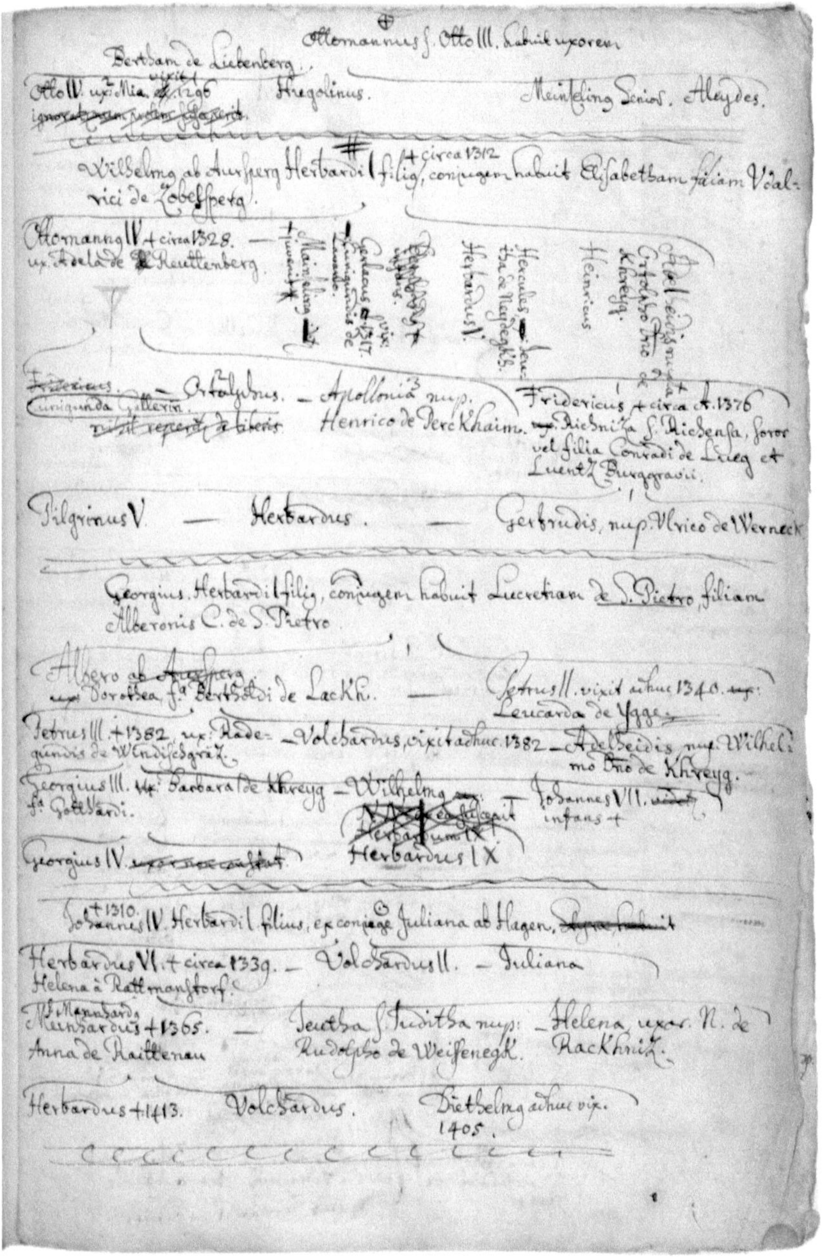

Back of the Auersperg table. Note the sign at the top of the page referring the reader back to the connector found on the front of the page, at the lower right edge (see previous figure). GNM, Généalogies A (vol. 6), unpaginated (Auersperg).

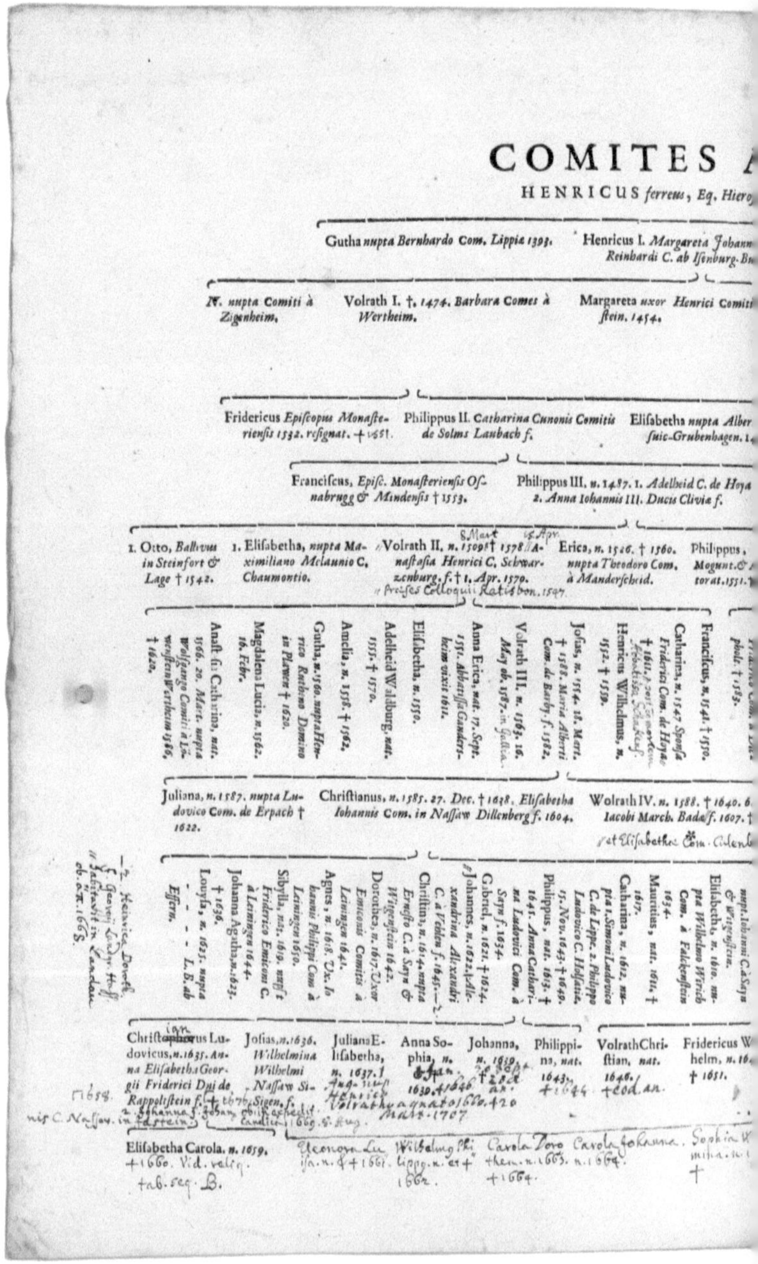

Imhoff's handwritten additions to a printed table from Rittershausen's 1664 edition of the *Genealogiae Imperatorum*. GNM, Généalogies W (vol. 22), unpaginated (Waldeck).

WALDECK.

…s, *Elisabetha Com. de Berg. f. 1370.*

Wisbaden *f. 2. Anastasia* …1440. | Adolphus, *Agnes Com. de Zigenheim 1406.*

Henricus. *1431.* | Otto, † *1454. Anna Mauritii C. de Oldenburg & Delmenhorst. f.* | Johannes. *1431.*

Otto † *1495. 1. Metta Imbrechti C. de Nevenar f. 1464. 2. Elisabetha C. Teckelnburgica.*

…n- | Philippus I. *nát. 1452.* † *1475. Johanna Com. de Nassaw Dillenberg. f. 1464.* | Eva, *moritur sponsa Bernhardi Comitis Lippiæ 1489.*

…gius, *n. 1483.* …Parisiis. | Catharina, *nupta Bernhardo Comiti Lippiæ.* | Henricus, † *1512. Anastasia Wilhelmi Com. de Runckel Isenburg f. 1492.*

…ncifcus, *N. Johannis Hoggrevii Cancell. …liacensis f.* † *1580.* | Johannes Pius, † *1567. Anna Simonis C. Lippiæ f. 1550.* | Mathildis. | Catharina, *Ux. Bernhardi C. Lippiæ 1550.* | Wilhelmus, *Canon. Coloniensis.* | Philippus IV. *n. 1493.* † *1574. 1. Margareta Ezardi C. Ostfr. f. 1523.* † *1537.*

Bernhardus, *Episc. Osnabrug. el. 1585. 25. Oct.* † *1591. 11. Mart.* | Margareta, † *1580. Uxor Guntheri Comitis à Waldeck.* | Franciscus, *n. 1553.* † *1597. Walburgis Christophori Dynastæ à Plessen f. 1582.* | Philippus *Can. Argentor. n. 1556.* † *1567.* | Henricus, *n. 7. Dec. 1531.* † *1577. 4. Oct.* | Daniel, *n. 1. Aug. 1530.* † *1577. 7. Jun. Barbara Philippi Landgr. Hassiæ f. 1568.* | Samuel, *n. 2. May 1528.* † *1570. 6. Ian. Anna Maria Henrici C. Schwarzenburg f. 1554. 8. Oct.* | Fridericus, *n. 1534.* † *1557.* | Margaretha, *nat. 1553.* † *1554.* | Elisabetha *n. 1525. nupta Reinhardo C. Isenburg.* † *1543.* | Magdalena. | Anastasia.

Magdalena, *n. 1558, nupta 1. 1576. 5. Feb. Philippo Ludovico Com. Hanaw. 2. Johanni Com. Nassaw-Sigen.* | Daniel, Samuel, *& alii* †† *infantes.* | Guntherus, *n. 1556. 20. Feb.* † *1585. 1. Margareta agnata. 2. Margareta Georgii Com. à Gleichen f.* 1582.

Culenburg.

Maria Magdalena, *n. 1606. nupta 1623. Simoni II. C. de Lippe.* | Volrath V. *n. 1625.* † *1657.* | Christianus. | Jacobus, *n. 1621.* † *1645.* | Georgius Fridericus, *n. 1620. Elisabetha Charolotta Wilhelmi Com. à Nassaw f. 1643.* | Johannes Ludovicus, *n. 1616.* † *1630.* | Philippus Theodorus, *n. 1614.* † *1645. Maria Magdalena Wilhelmi Com. à Nassaw Sigen. f. 1639.* | Josias Floris, *n. 1612.* † *1613.* | Maria Elisabetha, *nat. 1608. nupta 1634. Friderico Marchioni Badensi* † *1643.* | 2. Wilhelmus Ernestus, *nat. 1584.* † *Tubingæ 1598.* 16. Sept.

…us Wil… *nat.* † *1653.* | Louysa Amelia, *n. 1653.* | Charlotta Amelia, *n. 1654.* † *1657.* | Fridericus Wilhelmus, *nat. 1657.* † 1670 | Amelia Catharina, *n. 1640.* | Henricus Volrath, *n. 1642.* † 1664. | Floris Wilhelm, *nat. &* † *1643.*

Vide tab. seq. C.

later to indicate their birth order by superimposing arabic numerals. His information became more precise as his research wore on, and so Imhoff needed to update his table. An early statement about "many other children who seemed to have died young"[55] was accordingly replaced by these individuals' names. But now there was no longer enough space at the proper place, so Imhoff had to add the names of these children at the bottom of the sheet. All he could do to connect them to their father, mother, and siblings was add a number of shaky lines traversing much space. On other occasions, when additional information no longer fit on the main page after a first round of drafting, Imhoff moved it to supplemental pages and used graphical signs to mark the connections, as he did in the case of the Auersperg table. These observations illustrate how Imhoff transformed his many incoming sources into one coherent vision of a noble family by means of episodes of diagrammatic drawing and writing. The physical act of drawing quill over paper drew individuals together into families and dynasties.

Such sketches also demonstrate in physical terms how genealogical knowledge production was a protracted process, advancing step by step, as details were often added here and there in an haphazard sequence. The ongoing need to redraw previous tables reflects the unsystematic nature of genealogical research. A sketch like the Auersperg table highlights how genealogists continuously revised their conclusions, integrating new details into existing drafts. They provide material evidence that the rhetoric of preliminarity, discussed previously as a strategy to mitigate the fallout of premature or erroneous statements, was grounded in practical experience. Genealogical knowledge remained tentative for a long time during this process.

Even clean copies of family trees were, in effect, only ever provisional. They quickly became a material platform for the next round of annotations, reflecting further research and new family developments. Many clean copies of genealogical tables in Imhoff's papers contain secondary layers of manuscript writing, indicating corrections, additions, or alternatives to the presumably finished genealogy. Imhoff even inserted handwritten annotations into unique documents, for instance, adding marriages, deaths, and births of the 1670s to a rare original manuscript by Nicolaus Rittershausen.[56]

Imhoff and his genealogical peers also habitually used printed family trees as a material basis for keeping track of the most recent dynastic events, adding handwritten lines and brackets, along with names and dates, to update such tables as time moved on. Imhoff regularly used Rittershausen's tables from 1664 in such a way, casually expanding or altering them as he acquired new information. He even went so far as to disassemble a copy of Rittershausen's *Genealogiae Imperatorum* and commenced each of his own dossiers on the same noble families with the relevant tables from 1664. Rittershausen's work thus served as a foundation for Imhoff's own project even in a physical sense.[57] Imhoff subjected his own works to the same routine, transforming them into base texts on which later developments could be registered. This was a key practice of recording and managing genealogical information.

Sketching and resketching existing (preliminary) family trees were essential steps in the process of understanding dynastic history. Drawing diagrams brought fragmented information together—this was how genealogical compilation worked. Since it was a process of constant improvement on previous versions, however, it gave genealogy the appearance of instability. Genealogical works, including diagrams, were never truly finished. Their open-endedness made it necessary (indeed possible) for readers to personalize existing tables by reworking them according to their own, improved standard of knowledge. If few printed products of early modern Europe can be considered as fixed, stable, and definitive, this is true of genealogical works a fortiori.[58] Whether printed or in manuscript form: genealogies, and especially genealogical diagrams, were hardly ever finished and static.

Creating a Genealogy: The House of Este

A case study allows us to observe up close how Imhoff's working routines and research strategies came together to help him produce impressive comprehensive genealogies. In his 1702 book on the Italian nobility, *Corpus Historiae Genealogicae Italiae et Hispaniae* (Corpus of the historical genealogy of Italy and Spain), Imhoff dedicated an extensive discussion and several genealogical tables to the prominent house of Este from Modena, Italy: a total of four double-pages of diagrams and thirteen pages of attached commentary.[59] As in many other cases, he based his work on the Este on Nicolaus Rittershausen's table from 1664. It is by comparing Imhoff's tables (and narratives) against Rittershausen's versions, that Imhoff's approach and practices emerge most clearly.

The first thing to note is that Imhoff dispensed with Rittershausen's self-imposed chronological limits. In keeping with his practice of tracing genealogies back only to circa 1400, Rittershausen had commenced his Este genealogy with Duke Albert (1347–93). In his own "Tabula I," in contrast, Imhoff started his genealogy some three hundred years earlier, with Duke Alberto Azo II, who lived in the eleventh century. Rittershausen's point of departure,

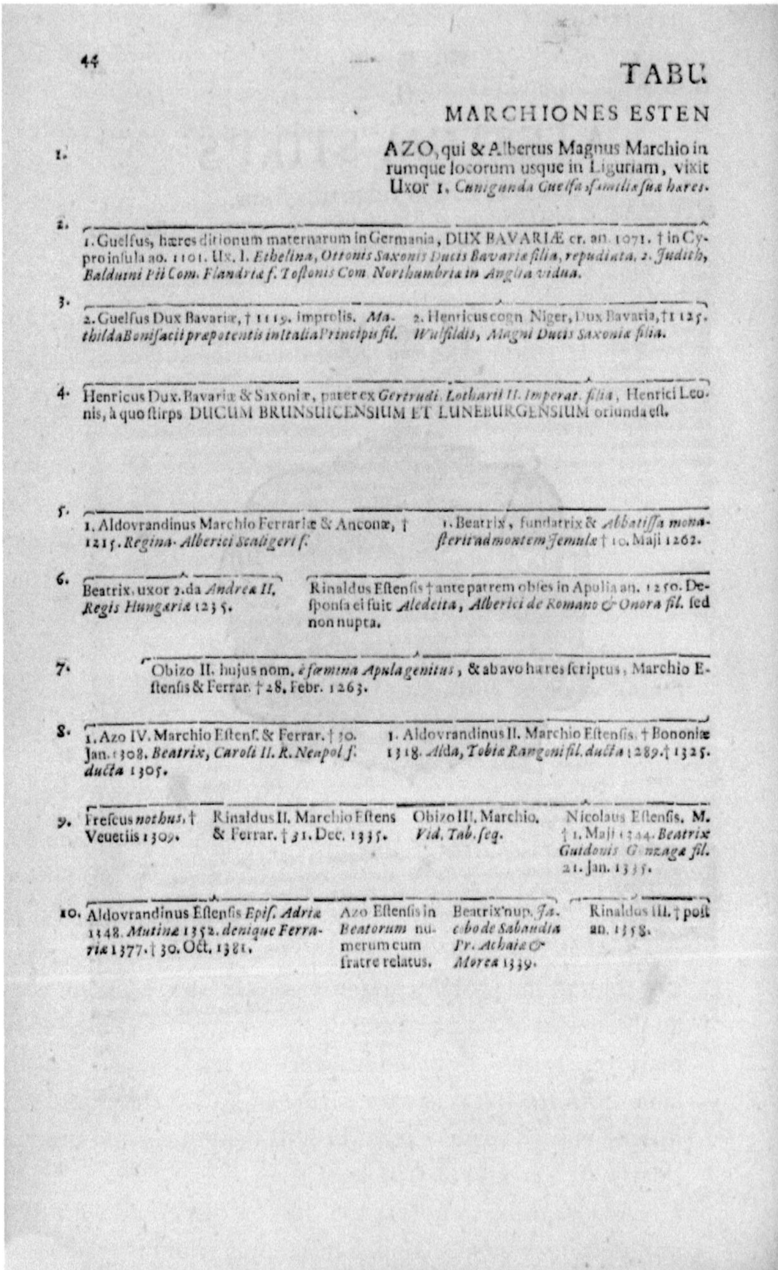
44

TABU

MARCHIONES ESTEN

1. AZO, qui & Albertus Magnus Marchio in rumque locorum usque in Liguriam, vixit Uxor 1. *Cunigunda Guelfa, familiæ suæ hæres.*

2. 1. Guelfus, hæres ditionum maternarum in Germania, DUX BAVARIÆ cr. an. 1071. † in Cypro insula ao. 1101. Ux. 1. *Ethelina, Ottonis Saxonis Ducis Bavariæ filia, repudiata. 2. Judith, Balduini Pii Com. Flandriæ f. Tostonis Com. Northumbriæ in Anglia vidua.*

3. 2. Guelfus Dux Bavariæ, † 1119. improlis. *Mathilda Bonifacii præpotentis in Italia Principis fil.*
2. Henricus cogn. Niger, Dux Bavariæ, † 1125. *Wulfildis, Magni Ducis Saxoniæ filia.*

4. Henricus Dux. Bavariæ & Saxoniæ, pater ex *Gertrudi, Lotharii II. Imperat. filia*, Henrici Leonis, à quo stirps DUCUM BRUNSUICENSIUM ET LUNEBURGENSIUM oriunda est.

5. 1. Aldovrandinus Marchio Ferrariæ & Anconæ, † 1215. *Regina. Alberici Scaligeri f.*
1. Beatrix, fundatrix & *Abbatissa monasterii ad montem Jemulæ* † 10. Maji 1262.

6. Beatrix, uxor 2.da *Andreæ II. Regis Hungariæ* 1235.
Rinaldus Estensis † ante patrem obses in Apulia an. 1250. Desponsa ei fuit *Aledeita*, *Alberici de Romano & Onora fil.* sed non nupta.

7. Obizo II. hujus nom. *è fœmina Apula genitus*, & ab avo hæres scriptus, Marchio Estensis & Ferrar. † 28. Febr. 1263.

8. 1. Azo IV. Marchio Estens. & Ferrar. † 30. Jan. 1308. *Beatrix, Caroli II. R. Neapol f. ducta* 1305.
1. Aldovrandinus II. Marchio Estensis. † Bononiæ 1318. *Alda, Tobiæ Rangoni fil. ducta* 1289. † 1325.

9. Frescus *nothus*, † Venetiis 1309.
Rinaldus II. Marchio Estens & Ferrar. † 31. Dec. 1335.
Obizo III. Marchio. *Vid. Tab. seq.*
Nicolaus Estensis. M. † 1. Maji 1344. *Beatrix Guidonis Gonzagæ fil.* 21. Jan. 1335.

10. Aldovrandinus Estensis *Epis. Adriæ* 1348. *Mutinæ* 1352. *denique Ferrariæ* 1377. † 30. Oct. 1381.
Azo Estensis in *Beatorum* numerum cum fratre relatus.
Beatrix nup. *Jacobo de Sabaudia Pr. Achaiæ & Moreæ* 1339.
Rinaldus III. † post an. 1358.

The first table for the Este family in Imhoff, *Corpus Historiae Genealogicae Italiae et Hispaniae*, 44–45. Munich, Bayerische Staatsbibliothek, 2 Geneal. 93#Beibd.1, urn:nbn:de:bvb:12-bsb10328176-7.

LA I.

SES ET FERRARIENSES.

Lombardia , DOMINUS DE ESTE , Calaone alio-
feculo decimo &undecimo, & deceflit pœne centenarius.
2. *Ermengardis, Hugonis Com. du Maine in Gallia fil.* 1.

2. Hugo Comes du Maine in Gallia, fed brevi tempore. Uxor ei fuit *filia Roberti, cogn. Guifcard. Com. Apuliæ.*

2. Fulco, Marchio, Dom. de Efte. 2.

Obizo five Opicius, Marchio, Dn. de Efte, Poteftas Patavii an. 1177. & 1178. Ferrariæ 1195. † 1196. *Sophia,* 3.

Azo II. Marchio Eftenfis & Ferrariæ an. 1196. Poteftas Paduæ 1199. Veronæ 1207. Marchio Anconitanus 1208. † 1212. Ux. I. *Leonora, Thomæ I. Com. Sabaudiæ f. 2. Marchefella Adelardi, qui Guilielmi Poteftatis Ferrariæ frater erat, fil. defponfata 1196. fed ante confummatum matrimonium mortua. 3. Elifa, Ludovici Com. de S. Bonifacio f.*

Agnes, uxor fecunda *Ezelini II. de Romano, qui Monachus factus & dictus eft.* 4.

1. Conftantia, † puella 1213.

3. Azo III. Marchio Eftenfis & Ferrar. † 13. Febr. 1264. *Alifa, Rinaldi de Chatillon ex Conftantia Princip. Antiochiæ filia.* 5.

Beatrix, *virgo religiofa in monafterio B. Antonii juxta Ferrariam.*

Cubitofa nupta *Isnardo Malafpinæ, Marchioni de Lunigiana, Maffa & Carraria.* 6.

Uxor. I. *Jacobina Flifca Genuenfis ducta 1263.* † *1287.* m. Dec.
2. *Conftantia Scaligera, Alberti Dom. Veronæ f. 1288.* 7.

1. Beatrix nup. 1. *Nino Scotto Pifano Judici, Dom. Gallura in Sardin. 2. Actio Vicecomiti Princ. Mediol. 1301.*

Francifcus Marchio Eftenfis occ. 23. Aug. 1312.
Thadaa, Johannis Barbiani f. 8.

Elifa nup. *Rinaldo Paffarino Bonacolfo, Mantuæ Domino.*

Jacobina ux. *Remei Pepoli Bononienf. 1325.*

Azo M. Eft. † 1318.

Bertholdus M. Eft. & Ferrar. 1. N. *Pandulfi Malatefta f. 2. Catharina, Richardi de Camino f.*

N. uxor. N. de *Scrovegnis Arenæ Patavinæ Domini.* 9.

1. Francifcus II. Marchio Eftens † poft an. 1358. *Catharina, Luchini Vicecomitis Mediol. f.* 2. Jun. 1342. 10.

Azo V. Marchio Eftenfis, † in Creta. Ux. N. de *Collalto.* 11

Thadæus de Efte, Tribunus milit. ap. Venetos, † 1448. 12.

Bertholdus de Efte, Dux milit. Venet. in Græcia, occ. in Morea an. 1463.

Helena ux. *Ludovici Com. de Turre.* 13.

therefore, appeared in Imhoff's second table. It was important to Imhoff to research the family's remoter past, and so he conspicuously broke with Rittershausen's presentist stance.

In doing so, Imhoff took up an ongoing debate that was as exciting to scholars as it was important to politicians and princes. At the time, the question of the origin of the house of Este was of the utmost importance not only to the Este themselves, but also to the Guelph family in Lower Saxony, one of the most prominent German dynasties (and soon to be kings of England). Both families insisted (correctly) that they shared a common medieval ancestor, but information about this ancestor had remained vague until recently.[60] A few years before Imhoff's book appeared in print, however, a decisive breakthrough had occurred. None other than Gottfried Wilhelm Leibniz had clarified the details of the Este-Guelph connection. In a brief publication from 1695, he showed that the Este and the Guelphs both descended directly from Count Alberto Azo II of Este (d. 1097)—the man at the top of Imhoff's "Tabula I"—by two different wives.[61]

Imhoff knew about other scholarship on the early Este genealogy, too, yet he relied primarily on Leibniz's groundbreaking work.[62] He adopted Leibniz's account of the medieval genealogy of the Este so completely that—just like the scholar from Hanover—he lost all interest in the history of the Este prior to Alberto Azo II.[63] He also lifted some of his own explanatory text almost verbatim from Leibniz.[64] By integrating Leibniz's history of the medieval genealogy of the Este, Imhoff did more than merely add factual depth to Rittershausen's limited treatment. He also made his work more cutting-edge, both by associating it with one of the most prominent contemporary intellectuals and by catering to contemporary scholarly and political interests.

But extending Rittershausen' genealogy back to the High Middle Ages was not Imhoff's only concern. He was equally interested in improving more recent branches of his Este genealogy, raising the value of his book as a source of information for understanding contemporary Europe. One area that Rittershausen had left only poorly developed was his treatment of the offspring of Cesare VI d'Este (1562–1628). Imhoff, in contrast, highlighted the importance of Cesare, who clashed with the papacy in a protracted dispute for possession of Ferrara, and his family by dedicating a new table ("Tabula III") to them—a clear example of highlighting a previously neglected branch. Imhoff was in a position to do this, because on January 12, 1694, he received a detailed dossier about this branch from Benedetto Bacchini in Modena. Imhoff grafted Bacchini's material more or less unaltered onto Rittershausen's rudimentary tree. This was far more than a trifling addition. In integrating Bacchini's information about this junior branch, Imhoff added an extremely important contemporary to Rittershausen's original table, namely Rinaldo III d'Este (1655–1737). Rinaldo had become a cardinal in 1686 and thus warranted mention in any genealogy of his house. Rittershausen might already have known about Rinaldo's birth, but he certainly could not foresee Rinaldo's impressive career. Imhoff, however, understandably wanted to include him in his family's genealogy to pay homage to a prince of the church.

New developments, however, rapidly invalidated Bacchini's information. Just months after the arrival of the dossier in Nuremberg, in September 1694, Cardinal Rinaldo d'Este renounced his ecclesiastical offices and became Duke of Modena after the death of his childless nephew, Duke Francesco II (1660–94). Rinaldo unexpectedly had to take over as sovereign. He also needed to ensure the continuity of his line by producing an heir. Hence, two years later, Rinaldo—now Duke of Modena—married a German princess, Charlotte Felicitas of Brunswick-Wolfenbüttel (1671–1710), the daughter of Leibniz's first patron. Bacchini could not have known about these events, but Imhoff duly supplied all the relevant information anyway. Given Imhoff's excellent connections to Lower Saxony, he could have learned about these developments from any number of sources.[65]

Imhoff also added two further individuals to his Este genealogy, Marquis Foresto d'Este (1652–1725) and his brother Cesare Ignazio (1653–1712), who had likewise eluded Rittershausen. Their dates of birth had perhaps been too recent for Rittershausen to know of them. Imhoff even double-checked twice with Muratori to ascertain whether the two newfound brothers Foresto and Cesare were still alive.[66] He updated other details as well. Imhoff was able to give the names of Rinaldo's four children to date, a piece of information he received from Muratori in September 1701, shortly before the book went to press.[67] Clearly, the Nuremberg patrician spent at least as much energy to keep abreast of contemporary events as he did to research the remote past. His success in both areas, to judge by the case of the Este, is what made his work such a reliable resource.

What Imhoff's informants had not covered were the generations between the twelfth century (Leibniz) and the fifteenth to seventeenth centuries (Rittershausen, Bacchini, Muratori). Imhoff himself was presumably responsible for establishing the Este genealogy of this period. He mentions several historiographical works that would have contained relevant information. Giovan Battista Nicolucci, known by the name Pigna (1529–75), was a key reference.

Pigna had published an authoritative *Historia dei principi d'Este* in 1570.[68] After digesting Leibniz's nuanced assessment of the work, Imhoff relied on it heavily, but he carefully double-checked Pigna's information by consulting other (published) sources.[69] He also culled information from Henniges's *Theatrum Genealogicum*, a standard work on genealogy that Imhoff quoted frequently on other occasions as well. In Henniges's work, he found a reprint of an extensive and accurate Este family tree created by Girolamo Falletti (d. 1564) in the 1560s at the request of Alfonso II d'Este.[70] Imhoff exhibited considerable versatility as a compiler not only by finding these books but also by critically cross-referencing them to create a homogeneous pastiche of information from these disparate sources.

There was one final addition to this assemblage, namely, Imhoff's "Tabula IV," dedicated to the Este di San Martini e Borgomanero, a branch not even mentioned by Rittershausen. This line descended from Sigismondo I d'Este (1433–1507) and endured until 1752.[71] For most of its existence, this secondary branch was easy to overlook—Rittershausen could be excused for the omission. Recently, however, in a reversal of fortunes that was not all too uncommon in early modern dynasties, this junior branch had suddenly become important because of the extraordinary political career of one of its members. Carlo Emanuele di San Martini e Borgomanero (1622–95) rose to prominence in the Spanish diplomatic corps, serving as ambassador of Spain in Vienna from 1681 to 1695. Carlo Emanuele and his entourage made sure that Imhoff, as a leading genealogical authority, would not overlook his family branch as Rittershausen had done. Shortly after his arrival at the imperial court, the ambassador started forwarding information about his family to Nuremberg.[72] In 1694, moreover, Bacchini sent Imhoff crucial additional information about the Borgomanero branch.[73] When he set out to compile his book shortly before 1700, Imhoff accordingly could draw on significant material from several sources to discuss the San Martini e Borgomanero branch, which he duly included despite Carlo Emanuele's death in 1695. In this way, Imhoff adjusted his genealogical encyclopedia to reflect the lasting recalibration of social standing among the many branches of the Este following the recent rise of the San Martini e Borgomanero.

Capitalizing on his ability to elicit genealogical information from the nobility and other informants, Imhoff delivered an unusually complete and up-to-date discussion of the house of Este. Drawing on multiple sources, he merged information about the past and the present history of the family. Through a complex process of genealogical bricolage, he integrated new fragments of knowledge into a coherent whole. This process entailed collating different data, identifying lacunae, and updating and cross-checking preliminary ideas. Although a draft of Imhoff's Este tables comparable to that for the Auersperg family has not come down to us, we may easily imagine the genealogist repeatedly sketching such diagrams. His work on the genealogy of the Este was certainly of an equally material and physical nature as his work on other family histories. For sure, Imhoff would have placed different memoranda, printed artifacts, letters, and fact sheets side by side, as he pondered how to fit them together. As he patiently reconciled these disparate fragments, a comprehensive survey of the house of Este gradually emerged.

The rich paper trail that Jakob Wilhelm Imhoff left behind as he produced his many publications on European genealogy permits scholars to follow his daily working routines step by step. As a genealogical compiler, Imhoff not only needed to make relevant information flow from every corner of Europe; he also had to find ways to transform incoming genealogical data into comprehensive narratives and visual representations. Collecting extant information and producing new knowledge were two distinct, yet closely related activities. As the preceding material analysis of many of Imhoff's extant working papers has demonstrated, the processes of collecting and writing were cyclical. Imhoff typically began a new round of inquiries in order to improve existing genealogical knowledge, while, at the same time, incoming information had to be integrated into existing surveys.

Imhoff relied on specific types of manuscript documents to communicate with correspondents and to compile information. Furthermore, a range of intermediary written artifacts were needed to transform collected materials into compiled information. Annalistic lists of dynastic events, whether culled from *notificationes* or newspapers are a case in point. Such fact sheets may be regarded as a preliminary way to store potentially relevant genealogical information. They neither resembled the original sources nor foreshadowed the subsequent narrative or diagrammatic results. Yet the importance of these intermediary products of scholarly writing practices can hardly be overstated. These written artifacts were the physical workhorses of genealogical bricolage in the service of encyclopedic projects.

It is important to highlight the physical, material side of genealogical production. Occasionally, it was necessary to prepare the writing surfaces used in the process of genealogical bricolage with the help of scissors and glue. A much more obvious material practice in Imhoff's case, however, is the role of drawing and sketching diagrams. Arranging notes about individuals spatially on (blank) pages and connecting them literally by means of visual indications were essential physical activities for genealogists. This practice fixed, if only on a preliminary basis, the familial relationships between individuals. Spatializing knowledge materially by inscribing it onto (empty) surfaces permitted genealogical knowledge to emerge.

The profoundly visible, material complexity of bricolage mirrors the epistemic challenge of constructing individual genealogies from initially disjointed fragments of information. The example of Imhoff's Este genealogy illustrates how the compiler utilized incoming material like building blocks, which he had to arrange and combine into a coherent genealogical edifice. In doing so, the genealogist had to normalize information, so that his coverage of all the individuals and branches of a noble house adhered roughly to a similar standard. Imhoff's consistent practice of asking his informants follow-up questions, of verifying and clarifying details and dates, facilitated this homogenization of his information base. As a result, his finished printed diagrams and narratives exhibit a remarkable degree of uniformity.

The physical features of Imhoff's working papers draw attention to the preliminary and often only rudimentary status of genealogical knowledge. On Imhoff's end, the documents illustrate how he came to understand individual families step by step, often taking a long time to determine how the members were related by piecing them together in a literal sense. The completeness—or incompleteness—of his information was yet another problematic issue: numerous family members and sometimes entire branches of a family were initially unknown to him and had to be inserted into genealogical narratives after they had been "found" or "(re)discovered."

Genealogical knowledge was also often at best fragmentary and partial on the end of Imhoff's informants. As the case study of the Este shows, the coverage of Imhoff's informants was indeed incomplete. Whether more complete genealogies were unavailable to them, or whether they felt responsible only for their share of family history is often difficult to decide. Whereas, as illustrated by Leibniz, some of Imhoff's informants withheld the full extent of their knowledge, other correspondents had only rudimentary information in the first place. The nobility itself was also rarely in a position to provide complete genealogies; the task required extensive preparation, including the help of genealogists.

Unfazed by these difficulties, Imhoff stubbornly went about collecting, conserving, and compiling genealogical information. By cultivating a specific repertoire of writing practices, which included drawing, listing, and the skillful physical manipulation of written artifacts at large, he managed to produce coherent and convincing genealogical diagrams and narratives. It was on these material scholarly practices that his growing fame among scholars, politicians, and readers rested.

Publishing and Reading Genealogy

Imhoff's family assiduously tracked Jakob Wilhelm's growing public fame. Several lists of quotations praising his genealogical works survive among the family papers.[1] Everyone knew about the genealogist's "reputation."[2] Obviously, Imhoff and his publishers created successful products. Imhoff was hailed as a "shrewd man, a lover of truth, and untouched by the bias of his peers,"[3] and his reviewers deplored the fact that Imhoff had not published *more* than his already impressive series of works.[4] Jakob Friedrich Reimmann, for instance, when surveying the history of genealogy in 1710, repeatedly mentioned the Nuremberg genealogist as the epitome of modern genealogy.[5] A few years later, an annotated bibliography of recent genealogical publications also singled out Imhoff as the key author of the "latest" generation of authors. His books epitomized what readers hoped to find in a genealogical compendium.[6]

Several times, the reviewers explicitly mentioned the extraordinary physical quality of his books, their splendid typesetting in particular, in addition to the high quality of information. The Leipzig-based *Deutsche Acta Eruditorum* praised his *Regum Pariumque Magnae Britanniae Historia* in June 1690 for its felicitous combination of diagrammatic and textual content.[7] In 1703, they called his work on Italy, the *Corpus Historiae Genealogicae Italiae et Hispaniae*, a "most beautiful book," only to praise his *Genealogiae Viginti Illustrium in Italia Familiarum* (Twenty genealogies of illustrious families in Italy) as a "splendid work" in 1710.[8] In contrast to many otherwise comparable products on the burgeoning genealogical book market, Imhoff's works distinguished themselves with their exquisite production quality. They set themselves apart from the many cheaply and hastily produced printed works on which many publishers seemed to focus. Like Imhoff's approach to genealogy in general, his response to genealogical publishing trends were both confidently avant-garde and yet also consciously traditional, a combination of a market-driven and a genteel approach to genealogy.

The success of a printed product never depended solely on its author or on the qualities of his or her text. Roger Chartier and several other scholars have reminded us of the importance of publishers, printers, and booksellers in the process of making literary successes.[9] Imhoff also worked closely with the genealogical publishing industry to ensure that his scholarship appealed so broadly.

Such collaboration was all the more important as the genealogical book and information market underwent rapid change circa 1700. Volker Bauer has called attention to this shift in a number of important publications.[10] New genres emerged, including genealogical calendars and universal or encyclopedic genealogical publications (Imhoff being a major exponent). This new literature was produced with an eye to profit, as publishers increasingly found genealogical publications to be a lucrative commodity. Moreover, the readership of genealogical literature diversified and became more heterogeneous. Genealogy expanded beyond the traditional sphere of courtly self-promotion and became a key resource for understanding contemporary politics. These developments brought new challenges, but also new opportunities to genealogists as they sought to publish their findings. Publishers now actively solicited manuscripts, anticipating general interest in new books. At the same time, however, competition grew significantly. Authors and publishers had to survive in an increasingly crowded market. As competition heated up, publishers and authors experimented with new forms of marketing. Capitalizing on past successes, they tried to transform best-selling works and authors into genealogical "brands."

Imhoff's works and research practices were closely related to this evolving editorial context. He tried quite deliberately (and successfully) to position his works in the expanding market for genealogical information. His strategies of compiling, for instance, clearly adhered to the conventions and requirements of the encyclopedic genre, while his initial self-promotion as a continuator of Rittershausen reflected the marketing advantages conferred by a "big name" brand. And yet, despite the fact that Imhoff clearly adapted to many of the most recent developments of the genealogical information economy, he was not exclusively driven by market logic. As we have seen in previous chapters, his method of producing genealogies originated in his social background and his overall cooperative approach toward the nobility as much as it did in an analysis of market forces. It was a balanced combination of all of these factors that lent Imhoff's genealogies their unique character.

This chapter surveys the presence of Imhoff's genealogies in the book market. I first discuss the features of his works that guaranteed them a leading place in the early modern book market. Imhoff's success serves as a case study in forms of cooperation between authors and publishers to create successful print products. The second section of this chapter then explores several examples that illustrate how Imhoff's readers used his works. Given the impact of his books across Europe, it comes as no surprises that a significant number of annotated copies and notes on Imhoff's works have come to light. Although these yield at best only fragmentary insights, careful analysis of these readers' material traces nevertheless permits me to elaborate a rough survey of responses to Imhoff's published works, even decades after their original publication. The third and final section of this chapter details how Imhoff's genealogical expertise transcended the realm of published texts. Although his printed books remain his most important legacy, Imhoff eventually also promoted and distributed genealogical knowledge in other, more informal and private ways. His study in Nuremberg, it seems, eventually became a major hub of dynastic information for visitors generally, and it was accessible even remotely, by letter.

Crafting Editorial Success

If we survey Imhoff's publications, we can discern two distinct periods in his career, representing two crucial steps in the author's rise to fame. Early in his career, Imhoff gained access to the book market by linking himself to a well-known genealogical brand, connected with the name of "Rittershausen," and cooperating closely with that brand's primary promotor, the firm of Brunn and Cotta in Tübingen. In his later career, after he had established his public reputation, Imhoff left this arrangement and cultivated his own signature style of presenting genealogical knowledge. In the process, he made his own name synonymous with high-quality information in the field.

Harnessing a Brand and Launching a Career

As a young man new to the fields of genealogical research and publishing, Imhoff took a well-trodden path to boost his projects. He associated them with one of the most prominent recent names in genealogical scholarship, that of the late Altdorf professor Nicolaus Rittershausen. Aligning himself with Rittershausen was, in part, a purely pragmatic choice for a Nuremberg patrician. Rittershausen had worked in Altdorf, and there was also a family connection between him and the Nuremberg patriciate by way of Imhoff's father. Rittershausen's legacy was an obvious starting point for Imhoff's scholarly career. Yet it was also a shrewd way to secure immediate public attention for his research. Rittershausen was a household name among interested readers—one that was associated with a recognizable publishing style and suggested reliable information.

Such "brands" became ever more important as the number of genealogical publications proliferated.[11] By connecting new works to a renowned brand name, authors and publishers courted attention and inspired confidence in potential buyers. Publishers were eager to keep successful brands flourishing by constantly churning out new books, even after the death of the original namesake. They launched reprints, updated editions, and spin-offs of successful genealogical publications and brought new "continuations" onto the market under the original author's name. The people who did the actual work on such spin-offs are often very hard to identify. Ghostwriters for brands like "Rittershausen" usually went unacknowledged. All that mattered was the brand name, while the actual editors or continuators were relegated to anonymity and remained "hidden helpers."[12]

The transformation of Rittershausen's family name into a recognizable brand was already well under way when Imhoff arrived on the scene. Even before Rittershausen's death in 1670, publishers used his name to market new products. Rittershausen's publisher Philibert Brunn and his successor Georg Cotta in Tübingen worked hard to create an ever-growing body of "Rittershausiana." Among the first spin-offs of Rittershausen's genealogical works were the so-called *Tübinger Tafeln*, which first appeared in Tübingen in 1656.[13] This was a collection of initially twelve genealogical tables dedicated to important German families that bore a strong resemblance to Rittershausen's original work. The idea for this by-product was not Rittershausen's, but his enterprising publisher Brunn quickly brought him on board.[14] It is impossible to determine how strong Rittershausen's influence on the *Tafeln* ultimately was. Be that as it may, the *Tafeln* were successfully linked to the Altdorf professor, and the series stuck. New editions appeared in 1660, 1670, 1685, 1692, and 1695, occasionally containing explicit references to Rittershausen's personal influence.[15]

The most important next step in establishing "Rittershausen" as a genealogical brand was taken in Nuremberg under the aegis of the Imhoff family in cooperation with Brunn and Cotta. Wilhelm Imhoff (Jakob Wilhelm's father) had been instrumental in producing the definitive third edition of Rittershausen's *Genealogiae* in 1664. When Rittershausen died in 1670, Imhoff Sr. took over some of Rittershausen's unfinished projects. Relying on this material under Cotta's guidance, Imhoff Sr. curated a posthumous spin-off, the *Brevis Exegesis Historica Genealogiarum Praecipuorum Orbis Christiani Procerum* (Brief historical exegesis of the genealogies of the eminent nobles of the Christian world, 1674). This addition to the corpus of Rittershausiana added historical narratives to the original book, which consisted only of tables.[16] As was typical, the elder Imhoff's important role in producing this work went unacknowledged in print.

Then Jakob Wilhelm took over.[17] Since at least 1676, he was widely known for his work on an appendix to Rittershausen's *Genealogiae* of 1664.[18] In 1683 and 1685, two volumes of Imhoff's *Spicilegium Rittershusianum* finally appeared. Again, the continuator's name was not mentioned in the book. Instead, both the publisher and the genealogist agreed to highlight the brand name, this time in the title. Even Imhoff's first magnum opus, his *Notitia Procerum* of 1684, was published within this "Rittershausian" framework. Sandwiched as it was between the two volumes of the *Spicilegium*, and adopting Rittershausen's slightly unusual signal-word *Procerum* in the title, it conjured the legacy of Imhoff's great exemplar.[19] The *Notitia Procerum*, however, appeared under Imhoff's own name. He was emerging from the shadow of commercial publishers and was about to leave the genealogical publishing world of Altdorf and Tübingen behind. Except for the many subsequent editions of the *Notitia Procerum*, Imhoff now abandoned the role model that smoothed the way for his initial steps as an author.

Imhoff agreed to work as an anonymous continuator of another author's popular work one more time. After the premature death of Georg Lohmeier in 1691, Imhoff produced a second edition of his genealogical textbook in 1695. This revised edition, however, was de facto a new work. Although the style and appearance of the original remained largely unaltered, the work had roughly doubled in size, and many individual tables were thoroughly rearranged. Imhoff's identity remained hidden behind the acronym ANPI (that is, "Auctor *Notitiae Procerum Imperii*," author of the *Notitia Procerum Imperii*), but he unmistakably poured his wealth of genealogical knowledge into remaking Lohmeier's handbook. Why he did so is not entirely clear. Some of his friends believed that Lohmeier's work should have been viewed alongside the recent 1693 edition of the *Notitia Procerum*.[20] Perhaps Imhoff wanted to add a volume of tables to the text-only *Notitia*? Perhaps the Nuremberg genealogist saw this project of revision as an experiment, exploring the possibility of publishing in German so as to address a different audience? Whatever his reasons, Imhoff took the Lohmeier project very seriously. Yet another improved edition appeared in 1701 (again shortly after a new edition of the *Notitia* in 1699), and rumors of yet another edition circulated throughout the early 1700s.[21]

What first emerges from these episodes is the preciousness of a successful genealogical brand. Everyone appreciated its enormous economic value. All the actors involved accordingly showed great deference to great names, even after the original author's death. Imhoff and his friends expressed relief that Rittershausen's works reputedly "only require[d] additions, not alterations."[22] The reworking of Lohmeier's handbook, although very substantial, likewise never resulted in outright criticism of the original publication. A second point to observe is that the cultivation of editorial "brands" was largely the publishers' doing. *Their* needs and *their* initiatives drove brand-continuation as a business model. From their perspective, doing so enabled them to survive in a book market that was increasingly saturated with genealogical publications. Their "hidden helpers" and ghostwriters, in contrast, including Imhoff *père et fils*, usually remained anonymous.

A third and final point to note is that although the publishers' interests drove genealogical brand-development, it also presented a great opportunity for upcoming writers like Imhoff. The contributions of knowledgeable ghostwriters may not have been publicly acknowledged, but they were nevertheless widely known. Accepting a role as an anonymous hidden hand may therefore have been an acceptable compromise for a young genealogist to establish a reputation. The editorial trend of promoting brand names in genealogy circa 1700 thus created significant opportunities for both publishers *and* authors. Imhoff profited greatly from his early association with the Rittershausen legacy.

A Trademark Style: Imhoff's Signature Combination of Text and Diagram

Once he had established his own reputation after 1685, Imhoff was confident enough to move on, eventually shedding his association with the Rittershausen name and devising his own trademark style of publication. His most recognizable products were his many books on noble houses outside of Germany, the layout and presentation of which differed significantly from his earlier works. By far the greatest difference consisted of the exquisite and crystal-clear combination of narrative and visual presentations of genealogical information. Imhoff's move away from the Altdorf-Tübingen editorial circle and the Rittershausen brand coincided with a forward-looking integration of diagrams and texts.

Diagram and text are the two basic forms of presenting genealogical knowledge. When Imhoff launched his genealogical projects, many of the most influential encyclopedic genealogists kept these two media formats largely separate. This meant in practice that they published books that consisted either entirely of genealogical tables or exclusively of narrative text. Rittershausen himself was a prime example. His *Genealogiae* contained hundreds of genealogical tables, but no text. The author published an accompanying volume of genealogical narratives posthumously, courtesy of Imhoff Sr.[23] Philipp Jakob Spener, Imhoff's second important role model, also regularly published tables and texts in different volumes.[24]

At first, Imhoff was unsure of his own preference. One of the key influences in his earliest career, the Zittau rector and playwright Christian Weise, insisted on the necessity of schemata, but Imhoff expressed his skepticism of Rittershausen's "naked tables" (*tabulas nudas*).[25] Diagrams were powerful tools to convey biological relationships and personal information, yet they could not accommodate information about "history and law," two important topics that Imhoff closely associated with genealogy. The Nuremberg patrician favored a broader understanding of genealogy, one that went beyond merely providing personal data. He initially solved the conundrum by providing tables and narratives in separate books, published roughly simultaneously. The *Spicilegium Rittershusianum* and the *Notitia Procerum* divided tables and text neatly between them, yet both volumes appeared almost simultaneously.

This approach, however, was not completely satisfactory to Imhoff. Starting in 1687 with his *Excellentium Familiarum in Gallia Genealogiae*, on French genealogies, Imhoff adopted a new formula. Instead of dividing text and tables between separate works, he now integrated them. This fusion was to become Imhoff's signature style, a highly satisfying combination of genealogical diagrams and narrative. It was this well-crafted arrangement that attracted Imhoff's readers. Bohuslav Balbín, praising Imhoff's book on France in 1687, noted how it broke with established layout preferences. "I adopted another, different logic from yours in my genealogies: I presented plain tables, without notes, in this *Tabularium*; I will present the family origins later in a book called *Stemmatographia* that I am currently preparing."[26] Another friend enthusiastically told Imhoff, "I am happy that you added historical notes, with which the French so especially like to illustrate their genealogies, to the tables."[27] Two decades later, reviewers still felt that this layout set Imhoff's works apart: "the individual stocks with their spreading branches are placed before inquisitive eyes in tables; while, so that the reader is not wearied by a pure recitation of names, the *historical exegeses* that he has added abundantly satisfy their thirst."[28] Here lay a major selling point of Imhoff's books.

There were a few precursors to Imhoff's sophisticated approach. Earlier genealogists had occasionally inserted large amounts of narrative directly into their diagrams, effectively creating genealogical tables that contained several lines of text for each individual. Hieronymus Henniges's *Theatrum* was a well-known example. Imhoff also experimented with this format, using it in his updates of Lohmeier. Overall, however, this approach resulted in cluttered pages that sacrificed the visual clarity of the diagram, as one anonymous reader of Imhoff's edition of Lohmeier experienced. In his private copy, he drew lines and added comments to highlight and repeat basic information that Imhoff's text already provided. In effect, this reader made visible yet again decisive pieces of information that had otherwise been buried in crowded columns of text.[29] Such examples illustrate why Imhoff presumably did not find this layout particularly effective.

Other works integrated genealogical text and image by printing them consecutively. In an early work from 1608, discussing numerous Christian dynasties, Antonio Albizzi had appended brief historical narratives to naturalistic family trees on the page after each image.[30] But this work did not set a lasting trend, it seems. The second edition (1677) of Spener's *Sylloge Genealogico-Historica*, in which the author added a few multipage tables to a book that had initially consisted only of text may have been more influential on Imhoff. Yet the differences between Spener's and Imhoff's works are more significant than their similarity in this regard. In Spener's updated edition, the added tables were tucked away in roughly nine hundred pages of small print. They were retrospective additions, not an integral part of a preconceived mode of presenting information.

Imhoff's masterful synthesis of tables and texts was a distinctive feature of his works. He followed best practices of diagrammatic *and* textual genealogy and combined both in a balanced way. His tables continued the aesthetically refined tradition of Rittershausen and the *Tübinger Tafeln*. These works featured well-arranged and amply spaced diagrams printed on double folio pages. Rittershausen's tables were pleasant to behold. Imhoff took up this format and refined it even further. His readers applauded his "clearer and more accessible, and indeed analytical and methodical" presentation of the material.[31] In

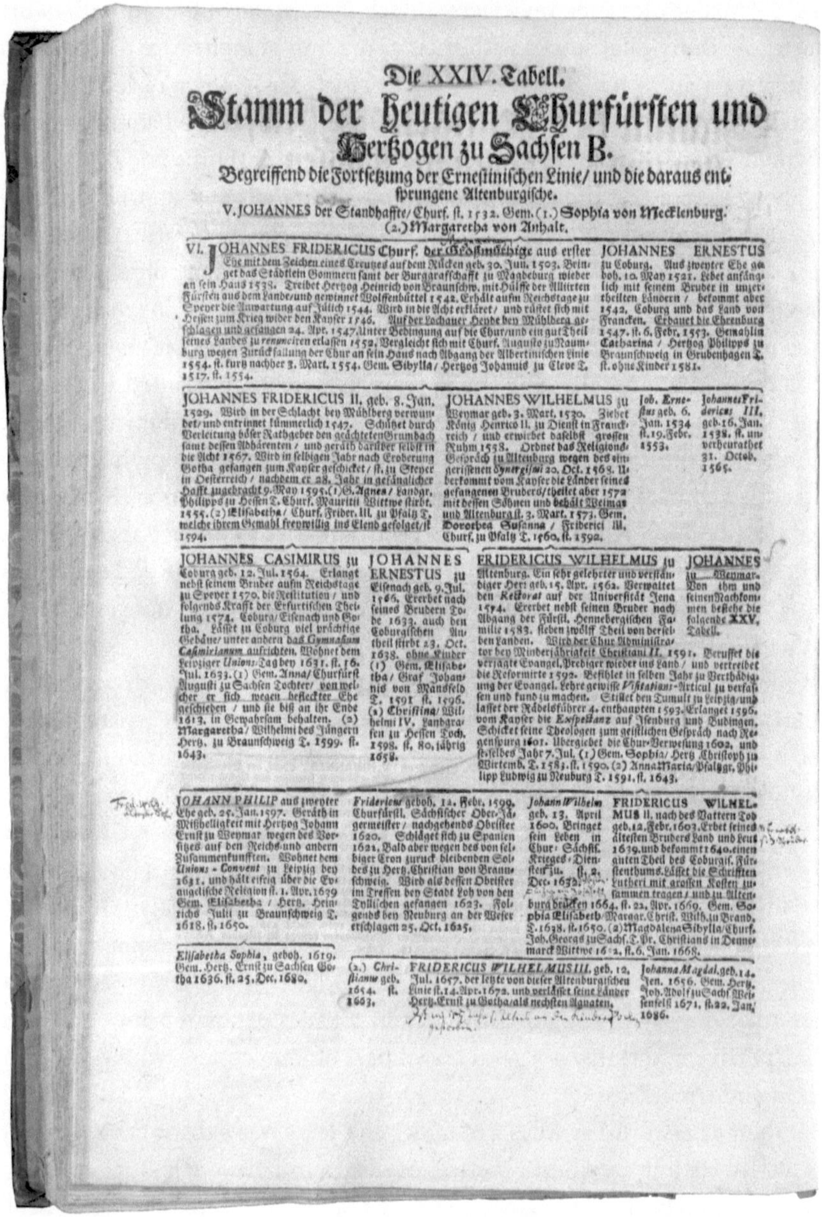

Die XXIV. Tabell.

Stamm der Heutigen Churfürsten und Hertzogen zu Sachsen B.

Begreiffend die Fortsetzung der Ernestinischen Linie/ und die daraus entsprungene Altenburgische.

V. JOHANNES der Standhaffte/ Churf. st. 1532. Gem. (1.) Sophia von Mecklenburg. (2.) Margaretha von Anhalt.

VI. JOHANNES FRIDERICUS Churf. der Großmüthige aus erster Ehe mit dem Zeichen eines Creutzes auf dem Rücken geb. 30. Jun. 1503. Bringet das Städtlein Gommern samt der Burggrafschafft zu Magdeburg wieder an sein Haus 1538. Treibet Hertzog Heinrich von Braunschw. mit Hülffe der Alliirten Fürsten aus dem Lande/ und gewinnet Wolffenbüttel 1542. Erhält aufm Reichstage zu Speyer die Anwartung auf Jülich 1544. Wird in die Acht erkläret/ und rüstet sich mit Hessen zum Krieg wider den Kayser 1546. Auf der Lochauer Heyde bey Mühlberg geschlagen und gefangen 24. Apr. 1547. Unter Bedingung auf die Chur/ und ein gut Theil seines Landes zu *renunciren* erlassen 1552. Vergleicht sich mit Churf. Augusto zu Naumburg wegen Zurückfallung der Chur an sein Haus nach Abgang der Albertinischen Linie 1554. st. kurtz nachher 3. Mart. 1554. Gem. Sibylla/ Hertzog Johannis zu Cleve T. 1517. st. 1554.

JOHANNES ERNESTUS zu Coburg. Aus zweyter Ehe geboh. 10. May 1521. Lebet anfänglich mit seinem Bruder in unzertheilten Ländern/ bekommt aber 1542. Coburg und das Land von Francken. Erbauet die Ehrenburg 1547. st. 6. Febr. 1553. Gemahlin Catharina/ Hertzog Philipps zu Braunschweig in Grubenhagen T. st. ohne Kinder 1581.

JOHANNES FRIDERICUS II. geb. 8. Jan. 1529. Wird in der Schlacht bey Mühlberg verwundet/ und entrinnet kümmerlich 1547. Schützet durch Verleitung böser Rathgeber den geächteten Grumbach samt dessen Adhärenten/ und geräth darüber selbst in die Acht 1567. Wird in selbigen Jahr nach Eroberung Gotha gefangen zum Kayser geschicket/ st. zu Steyer in Oesterreich/ nachdem er 28. Jahr in gefänglicher Hafft zugebracht 9. May 1595. (1) G. Agnes/ Landgr. Philipps zu Hessen T. Churf. Mauritii Wittwe stirbt. 1555. (2) Elisabetha/ Churf. Frider. III. zu Pfaltz T. welche ihrem Gemahl freywillig ins Elend gefolget/ st. 1594.

JOHANNES WILHELMUS zu Weymar geb. 3. Mart. 1530. Ziehet König Henrico II. zu Dienst in Franckreich/ und erwirbet daselbst grossen Ruhm 1558. Ordnet das Religions-Gespräch zu Altenburg wegen des eingerissenen *Synergismi* 20. Oct. 1568. Uberkommt vom Kayser die Länder seines gefangenen Bruders/ theilet aber 1572 mit dessen Söhnen und behält Weimar und Altenburg st. 3. Mart. 1573. Gem. Dorothea Susanna/ Friderici III. Churf. zu Pfaltz T. 1560. st. 1592.

Joh. Ernestus geb. 6. Jan. 1534 st. 19. Febr. 1553.

Johannes Fridericus III. geb. 16. Jan. 1538. st. unverheurathet 31. Octob. 1565.

JOHANNES CASIMIRUS zu Coburg geb. 12. Jul. 1564. Erlangt nebst seinem Bruder aufm Reichstage zu Speyer 1570. die Restitution/ und folgends Krafft der Erfurtischen Theilung 1574. Coburg/ Eisenach und Gotha. Lässet zu Coburg viel prächtige Gebäue/ unter andern das *Gymnasium Casimirianum* aufrichten. Wohnet dem Leipziger *Unions*-Tag bey 1631. st. 16. Jul. 1633. (1) Gem. Anna/ Churfürst Augusti zu Sachsen Tochter/ von welcher er sich wegen befleckter Ehe geschieden/ und sie biß an ihr Ende 1613. in Gewahrsam behalten. (2) Margaretha/ Wilhelmi des Jüngern Hertz. zu Braunschweig T. 1599. st. 1643.

JOHANNES ERNESTUS zu Eisenach geb. 9. Jul. 1566. Ererbet nach seines Brudern Tode 1633. auch den Coburgischen Antheil stirbt 23. Oct. 1638. ohne Kinder (1) Gem. Elisabetha/ Graf Johannis von Mansfeld T. 1591 st. 1596. (2) Christina/ Wilhelmi IV. Landgrafen zu Hessen Toch. 1598. st. 80. jährig 1658.

FRIDERICUS WILHELMUS zu Altenburg. Ein sehr gelehrter und verständiger Herr geb. 15. Apr. 1562. Verwaltet den *Rectorat* auf der Universität Jena 1574. Ererbet nebst seinen Bruder nach Abgang der Fürstl. Hennebergischen Familie 1583. sieben zwölff Theil von derselben Landen. Wird der Chur Administrator bey Minderjährigkeit Christiani II. 1591. Beruffet die verjagte Evangel. Prediger wieder ins Land/ und vertreibet die Reformirte 1592. Befihlet in selben Jahr zu Verthädigung der Evangel. Lehre gewisse *Visitations*-Articul zu verfassen und kund zu machen. Stillet den Tumult zu Leipzig/ und lasset drey Rädelsführer 4. enthaupten 1593. Erlanget 1596. vom Kayser die *Exspectanz* auf Isenburg und Büdingen. Schicket seine Theologen zum geistlichen Gespräch nach Regenspurg 1601. Ubergiebet die Chur-Verwesung 1602. und st. selbes Jahr 7. Jul. (1) Gem. Sophia/ Hertz. Christoph zu Wirtemb. T. 1583. st. 1590. (2) Anna Maria/ Pfaltzgr. Philipp Ludwig zu Neuburg T. 1591. st. 1643.

JOHANNES zu Weymar. Von ihm und seinen Nachkommen bestehe die folgende XXV. Tabell.

JOHANN PHILIP aus zweyter Ehe geb. 25. Jan. 1597. Geräth in Mißhelligkeit mit Hertzog Johann Ernst zu Weymar wegen des Vorsitzes auf den Reichs- und andern Zusammenkunfften. Wohnet dem *Unions-Convent* zu Leipzig bey 1631. und hält eifrig über die Evangelische Religion st. 1. Apr. 1639 Gem. Elisabetha/ Hertz. Heinrichs Julii zu Braunschweig T. 1618. st. 1650.

Fridericus geboh. 12. Febr. 1599. Churfürstl. Sächsischer Ober-Jägermeister/ nachgehends Obrister 1620. Schläget sich zu Spanien 1621. Bald aber wegen des von selbiger Cron zuruck bleibenden Soldes zu Hertz. Christian von Braunschweig. Wird als dessen Obrister im Treffen bey Stadt Loh von den Tyllischen gefangen 1623. Folgends bey Neuburg an der Weser erschlagen 25. Oct. 1625.

Johann Wilhelm geb. 13. April 1600. Bringet sein Leben in Chur-Sächsis. Krieges-Diensten zu. st. 2. Dec. 1632.

FRIDERICUS WILHELMUS II. nach des Vattern Tod geb. 12. Febr. 1603. Erbet seines ältesten Bruders Land und Leut 1639. und bekommt 1640. einen guten Theil des Coburgis. Fürstenthums. Lässet die Schrifften Lutheri mit grossen Kosten zusammen tragen/ und zu Altenburg drucken 1664. st. 22. Apr. 1669. Gem. Sophia Elisabeth/ Marggr. Christ. Wilh. zu Brand. T. 1638. st. 1650. (2) Magdalena Sibylla/ Churf. Joh. Georgs zu Sachs. T. Pr. Christians in Dennemarck Wittwe 1652. st. 6. Jan. 1668.

Elisabetha Sophia, geboh. 1619. Gem. Hertz. Ernst zu Sachsen Gotha 1636. st. 25. Dec. 1680.

(2.) *Christianus* geb. 1654. st. 1663.

FRIDERICUS WILHELMUS III. geb. 12. Jul. 1657. der letzte von dieser Altenburgischen Linie st. 14. Apr. 1672. und verlässet seine Länder Hertz. Ernst zu Gotha/ als nechsten Agnaten.

Johanna Magdal. geb. 14. Jen. 1656. Gem. Hertz. Joh. Adolf zu Sachs. Weisenfels 1671. st. 22. Jan. 1686.

Improving Lohmeier's visual clarity by drawing highlights and marginalia. The diagonal line in the bottom third of the page emphasizes intergenerational links, which were indicated inconspicuously in the original by a printed bracket. Lohmeier, *Der Europäischen Reiche und Fürstenthümer*, 1701, unpaginated. Bamberg, Staatsbibliothek, Gen.f.6.

contrast, for instance, to the dense tables of Spener's *Theatrum* and Henniges's and Lohmeier's crowded diagrams, Imhoff's genealogies were visually "accessible."

In addition, Imhoff carefully balanced text and diagrams. Both media were of equal importance, neither being subservient to the other. In keeping with his previous practice in the *Notitia Procerum*, Imhoff fleshed out his narratives where it mattered, yet did so without exhausting his readers. In effect, he kept his texts to a few manageable pages per family, providing succinct surveys of family history together with more focused paragraphs about select individuals, events, and branches. These texts were not exhaustive treatments but rather reference works, roughly comparable to the biographical dictionaries, which came into vogue in the early eighteenth century. Without overwhelming readers, Imhoff's texts helped contextualize individual nobles and families in greater historical detail than a plain table could have ever done. Finally, Imhoff carefully related his texts to the accompanying tables. His narrative always referred to specific tables (for example, "Ad tab. I") and was divided into individual numbered paragraphs in which important individuals were often highlighted in italics. A reader could easily find individuals mentioned in the text in the tables and vice versa. For all the research they contained, the narratives were still recognizable as commentary on the tables.

Imhoff thus created and perfected a successful formula for presenting genealogical information in print. The new style of his publications after 1687, while perhaps not entirely unprecedented, was nevertheless unusually clear and elegant. His publications retained the visual appeal of the best tables-only books, while also providing in-depth information about the past and present of each dynasty to satisfy readers who wanted to know more than merely the dates of births, marriages, and deaths. Few products on the German book market at the time could compete with Imhoff's refined combination of visual and textual genealogy. On the contrary: most other contemporary works fell short of Imhoff's, either because they did not integrate text and diagrams at all, or did so in a less appealing visual form, or because they could not match the high quality of Imhoff's information.[32] The visual appeal and artisanal perfection of his books, along with the reliability of their information translated into resounding success.

Imhoff and His Publishers

Creating such elaborate books required the cooperation of competent partners. Imhoff worked exclusively with elite publishers virtually from the start.

His early works in the Rittershausen tradition all appeared in Tübingen (later Stuttgart) with Brunn-Cotta, who were building a strong portfolio of genealogical products at the time.[33] Later, he moved his projects to Nuremberg, where he cooperated with Georg Andreas Endter (1654–1717) until the firm lost interest after 1702. After a brief period of adjustment, Imhoff published in Amsterdam with Zacharie Châtelain, the famous printer of maps and atlases.[34] Only for his last major work of 1712, *Genealogiae Viginti Illustrium in Hispania Familiarum*, did he change publishers once more, producing this final opus with Gleditsch in Leipzig. As this list of partners demonstrates, Imhoff worked only with the best and most successful printers and publishers of his time. A few minor incidents notwithstanding, the Nuremberg patrician never struggled to get his works published.[35] On the contrary, he liked to report how publishers "asked" him to prepare new editions of his works.[36]

Insofar as he was in a comfortable position, Imhoff could afford to be fussy about the precise terms of cooperation. He knew the rules and played hard to get his way. For example, he once turned down an offer from Jean Anisson (1642–1721), the official royal printer of Louis XIV.[37] Anisson had set only two conditions for their cooperation: First, Imhoff would exploit his contacts with the German community in Paris to recruit skilled proofreaders who would check all the names in the final proofs. Second, Imhoff would buy three hundred copies of his own work for two *écus* a piece.

These two conditions reflected Anisson's pragmatic perspective on genealogical books as a publisher. His first requirement resulted from a superior artisan's pride in perfection. "I do not want to publish any work that is not beautiful and correct," he told Imhoff when asking for competent proofreaders.[38] Genealogy, with all its names and details, was a challenging business. Anisson's second stipulation, meanwhile, highlighted the risk that genealogical works posed for publishers. Despite the fact that Imhoff's books were popular among aficionados, savvy publishers like Anisson nevertheless hedged their bets. Imhoff, however, was outraged by Anisson's condition. Seconded by d'Hozier, he viewed the mandate that he must purchase hundreds of copies as indicative of publishers' well-known greed. "Book merchants—they want the work, the toil for free . . . and find nothing unjust about this conduct," d'Hozier fumed.[39] Imhoff agreed and turned Anisson down. Clearly enough, he felt he was in a position to dictate his own terms.

Occasionally, the times were ripe for a genealogical work. Just as Imhoff's decade-long research on Spanish (and Italian) genealogy neared its end around 1700, the Spanish succession crisis heated up. When the War of the Spanish

Succession finally broke out in 1701, it was easy to anticipate a general surge in interest in Spanish genealogy. Suddenly, Imhoff's forthcoming book about Spain was a "contemporary history," as he proudly announced.[40] His publishers were now also especially eager to proceed and rushed his book to press.[41] Imhoff's *Recherches historiques et genealogiques des grands d'Espagne* (Historical and genealogical research on the grandees of Spain), a short handbook in French about the Spanish nobility in 1707, also reflected recent events. Imhoff had gathered his information for this work during his larger project on Spain but then significantly condensed it to make for easy reading. Opening with a twenty-page introduction to the Spanish high nobility that highlighted their prerogatives and legal status, the small book almost reads like a French invader's guide to a foreign aristocracy, to be consulted for orientation before conversations with dukes, marquises, and counts.[42] Châtelain in Amsterdam clearly anticipated a need for easily digestible and ready-to-use forms of genealogical knowledge.

The fact that Imhoff had a very clear idea of where he wanted to position his books in the market can also be glimpsed, *ex negativo*, by his reluctance to adopt some of the latest methods of commercializing genealogical knowledge. Several friends tried to force media practices on Imhoff that were characteristic of the most market-driven sectors of the genealogical publishing business. Should not *separata* be extracted from Imhoff's encyclopedic works, so that individual families could buy only "their" particular genealogies?[43] Printing *separata* by themselves would break down the encyclopedic model into single-family genealogies, allowing for very targeted marketing. Interested dynasties could acquire Imhoff's products in fragments without having to buy entire books. Johann Seifert, a near contemporary genealogist from Regensburg, was a pioneer of this approach. Imhoff would have known about Seifert and about the practice in general, yet he never showed any interest in it.

The same somewhat conservative reserve also appears in Imhoff's reluctance to take up a second suggestion. Instead of constantly putting new editions of *Notitia Procerum* on the market, could Imhoff produce addenda and updates that contained only the dynastic events that had occurred between the last edition and the present?[44] Again, this idea derived from recent media developments, as genealogical encyclopedias were gradually transformed into genealogical calendars that covered only the most recent events. Serializing genealogical information was a forward-looking strategy, as the subsequent rise of periodical works on European noble houses, such as the famous *Gotha*, an annually updated handbook of noble families, demonstrates.[45] Imhoff, although focused on staying up to date about genealogical developments, never adopted this cutting-edge genre. In contrast, he retained the traditional approach and relied on large-scale, high-quality books that combined the visual aura of genealogical tables with informative, well-balanced, and updated, mid-length historical introductions to individual dynasties and nobles. His works thus steered a middle course, broadly following the recent trend toward encyclopedic genealogy, but without wholly embracing the latest innovations in the field. Imhoff was receptive to market forces but not driven by them.

Surviving Competition

The growing dynamics of the genealogical knowledge economy had unforeseen social consequences in Imhoff's lifetime. These dynamics created new forms of competition among genealogists. Courtly genealogy, largely controlled and paid for by individual nobles and dynasties as it was, had known little competition among authors. In most cases, there was—at best—one genealogical writer per family and generation. If competing versions of a family's genealogy appeared, it was predominantly a reflection of social conflict or political change. Competition resulted in alternative views of family history. With projects like Imhoff's, this changed. Once genealogy moved out of the closely monitored courtly production sites, it became hard for authors to avoid the struggle for market dominance. Several of Imhoff's European acquaintances indeed viewed the publication of his encyclopedic works as a threat to their own projects.

Since encyclopedic genealogies, by the very logic of their genre, touched on any number of dynasties, they might unwittingly compete with ongoing projects. Authors who had dedicated their lives to a particular genealogical topic occasionally considered Imhoff an intruder in their territory. In France, the antiquarian and manuscript collector Claude de Guénegaud, sieur de Bresse (1640–1720), and since 1696 grand chancellor of the Order of St. Lazare, is a case in point. Guénegaud was working on a history of his order, and some of Imhoff's requests about French nobles affiliated with this institution made it seem as if the German intended to cover some of the same material, thus preempting his own work.[46] The French genealogist accordingly responded coolly to Imhoff and was reluctant to give support.

In Naples, Imhoff's unexpected appearance on the local genealogical scene also prompted rivalry. Imhoff had correctly placed high hopes in Biagio Aldimari, yet the courtly genealogist of the house of Carafa guarded his research just as "secretively" as Gottfried Wilhelm Leibniz had in Hanover. "Aldimari never shows anything to anybody," as locals complained.[47] There were probably several reasons behind his reluctance to share his findings, but concern for the editorial success of his work was definitely among them. Aldimari had a very market-savvy publisher, the above-mentioned Antonio Bulifon.[48] While he helped Imhoff gain an epistolary foothold among Neapolitan genealogists, Bulifon also tried to delay Imhoff's publication until after Aldimari's work had appeared in print. Through Magliabechi, Bulifon explicitly suggested to Imhoff that he should postpone any publications about the Carafa until after he had thoroughly digested Aldimari's forthcoming magnum opus.[49] Clearly, Imhoff's reputation had made encyclopedic genealogy serious competition for works of dynastic genealogy that had previously faced very little.

Rivalries were also rampant among encyclopedic genealogists, who all served the same market. Successful products and promising editorial formulas were quickly copied. Since there seemed to be significant demand for encyclopedic genealogy, authors and publishers had incentive to deliver competitive products. This was particularly obvious at the beginning of Imhoff's career, when he faced a range of competitors for Rittershausen's legacy, including Christian Weise and Philipp Jakob Spener.[50] Spener, in particular, would have been a fine choice as a continuator. He had made numerous additions and corrections in his own copy of Rittershausen, laying the groundwork for an improved edition. Yet, as Spener himself acknowledged, his career had started to shift away from genealogy toward theology and church politics. He found neither time nor leisure for further genealogical publications. In 1683, therefore, Spener decided to sell his annotated copy of Rittershausen and many additional papers on French genealogies for two hundred florins. Tellingly, Spener approached a publisher—Cotta in Tübingen—and not a specific author with this offer.[51] As we have already seen in the case of Antonio Bulifon, publishers emerged as powerful brokers in the genealogical information and book market.

In 1686, the prospect of competition with Imhoff came up for Spener once again and was centered even more pointedly on the subject of French genealogy. Spener attempted a final push to complete a long-planned work on France.[52] Now, however, the tables had turned: Imhoff was a well-established author, and his publisher Endter had already announced his new book on France, his *Excellentium Familiarum in Gallia Genealogiae*, in the company's catalog of forthcoming works. Spener trailed Imhoff. The theologian tried to remedy the unfortunate coincidence of near-identical book projects. They needed, he wrote to Imhoff, to avoid direct competition in the book market "so that we do not hurt each other's work."[53] Spener inquired whether the two planned books on

France would either complement each other or overlap. In the first case, both works could appear together in print; in the second case, however, something had to be done. Since Imhoff's work was already in press in 1686, Spener agreed to omit the noble houses discussed by Imhoff from his own work.[54] That is exactly what happened: Imhoff had passed over enough genealogies to make a second book viable on the book market.[55] Spener's *Illustriores Galliae Stirpes* (Illustrious French lineages) thus appeared in 1689 explicitly as a "supplementum" to Imhoff's earlier volume.[56]

Encyclopedic genealogies like Imhoff's and Spener's were competitors not so much because of alternative ideological standpoints or historiographic visions but rather because they shared a similar outlook and pursued parallel projects. Even though this particular conflict came to a happy ending, the episode nevertheless demonstrates just how much competition there was in the increasingly crowded genealogical book market circa 1700. Imhoff, Pregitzer, Spener, and a host of other authors all vied for the public's attention.[57] In practice, publishers and authors divided the market so as not to hurt each other directly by producing too many similar books. Encyclopedic genealogists had to situate their work carefully vis-à-vis a number of parallel projects. Imhoff mastered this difficult art very successfully.

Reading Imhoff

The books of Jakob Wilhelm Imhoff had many readers, and they came from many strata of society. As genealogy diversified and burst into the open information and book market, individual publications were attracting more and more diverse readerships. By 1700, it was hardly any longer possible to predict exactly who was going to study a genealogical encyclopedia and for what purposes. Imhoff's publications, too, sitting at the crossroads of celebratory, scholarly, and political genealogy, recommended themselves to many readers as they successfully integrated a range of different perspectives on the field. As long as they were able to read Latin, individuals interested in genealogy could, and did, access his works for very different reasons. Looking at traces of actual readership, we encounter a host of different approaches to Imhoff's works. Several annotated copies of Imhoff's books, together with a few excerpts made by individual readers, allow us insight into the reception of his works.

"Living Texts"

As Léopold Genicot wrote in his survey of medieval genealogical literature in 1975, works of genealogy are "living texts."[58] By that he meant, naturally, that no genealogical text or diagram is ever finished. Unless a family dies out, new members continue to be born and must be added to existing genealogies. Hence, existing works are constantly annotated, updated, and expanded. The eighteenth century eventually found a solution for this problem by creating serial publications, focused especially on recent news and updates about dynastic events, in the form of annually published genealogical calendars. Before the arrival of calendars and genealogical journals, however, existing texts constantly had to be expanded and updated by hand. Genealogies could never be fixed or static oeuvres. On the contrary, they were a powerful example of what Jeffrey Todd Knight calls the "malleable, mutable book."[59] They were "living" media.

Readers frequently personalized their copies of Imhoff's works, using them to record information about current genealogical events. All over Europe people modified Imhoff's tables—but rarely his narratives—and adorned them with updated genealogical information. One copy of Imhoff's *Genealogiae Viginti Illustrium in Italia Familiarum* of 1710, for instance, was interleafed so that an unidentified reader who read both Latin and French could update Imhoff's information on the houses of Cybo, Massa, and Pico della Mirandola.[60] The first part of a copy of Imhoff's *Excellentium Familiarum in Gallia Genealogiae* of 1687, meanwhile, was heavily annotated, corrected, and updated where Imhoff had discussed the French royal dynasties in several dozen tables.[61] A copy of his genealogy of the counts of Reuss, *Genealogia Ruthenorum Comitum et Dominorum in Plauen* of 1715, was updated with three entirely new handwritten tables (XI to XIII) on blank pages at the end of the original publication, adding several generations to Imhoff's original tables.

These annotated copies of Imhoff's works illustrate a very common form of continuous manuscript engagement with genealogical publications, made necessary by the nobility's inherent evolution. Imhoff worked like this himself, as we have seen, constantly annotating a copy of Rittershausen's 1664 edition, which served frequently as a starting point for his own more advanced genealogies. In fact, he also constantly annotated his own works, using marked-up copies of the *Notitia Procerum* as springboards for new editions. Although none of Imhoff's personal copies have survived, the practice is explicitly attested.[62] Thus, by Imhoff's time, keeping genealogies "alive" required constant transgression of the porous boundary between print and manuscript media. Prints were necessarily converted back into manuscripts so they could be kept up to date, and handwritten annotations in turn were eventually transformed into print. Cultivating genealogical knowledge required not only genealogists like Imhoff but also his readers to navigate and combine multiple media in highly sophisticated ways.[63]

Reading Imhoff's Books

Whereas updating genealogical diagrams was a rather generic form of engagement with (printed) books, some copies of Imhoff's works enable us to pinpoint certain readers' specific interests. Some evidence, for instance, can help us envision how nobles might have approached Imhoff's books in practice. In 1710, one of Imhoff's contacts had assumed that his works would be of great service to nobles in the process of "drawing up their proof of nobility or explaining their coats of arms."[64] Whenever nobles needed quick information about their families, he seems to suggest, they would turn to Imhoff's works.

It is not easy to tell how helpful genealogical encyclopedias actually were for compiling one's proof of nobility, yet members of the imperial nobility clearly found Imhoff's works to be highly relevant in difficult times. Albert Christian Ernst (1720–99), Count of Schönburg, for instance, turned to his works in 1758 at a moment of dynastic insecurity. The count and his dynasty belonged to a tangled cluster of interrelated noble families in Thuringia.[65] Ernst took great pains to maintain the preeminence of his own branch in this context. Various genealogical projects featured among his wide-ranging activities to protect his family's splendor, and the count read widely in available specialized literature. His readings naturally included Imhoff's forty-three-year-old booklet about the closely related counts of Reuss, published on behalf of Heinrich XIII of Reuss-Untergreiz (1672–1733) in 1715.[66]

Ernst annotated his personal copy heavily. He used the many empty pages of Imhoff's slim book to write down additional historical information about his family and its history. The count's marginalia illustrate his reading practice, as he records parallel passages from other authors. Ernst used Imhoff's book almost as a notebook. The annotations show how the count tried to keep track of the many different branches, many of them now extinct, and the division of territories and titles among them.[67] While most of this information did not serve immediate political or legal purposes, his notes nonetheless show that the count was actively engaged in studying his family's past, compiling an arsenal of information that he could muster to defend his family's legal and social preeminence. Imhoff's original genealogical table provided a workable matrix for investigating the dynasty's past; published without a historical narrative, however, the booklet from 1715 obviously lacked the necessary historical context. The count's annotations offer a rudimentary, self-made equivalent to the historical narratives that Imhoff otherwise provided in his mature works. In a moment of dynastic tension, even a relatively dated book might be an important item on a nobleman's politically motivated genealogical reading and research agenda.

The many annotations an anonymous reader entered into a copy of Imhoff's new edition of Lohmeier from 1695, today housed in Regensburg, are of a very different nature.[68] The unidentified owner had this exemplar interleaved, so that every folio, which featured a table each on recto and verso, is followed by one or two additional sheets covered in handwritten notes about the material printed on the opposite page. Since the genealogical tables in Lohmeier were packed with narratives, there was much to comment on. Subsequent additions are piled up in the margins, as the annotator was prone to comment on his own commentaries. In order to link his manuscript notes to the relevant printed passages, and later notes to earlier ones, the author devised a system of signs and utilized numerous handwritten lines and arrows.[69]

While some additions merely clarify details of Imhoff's narrative, most of the handwritten passages do much more than expand on the existing text. They provide new material and significantly enrich Imhoff's original, adding many new dimensions to the narrative. One distinctive feature of the annotations is their author's obsession with royal or princely mottos. The annotator enthusiastically records the emblematic and moralizing maxims chosen by various kings or princes as personal mottos, and he also not infrequently cites poetry, epitaphs, and various *bon mots*. These handwritten notes usually concern individual princes and nobles and their immediate family and often revolve around a key episode or anecdote, occasionally lending them a gossipy and entertaining flair. Only rarely, however, are there annotations strictly genealogical in nature. Once, for instance, the annotator provides a list of illegitimate children born to the English king James II (opposite Tabula XXXIX). In some places, he adds recently born children or dates of death. Annotations of a broadly "historical" nature, however, predominate by far. These notes contain bits of curious or (not so) relevant information about individuals mentioned by Imhoff.

While there is currently no way to establish the authorship of these marginalia, it is clear that they represent a readership very different from the courtly realm represented by Count Albert Christian Ernst.[70] Although the count also took notes on the fruits of his genealogical reading, his annotations are highly focused, concern a single family, and are largely restricted to legal and dynastic matters. The anonymous annotator of the Regensburg copy of Imhoff-

Lohmeier, in contrast, had a very broad understanding of genealogy. He was, in effect, interested in everything even remotely connected to the genealogies detailed by Imhoff. His additions read like the extensive contents of a fellow historian's or genealogist's notebook.[71] The anonymous reader used Imhoff-Lohmeier as a repository for his own wide-ranging readings, turning the interleafed print-publication into a commonplace book for his excerpts. Imhoff's tables—or, rather, the individuals mentioned in them—served as loci under which he subsumed all kind of material.

Since there are few indications of original source work, this particular annotator may not have come from a scholarly background. He was not primarily interested in philological or archival details. There is no doubt, however, that scholars with a more antiquarian bent also appreciated Imhoff's books and actively utilized them in their work. One erudite historian who left behind significant traces of a thorough examination of Imhoff's texts was an Augustinian monk from Bohemia, Valentin Weidner (1658–?).[72] A prolific church historian influenced by the nascent Maurist and Bollandist traditions, Weidner investigated the history of several Bohemian monasteries, including their founding families and benefactors. To lay a foundation for his genealogical research, he carefully read and excerpted the third and fourth editions of the *Notitia Procerum* and several other publications by Imhoff. One hundred twenty pages of notes from his reading survive today.[73]

Weidner's excerpts follow Imhoff's text very closely, even with respect to omissions and biases. Occasionally, he reads the Nuremberg Lutheran against the grain, as, for instance, when he transforms Imhoff's report about the counts of Mansfeld (in whose home Martin Luther passed away in 1546) into a biographical note about the death of the Reformer.[74] Weidner also regularly updated and improved on Imhoff's information, thereby keeping the original text "alive." Only rarely do his notes exhibit a markedly Catholic standpoint.[75] The Augustinian certainly read selectively, condensing sections of minor importance to himself and passing over details that he already knew.[76] Since Weidner left few printed works of his own behind, it is hard to discern how he worked with these excerpts from Imhoff and how he integrated Imhoff's information in his own scholarship. The rigor of this impressive collection of notes, however, attests to in-depth engagement with the Nuremberg genealogist's works by a dedicated scholar.

A final group of readers worth mentioning consists of politicians and administrators. The private libraries of such men were often well-stocked with genealogical works, among which Imhoff's publications frequently appear.[77]

A humble file on the tiny nearby territory of Blankenburg in the ducal archives of Wolfenbüttel nicely illustrates how politicians made use of Imhoff's books in everyday political affairs.[78] The dukes had fought for decades with the electors of Brandenburg for control of the miniature county.[79] Much ink, albeit little blood, was spilled in these long quarrels. At some point around 1700, officials in Wolfenbüttel compiled yet another dossier on the complex history of the territory and the dukes' rights, presumably in anticipation of a new round of clashes. The file consisted of several parts. It included older administrative documents from 1635, 1662, and 1666, surveying the territory's potential, along with a verbatim copy of the relevant pages from Imhoff's *Notitia Procerum*. Sitting side by side with administrative memoranda in this file, Imhoff's genealogies thus were seamlessly factored into the political decision-making processes.

If we pull all these examples of readers' responses together, it is obvious how genealogical knowledge had become valuable in so many ways by 1700. There were numerous sites of genealogical knowledge, all of them shaped by different perspectives on the field. Readers did very different things with Imhoff's works and the information they contained. Some even incorporated material from Imhoff into entirely unrelated contexts, adapting his historical narratives, for instance, for theological and polemical purposes.[80] As has been argued throughout this book, Imhoff worked hard to maintain a genealogical vision broad enough to please many audiences; and as this section has demonstrated, he largely succeeded: nobles, scholars, and politicians all turned to Imhoff's works with confidence. His books seemed to span the entire, rapidly diversifying field of genealogy.

Beyond the Books

Imhoff's success eventually transcended his books. Although he never used the popular early modern metaphor himself, he became something like a "bureau d'adresse" for all things genealogical. While much of the information that he cultivated so meticulously flowed more or less directly into his publications, a less focused interest in genealogical communication is evident as well: Imhoff wanted to learn about the nobility whether he was working on a new book or not. His correspondence with d'Hozier, in particular, concerned ongoing projects only in part, the rest serving rather as a means of staying up to date. Similarly, the letters Imhoff exchanged with Moritz Wilhelm, Duke of Saxony-Zeitz, were often only loosely connected to specific research projects.

A significant amount of the knowledge that Imhoff acquired accordingly circulated exclusively in his correspondence. In addition to reading his works, people also addressed letters to Imhoff in search of genealogical information for a variety of reasons. This started with his own family. Although the history of the Imhoff family was not Jakob Wilhelm's priority, his genealogical expertise nonetheless attracted inquiries from various relatives. They frequently approached him for "knowledge about our ancestors."[81] They also occasionally asked Imhoff for logistical help when they were required to prepare proof of their nobility.[82] Johann Joseph von Imhoff, the former administrator of Marlborough's German territory, contacted his cousin in Nuremberg several times to clarify the "origins of the Imhoff family."[83] In April 1721, Johann Joseph hoped that Jakob Wilhelm could help him with another difficulty pertaining to the world of academics and scholars. He needed an old legal opinion, given in 1583 by the law faculty of the University of Tübingen, and asked Imhoff for his help in obtaining it.[84] Clearly, family members considered "the erudite Imhoff" a fixer of sorts in matters connected to genealogy, history, and the scholarly world in general.

Other family members appropriated Imhoff's knowledge and authority for political reasons. His distant cousin Andreas Lazarus, a leading official at the court of nearby Sulzbach, was heavily involved in imperial politics. Genealogy often played a prominent part. In late 1686, Sulzbach became tangentially involved in an ongoing legal battle between Julius Franz (1641–89), (last) Duke of Saxony-Lauenburg, and the city of Hamburg. The dispute concerned lordship over the northern German town of Bergedorf, located between Hamburg and Lauenburg. Given that his employer in Sulzbach was the father-in-law of the Duke of Lauenburg, Andreas Lazarus Imhoff sided with the Lauenburg side.[85]

Such legal disputes usually had a historical angle, and since history entailed "lots of genealogy," the Sulzbach politician called on his cousin Jakob Wilhelm to clarify the complicated genealogical and historical context of northern Germany.[86] Had there been any discussion of lordship over Bergedorf at the Diet of Nuremberg in 1431? Could Jakob Wilhelm please check in Nuremberg's city archives?[87] As interest in northern Germany grew at the court in Sulzbach, Andreas Lazarus also discussed other conflicts with Jakob Wilhelm.[88] Although Imhoff did not proactively become involved in politics, others clearly tried to instrumentalize his genealogical expertise.[89]

Many more contemporaries wrote to Imhoff only occasionally. Some asked for his patronage in a general capacity.[90] Others needed help in bibliographical matters. The young Johann Peter von Ludewig (1668–1743), for instance, later a famous historian at Halle, once approached the Nuremberg patrician in 1694 because he needed some rare Spanish books.[91] A letter Imhoff received in 1714 from a little-known ex-Jesuit author and Bohemian noble, Michael Adam Franck von Frankenstein (1675–1728), was more closely related to Imhoff's field of expertise.[92] Frankenstein had recently taken an interest in genealogy and was compiling the lineages of regional families, including the house of Waldenstein, in the tradition of Bohuslav Balbín. Perhaps he had heard about the Jesuit's good relationship with Imhoff. In any case, the Bohemian noble asked the Nuremberg patrician to "censor" his own historical research, thus saving him from gross misrepresentations and, at the same time, sanctioning his findings. Similarly, a former Prussian official contacted Imhoff for advice about a fantastic plan: Heinrich Hartwig Knorn intended to find no fewer than sixteen thousand ancestors of the Prussian king Frederick I.[93] As with Frankenstein, Knorn asked Imhoff for advice and help. Similar requests continued to arrive until late in Imhoff's life. In 1720, an obscure friar from Tyrol, Justin Redn, OFM, (dates unknown), wrote to Imhoff for information about details of the *Stemma Desiderianum*.[94] Again, it seems that Imhoff dutifully replied. Even years after he had ceased all genealogical activity of his own, lovers of genealogy, nobles, and other people hoping for the patronage of a respected scholar all turned to Jakob Wilhelm Imhoff for help.

For five years, from 1696 to 1701, Charles-René d'Hozier incessantly urged Imhoff to have his portrait engraved. The French genealogist felt that this was the most appropriate way for Imhoff to accept his growing public fame.[95] Since Imhoff enjoyed a "considerable reputation" among Parisian genealogists, many people had requested his portrait. Assembling collections of such portraits became a widespread practice among genealogists and other scholars and was a highly effective means of establishing communities of scholars, who thus reassured one another of their reputation and cultural influence.[96]

After considerable hesitation, Imhoff finally had his portrait made in 1702 by the well-known Leipzig engraver Martin Bernigeroth (1670–1733). The images created by Bernigeroth circulated widely—some of his portraits reached print runs of 1,500 copies.[97] Imhoff distributed his likeness throughout the European republic of genealogists. Copies were available in Italy by 1703.[98] His image remained in high demand elsewhere, as well. Eventually, even Imhoff himself ran out of copies. In France, d'Hozier requested yet another batch in 1709, shortly after Imhoff's (only) work in French had appeared. Imhoff appears to have ordered an updated version of his portrait from Bernigeroth and sent several copies to Paris yet again.[99] This new portrait also appeared

Martin Bernigeroth, first version of his portrait of Imhoff, ca. 1702. StadtA Nürnberg E 8 no. 5156.

in *Deutsche Acta Eruditorum oder Geschichte der Gelehrten*, the well-known journal based in Leipzig, in 1712.[100] It thus was when Imhoff shut down his genealogical operation for good that his public image as a master genealogist potentially reached its widest circulation.

Martin Bernigeroth, second version of his portrait of Imhoff, ca. 1709. StA N, Mss 286. Also reprinted in *Deutsche Acta Eruditorum oder Geschichte der Gelehrten*, 1712.

Imhoff's enduring fame rested on his genealogical knowledge and, more specifically, on the books he created in collaboration with his publishers. His works responded positively to new literary trends in many ways. As genealogical knowledge expanded beyond courtly and dynastic circles, the public circulation of relevant material grew apace. The very genre of encyclopedic or universal genealogy, which Imhoff helped found, was one of the most obvious products and simultaneously a major catalyst of this shift. Although Imhoff contributed to and participated in such editorial innovations, he was never driven exclusively by these dynamics. In contrast to the works of more market-oriented genealogists, Imhoff's books retained a modicum of traditionalist exclusivity. His publications, especially his mature works on the noble houses of Europe, were products of exquisite craftsmanship, composed in Latin. The very materiality of these books made his works acceptable vehicles of courtly and noble representation, in addition to satisfying the informational needs of politicians, journalists, and the general reader. Imhoff's approach was receptive to recent editorial trends, not subservient to them.

With these complementary features, Imhoff's projects achieved great overall success. His books were well-respected and received positive reviews. Contemporaries considered them outstanding examples of "modern" genealogy. Accordingly, Imhoff had the privilege of cooperating with a distinguished group of publishers in different cities of Germany and beyond. While his letters reveal that he, too, sometimes felt the pressure of an increasingly crowded information market, he still usually had the upper hand in his dealings with publishers. As a successful and confident patrician author, Imhoff effectively controlled his own public profile. No publisher ever attempted to turn "Imhoff" into a brand name, as had happened to "Rittershausen."

Given the largely positive reception of his works, it is hardly surprising that Imhoff found many readers. More significantly, he found many readers from very different contexts. Judging from surviving marginalia and manuscript annotations, nobles, scholars, general readers, and politicians all used his texts and tables. The kind of genealogical knowledge that Imhoff provided obviously suited many purposes and many different agendas. While some read genealogical books for entertainment, others consulted them in moments of dynastic crisis. Readers interested in politics also found much to recommend such works. Imhoff's books were invaluable tools for making sense of contemporary politics; they provided a powerful interpretive framework for understanding international and domestic affairs. Moreover, genealogical works could also serve effectively as notebooks for storing and organizing new information.

Largely on the strength of his books, "the erudite Imhoff" gained a reputation not only as a successful author but also as a living archive of genealogical information—just as his father had been celebrated as a "living library" of knowledge concerning Nuremberg. Imhoff's authority as a genealogical expert is reflected in the many letters he received from people seeking his advice on their own projects. In his responses, he drew on his extensive, personal knowledge of European genealogies, which was significantly broader than what was contained in his published works. Imhoff's books may represent his greatest contribution to the public circulation of genealogical knowledge, but his informal and private transmission of information about the nobility must not be overlooked. It was not least through the letters that he exchanged independent of his research projects that he helped accelerate the growing public awareness of the nobility of Germany and Europe as a distinct class.

Conclusion

It had not been easy for Giulio d'Acquaviva (d. 1691), tenth Duke of Nardò, and Francesco Carafa, brother of Giovanni, the Duke of Nola, to find a government that would permit them to carry out their duel. Nowhere in Italy were duels allowed, since most authorities enforced the pope's numerous bans.[1] Hence, the two noblemen traveled all the way from southern Italy across the Alps to Protestant Nuremberg. This was apparently the nearest city where their duel could be held "freely and without giving offense." On November 15, 1673, the two Italians settled their dispute once and for all. More than one thousand people stood by and watched as the two combatants drew their swords. Battle was joined, blood was spilled: Carafa received a wound on his hand that made it impossible for him to continue. With that, the business was settled. The protagonists and their families had regained their honor and parted in "friendship." Jakob Wilhelm Imhoff published a brief account of these spectacular events in his book about the Italian nobility, his *Corpus Historiae Genealogicae Italiae et Hispaniae* of 1702.[2]

The sources insist that the public greatly "marveled at the valor and courage of the Neapolitan gentlemen."[3] A pamphlet about the events appeared in Nuremberg shortly after the incident, publicizing what was obviously a highly memorable occurrence.[4] There is no question that Imhoff also appreciated, and perhaps even admired, the Italians' values and mores, even though they differed dramatically from the social habitus of his own class, the Nuremberg patriciate. Throughout his works and in much of his preparatory research activities, he consistently showed a great deal of sympathy for the nobility,

Fürtrefflicher

Ritter-Kampf/

Der zweyen Neapolitanischen Fürsten

D. Francesco Carafa de Noja.

und

D. Guilio Antonio Aquaviva de Aragona.

am 5. 15. Novemb. 1673.

DJe Wolmeinung der unpartheyisch-verständigen ist jederzeit dahin gegangen/ daß die zwischen beyden hohen Neapolitanischen Häusern de Conversano und di Noja, biß auf zwey Jahr und etliche Monat ausgeschlagene Tod-Feindschafft/ am besten durch ein Duell mit Ritterlichen Gewehren könten ausgetragen und beygelegt werden: Zudem End dann gleich Anfangs ein bequemer Platz zu Vollziehung dessen gesucht/ niemals aber gefunden worden/ als am 5/15 Novemb. dieses 1673 Jahrs/ da auf ordentliche Ausforderung/ von D. Aquaviva geschehen/ D. di Noja, mit beyderseits erbetenen Secunden erschienen/ die zweyfache Gewehre/ als Dolchen und Degen gegeneinander gemessen/ und darauf in Beyseyn und Ansehen etlich 1000 hoher und niderer Personen aufeinander gegangen. Der Kreiß war von Pferden und Reutern umschlossen. D. Aquaviva, so der erste/ als Forderer/ auf dem Kampf-Platz gewesen/ erwartete/ mit besonders verspürtem Eifer / biß D. Carafa sich zum ersten Gang in Postur stellen würde. So auch nicht gesäumet. Sie stunden beyde mit subtilen von Silber und Gold ausgemachten Camisolen zugegen/ die noch vor dem Kampf von denen Secunden eröffnet/ und ob kein Vortheil darunter verborgen/ besichtigt worden. Vorhero aber liesen sie ihre Haare aufbinden/ und erwarteten einander mit blosen Häuptern. Der Angriff geschahe mit solcher Annehmlichkeit der Geberden gegē die Anwesende/ daß sie gleichsam/ als ihnē für die Beywohnung zu danckē/ sich darstelleten/ so auch bey jedes Gangs Anfang wiederholet wurde. Hochgefährlich schien der erste Gang durch ohne Unterschied angebrachte Hieb und Stich/ daß es scheinbarlich für eine Unmögligkeit gehalten wurde/ es müste dann einer der Kämpfer/ sonderlich auch wegen der anhaltenden Zeit-Länge/ im Stich und auf dem Platz bleiben. Der andere Gang war zwar etwas kürtzer/ weilen sie von dem ersten noch etwas ermüdet schienen: desto hitziger aber die folgenden/ da nicht allein Herr Aquaviva/ durch das Hembd und an das Knie gestossen worden/ sondern D. Carafa auch einen ungefehren Fall thate / und wiewol man vermeinte / sie dardurch zu Aufhebung deß Duells zubewegen/ mit Zusprechung/ daß ein jeder das seinige Ritterlich gethan / und also der Sach ein Genüge geschehen sey/ wollten sie sich doch nicht eher bequemen/ es hätte dann einer derselben eine blutige Wunde empfangen/ welches erst im siebenden Gang geschehen / da D. Carafa in den Arm einen Stich bekommen/ daß er davon geblutet/ und sich auf dem Platz müssen verbinden lassen. Er ließ nichts destoweniger eine Begierde weiter zu fechten spüren / so aber von denen Herrn Secunden verwehret wurde/ die sie zu Ablegung der Waffen beredet/ auch endlich dahin gebracht/ daß sie nicht allein die Händ einander freundlich gereichet/ sondern sich auch umhalset und geküsset/ welches dann bey den Zuschauern eine sonderbahre Bewegung der Gemüther verursacht. Worauf sie sich zusam in die Karete gesetzt/ und einander in die vorher ingehabte Herberge begleitet. Der Tag hiese nicht umsonst Blandina/ an welchem so hefftige Feinde wieder dergestalt zu Freunden worden/ daß sie einander geliebkoset.

Kein Löw fällt Löwen an/ kein Schlange ein andre sticht/
Kein Rab den andern hackt/ kein Wasser-Thier auch nicht.

Der Mensch nur suchet/ sich in Menschens Blut zu baden/
Und will dem andern mehr/ als Löw und Schlangen/ schaden.

Pamphlet about the duel between the princes of Carafa and Acquaviva in Nuremberg, 1673. Universitätsbibliothek Frankfurt am Main, Einblattdr.G.Fr. 335a.

their social standards, and their genealogical culture. Imhoff was a fascinated observer of the nobility.

In order to give an accurate account of the events of 1673, Imhoff had researched the incident in detail. There were many ways in which he might have learned about the duel. He may have personally been among the spectators, who witnessed the fight at first hand. In late 1673, he had just returned home from his extended grand tour, which had included a (brief) visit to Naples. There, he almost certainly would have heard about the two prominent families. It is even more likely that Imhoff knew of the local pamphlet produced after the duel, although no copy of it survives among his papers. At some point in the twenty-nine years between the event and the publication of his book in 1702, Imhoff visited Nuremberg's administrative archives and copied some of the official letters concerning the affair for his private collection—one of the very few documented cases in which Imhoff conducted (limited) archival work for his genealogical projects.[5] Perhaps he even wrote to Naples in retrospect for further details—if so, he would in all likelihood have received valuable information, since some of his contacts there clearly knew all about the affair.[6] The most probable way in which Imhoff acquired information about the case, however, was oral communication in Nuremberg itself. Duke Acquaviva had resided in the city for weeks before Carafa finally arrived and the fight could take place, and it is easy to imagine that the patriciate—including young Jakob Wilhelm, who was fluent in Italian and just returned from Naples—might have had numerous encounters with the Italian grandee.[7] Some of the information provided by Imhoff in his book—especially his recounting of the prehistory of the duel—appears in no other source. Personal conversation with Acquaviva (and, perhaps, also with Carafa) was probably their origin. There thus were many ways to acquire genealogically relevant information, and Imhoff was a master of most of them, including the difficult social art of oral communication.

This final case study underscores many of the broad points made in this book. Knowledge about the nobility was sometimes easy to come by in early modern Europe. Vast amounts of information concerning the aristocracy circulated in print, unpublished documents, and, especially, orally. To a certain extent, almost everyone in early modern Europe could potentially contribute something to a study of the ruling class. Since most people casually knew—or knew about—a few noble individuals and some of their important life-events, genealogically meaningful data were never far from the surface. Nevertheless, as Imhoff experienced on a daily basis, this experiential, locally circulating information, though abundant, was highly fragmentary in nature. People primarily knew about the nobles they saw or were in contact with, and they usually focused on information that was relevant to them personally or was otherwise notable. This was also true of nobles themselves, whose knowledge about their own and their peers' families must not be overestimated. Often enough, they were ignorant of even the most basic information. One or two prominent children of a prince or noble may have been well known, but less important siblings were likely to slip from memory, especially if they were female or died as infants.

Genealogists, however, needed to know all these things. Imhoff could not be content with incomplete family knowledge. How to proceed? He devised a plethora of strategies to attain complete genealogical information, whether through correspondence with local specialists, by tapping into courtly information channels, by accessing noble caches of dynastic information, or by systematically consulting contemporary printed materials. As the example of his research on the duel of 1673 indicates, Imhoff established many contacts with noble families and a wide range of other knowledgeable individuals, including priests and monks, scholars, administrators, and diplomats from his well-connected base of operations in the largely independent and culturally vibrant city of Nuremberg. Imhoff overcame innumerable practical difficulties, ranging from the loss of letters en route to the premature death of important informants, setbacks that occasionally threatened to undermine his entire project, to build an outstanding network of epistolary and oral communications, effectively linking numerous local and regional clusters of informants into a unique pan-European brotherhood of genealogists.

The transregional character of Imhoff's connections in particular warrants attention, since it appears to have been a pioneering achievement at the time. In fact, although he leaned heavily on popular scholarly practices associated with the Republic of Letters at large, Imhoff could rely on few conventions dedicated exclusively to the exchange of *genealogical* information. Tapping into the many regional reservoirs of such specific data was often an extended process of trial and error, as intermediaries were unknown, specialized research institutions were nonexistent, and few role models were available.

The successful exchange of knowledge resulted, in material terms, in heaps of manuscript texts that transported knowledge to Nuremberg. Out of these written artifacts, through a complex process of compilatory bricolage, Imhoff fashioned his genealogical diagrams and narratives. Bricolage in this book has meant two things. On the one hand, it refers to the epistemic operation of joining fragmentary pieces of knowledge together, including normalization of data and information from disparate origins. On the other hand, the term also evokes the many physical and material activities involved in effecting that assemblage. The epistemic aggregation of information relied on and, in turn, created specific material practices for handling paperwork. Building genealogical tables and narratives entailed, among other things, the physical and material manipulation of papers, manuscripts, and prints. Genealogical bricolage manifested itself materially in piles of preliminary documents. There were, for instance, fact sheets used to store personal data, and collections of excerpts from literature and journals. In Imhoff's case, drafts of visual representations of genealogical knowledge were especially prominent. The genealogist constantly sketched diagrams of kinship groups and continuously reworked them as his research progressed, to the point that the pages were no longer legible. These material practices of writing and rewriting, drawing and redrawing genealogies reflected the open-ended, unsystematic nature of genealogical research itself.

Yet despite such skills, genealogical research often remained a frustrating affair, full of obstacles, setbacks, and failure. As this book has demonstrated, genealogical knowledge was elusive and hard to come by. Information was scattered, fragmentary, and unstandardized. Although no member of Imhoff's genealogical brotherhood would have doubted the legitimatory potential of genealogy, all of them would have understood the difficulty of activating that potential. Despite genealogy's public role as a robust code for expressing social, cultural, and political claims to power, it was characterized de facto by vulnerability and instability.[8] Little wonder then, that many of the nobles who corresponded with Imhoff exhibited considerable anxiety and discomfort with respect to genealogy. This was an exciting but exacting field of knowledge.

This was true not least because "genealogy" meant different things in different contexts. There were numerous sites of genealogical knowledge, and all of them observed their own routines and applied their own standards as to what counted as proper information. We have encountered genealogy at noble courts, in bureaucratic offices, in courts of law, on the information market, in scholarly circles, and in the early modern city, to name only the most prominent contexts. There were significant differences between these sites of genealogical knowledge production and the epistemic standards and knowledge practices that shaped them, such as different attitudes toward oral knowledge and origin stories. In many ways, "genealogy" was an umbrella term in early modern Europe, extended to cover alternative notions of the field only with difficulty. Nevertheless, the genealogical contexts of the noble court, the university, the courtroom, and the city were by no means utterly cut off from one another. Practitioners such as d'Hozier and Leibniz, for example, integrated different perceptions of the field into their own activities. Imhoff also excelled at bringing together genealogists of different stripes and genealogical knowledge from different milieus, negotiating different standards of truth and homogenizing disparate types of information in the process.

One of the most important consequences of the diversification of genealogical knowledge circulation in early modern Europe was the diffusion of genealogy outside noble families and beyond their direct control. In addition to ongoing auto-genealogical activity by and for the nobility, hetero-genealogy grew in importance. A "genealogical gaze" shaped not only noble perspectives on the world; it also shaped perceptions of the nobility by government officials, publishers, and indeed the general public.[9] Genealogy increasingly turned the nobility into an object of outside scrutiny and study.

Judging by Imhoff's experiences, the relationship between auto- and hetero-genealogy was highly dynamic, at times mutually reinforcing, at times contentious and conflictual. There was no simple and self-evident alignment between what families thought about themselves and what outsiders saw in them. On the one hand, the external production and control of genealogical knowledge were occasionally helpful resources for safeguarding memory and protecting social status. We have encountered numerous nobles who cherished Imhoff's outside endorsement of their own positions and ideas—not for nothing did the imperial counts view his *Notitia Procerum* as a precious affirmation of their status within the complex framework of the Holy Roman Empire. For some families, encyclopedic genealogies, along with journals and other news media, were effectively the only affordable genres of genealogical representation at all—a situation that was not ideal, but nevertheless better than nothing. There were clear benefits to hetero-genealogy and its evolving media.

On the other hand, powerful hetero-genealogical initiatives could easily become a constraint or obstacle. Despite its enormous legitimatory potential, genealogy was often experienced as a burdensome obligation and even a trap. Constant outside scrutiny applied additional pressure on familial genealogical knowledge management. Hetero-genealogy promoted new standards and routines of epistemic control and publishing that lay outside the nobility's direct control, at least in part. The newly popular genre of universal encyclopedic genealogy also confronted noble families with unprecedented forms of external scrutiny. It was nearly impossible to ignore a reputable genealogist's request of information. Imhoff's works shaped the public perception of noble families, and they did so according to the genealogist's understanding of the facts. Imhoff presented his take on noble genealogy, and there was little that nobles could do to change it. Living and writing in Nuremberg as he was, Imhoff hardly could be coerced to do their bidding. Many noble families had to accept willy-nilly that genealogical authority was slipping from their grasp and falling into the hands of genealogical experts backed by a thriving information market. This transition altered power relations between genealogists and the nobility in subtle ways, especially if we compare this new arrangement to the traditional role of court genealogists employed by noble patrons. We have had numerous occasions to observe how it was Imhoff who set the agenda for his noble informants, as he gently pushed them to satisfy his own informational needs, yet without giving any guarantee that he would utilize their disclosures to their complete satisfaction.

This shift raises a broader question: How did all of these changes in the culture of genealogical knowledge influence the nobility's self-perception and social role prior to the end of the ancien régime? On the one hand, the European nobility easily reconciled genealogy's changing features with traditional approaches to the field. On numerous occasions, individual nobles openly supported the new critical methodologies developed by humanism and antiquarian philology. Not a few members of the nobility endorsed, even demanded, a modicum of methodological rigor, as we have seen in the case of the counts of Nassau. The nobility clearly appreciated critical methodologies as a valuable epistemic insurance against ridicule and naïveté, as the knowledge culture around them changed. Individual families welcomed new literary genres created with an eye to the emerging information market and appropriated them as useful means of self-representation. Thus, while the long-term changes in epistemic frameworks did not leave genealogical thinking unaltered, innovations in the field were incorporated into the nobility's self-image in numerous ways. There were no reasons why market-oriented or newly critical genealogists could not contribute to creating and promoting noble identities.

On the other hand, though, the shifting nature of genealogy potentially contributed to a long-term reassessment of the nobility and its social role. The rise of hetero-genealogy may have gradually permitted observers to develop new perspectives on the nobility. Imhoff's books not only collectivized countless noble houses into a single, normalized social body, but they also sketched a vivid portrait of the everyday life and social influence of "the" nobility. Stories like that of the duel between Acquaviva and Carafa in 1673 and thousands of other brief accounts of individual acts of valor and glory or vanity and pride would have fleshed out for readers what "typical" nobles, namely, those included in Imhoff's books, did and what they contributed to everyday life. Imhoff's works could shape his readers' social imaginaries, leading them to view the nobility plausibly as a coherent social class, an *état*, that wielded enormous cultural and political power and lived according to a set of shared conventions. Readers of Imhoff's works would have associated the nobility closely with titles, dynastic thinking, military valor, and a strong commitment to monarchical power and Christian piety, albeit also with wrangling over inheritances, precedence, and status. In this way, the genealogical encyclopedia encapsulated what later came to be known as the culture of the ancien régime in early modern Europe.

In this book, I have argued that genealogy should be considered a key area of Europe's history of knowledge. A history-of-knowledge approach not only supports research on the social and political function of ideas and modes of thought but also opens up new analytical perspectives on the social imaginary of Europe. Genealogy was more than a set of ideas, a "mentality," a "perceptual grid," or "pattern of thought" for understanding and ordering kinship relations and social life; it was also a sophisticated and diversifying knowledge practice.[10] By foregrounding such an approach, this book has highlighted how it was simultaneously both crucial and challenging for noble families to curate relevant information about themselves. Only a history that extends beyond "genealogy" and "genealogical literature" to encompass genealogical *knowledge* alerts scholars to the fact that such knowledge was available in wide variety of degrees and forms. It also reveals that such variations in accessibility profoundly influenced what "family" or "pedigree" actually meant to individuals in everyday life.

As has become evident in the preceding chapters, noble families should be conceptualized as *communities of knowledge* and, more precisely, as *communities of who's-who knowledge*. From the perspective of an individual noble, "family" was first and foremost a group of kinsmen that he or she might, or might not, be able to name and identify. Knowing one's family meant familiarity with its members, both past and present, including most prominently familiarity with basic genealogical facts—names, parents, siblings, dates of birth, marriage, or death.

Some individuals and families had better who's-who knowledge available to them than others. Sophisticated knowledge could prove to be a significant advantage. As a specialized field of intellectual inquiry that boasted its own literary genres, experts, and epistemic routines, genealogy was a key practice for enhancing and authenticating that body of family-related who's-who knowledge, for curating and improving the self-knowledge of noble families, whether for legal purposes, for self-promotion, or for the sheer joy of it. Genealogy, as a method for researching and memorializing ancestors, helped noble houses to expand their family knowledge beyond the three or four generations that people casually remember.[11] Having at hand detailed, relevant, and ideally complete information about one's family was not normal or natural. On the contrary: sophisticated research methods were needed to accumulate such knowledge constructively. Jakob Wilhelm Imhoff, as many noble families discovered in the years around 1700, applied these methods with extraordinary finesse.

Abbreviations

AA I	Leibniz, Gottfried Wilhelm. Akademie Ausgabe. Reihe I, *Allgemeiner, politischer und historischer Briefwechsel.* Currently 25 vols. Darmstadt/Berlin, 1923–.
ADB	*Allgemeine Deutsche Biographie*, 56 vols., Leipzig, 1875–1912.
GNM	Germanisches Nationalmuseum, Nuremberg
Généalogies	Series of Imhoff's genealogical ledgers, bound into 17 volumes, in GNM, Nachlass Imhoff VII, vols. 6–22. The volumes are cited with the letters of families they hold and the volume number. The dossiers do not have consistent pagination. I cite them by referring to the name of the family under discussion. All material belonging to individual families is collected together within the volumes.
HA	Historisches Archiv
Letters	Letterbooks (I and II) of Jakob Wilhelm Imhoff, GNM, Nachlass Imhoff VII, vols. 1 and 3. The letters are alphabetically ordered. There is no consistent pagination.
NDB	*Neue Deutsche Biographie.* Currently 27 vols. Berlin, 1953–.
Will 1–4	Will, Georg Andreas. *Nürnbergisches Gelehrten-Lexicon oder Beschreibung aller Nürnbergischen Gelehrten beyderley Geschlechtes nach Ihrem Leben, Verdiensten und Schrifften.* 4 vols. Altdorf: Schüpfel, 1755–58.

Introduction

1. Imhoff to Moritz Wilhelm, December 27, 1700, SächsHStA, loc. 8224, no. 10, fols. 81r–v.

2. For other early modern attempts to view the French dynastic crisis of 1711–12 in historical perspective, see Tölle, *Heirs of Flesh and Paper*.

3. Ketelaar, "The Genealogical Gaze." Kellner, *Ursprung und Kontinuität*, 13–131, discusses genealogy in a similar vein as premodern Europe's dominant "mental structure."

4. For the "society of princes," see Bély, *La société des princes*.

5. Kellner, *Ursprung und Kontinuität*; Melville, "Vorfahren und Vorgänger," 203–310.

6. Coote, "Prophecy, Genealogy, and History," 28.

7. Important examples include Heck, *Genealogie als Monument und Argument*; Trevisan, "Genealogy and Royal Representation," 257–75; Baker, "Tudor Pedigree Rolls and Their Uses," 125–65. More generally, see Andenna and Melville, *Idoneität*; Ingledew, "Book of Troy," 665–704; Radulescu and Kennedy, *Broken Lines*.

8. For a wide-ranging survey of the literature, see Barry, *La parenté*.

9. Duindam, *Dynasty*; Duindam, *Dynasties*; Geevers and Marini, *Dynastic Identity*; Pieper, *Einheit im Konflikt*, 32–34, 38–43. Schuster, "Familien- und Geschlechtsbewusstsein," 13–36, is a very perceptive example. Pointedly also, Hummer, *Visions of Kinship*.

10. Piétri and Luciani, "Généalogie, construction du présent," 6: "lieu de la genèse d'une identité."

11. Teuscher, "Bilatéralité," 1–18.

12. For a wide-ranging case study, see Berghorn, *Verwandtschaft*. Cf. also Teuscher and Lanzinger, "Editorial," 1–3.

13. E.g., Tucker, *City of Remembering*.

14. Shapin and Ophir, "Place of Knowledge."

15. Bouquet, "Généalogies imperiales," 146–178.

16. McDowell, *Invention of the Oral*.

17. Slauter, "Periodicals," 128–51. Frow, "Commodification," 374–78. For a new case study concerning France, see Schuwey, *Un entrepreneur des lettres*. See also Dover, *Information Revolution*.

18. Weduwen and Pettegree, *Dutch Republic*, 110–11.

19. On "public circulation," see Östling, "Circulation." On manuscripts, see Millstone, *Manuscript Circulation*.

20. Bauer, "Monographizitaet," 393–410.

21. For the genre, see Bauer, "Scope, Readership and Economy," 287–301.

22. Becker and Clark, *Little Tools*; Blair et al., *Information*.

23. The few (highly important) exceptions include Grafton, *What Was History?*, 142–58; Woolf, *Social Circulation*, 99–139; Broadway, *"No historie,"* 150–78; Kagan, *Clio and the Crown*. See also Williams, *First Scottish Enlightenment*, 215–43; and Ostenfeld-Suske, *Official Historiography*. Cf. Burkardt, *Die historischen Hilfswissenschaften*. Genealogy does not feature prominently in the following standard works: Turner, *Philology*; Woolf et al., *Oxford History of Historical Writing*; Woolf, *Global History*.

24. Völkel, *"Pyrrhonismus historicus"*; Dooley, *Social History of Skepticism*. Cf. Grafton, "Jean Hardouin," 241–67.

25. Dooley, *Dissemination of News*; Landwehr, *Geburt der Gegenwart*; Clark, *Time and Power*. More generally, see Levine, *Between the Ancients and the Moderns*; Lecoq et al., *La querelle des Anciens*. For the most recent entry into the bibliography, see Wallnig and Peper, *Central European Pasts*.

26. Schuwey, "Naissance," 303–19.

27. Pettegree, *Invention of News*.

28. Bauer, "Jetztherrschend." See also Jettot, "Intelligible to the Mind."

29. Caiani, "Re-inventing the Ancien Régime," 447; Peronnet, "L'invention de l'ancien régime," 49–58. See also, more broadly, Bann et al., *History of the European Restorations*, 2:15–60.

30. Mulsow, "History of Knowledge," 159–87; Daston, "History of Science," 131–54. See also the statements in the inaugural volume of the new *Journal for the History of Knowledge* 1, no. 1 (2020), 60, https://journalhistoryknowledge.org/issue/view/497, as well as the recent thematic issue of *History and Theory* 59, no. 4 (2020). See also Bauer et al., *Genealogical Knowledge*.

31. Woolf, *Social Circulation*, 113–14, argues that in England by the time of the Restoration public concern with genealogy had become dissociated from any real social impact.

32. For a survey of recent scholarship, see Lezowski, "Ramificazioni della storia." For a study of the production process of genealogical knowledge, see Friedrich, "Genealogy as Archive-Driven Research Enterprise." For contemporary genealogists, see Yakel, "Seeking Information"; Duff and Johnson, "Where Is the List"; Darby and Clough, "Investigating the Information-Seeking Behaviour." See also Duff and Johnson, "Accidentally Found"; Timm, "Grounding the Family."

33. Dewald, *Status, Power*, 32.

34. Schuwey, "Les périodiques."

35. The phrase "der gelehrte Imhoff" appears numerous times in GNM, Nachlass Imhoff I (Archiv der Gesamtfamilie), fasc. 2, unpaginated (ledger 1c, "Fremde Imhoffs betreffend").

36. The major exception is Imhoff's correspondence with Fabricius, which contains some private and personal information.

37. The best biography is Köhler, *Der historischen Münz-Belustigung*, 401–8. This is mostly also the basis for the entry in Will 2:241–44, and for the later article in *ADB*.

38. Retrospectively, Imhoff claimed to have met a knowledgeable Portuguese genealogist in Venice, namely, Rodrigo Mendez Silva, who died in 1670; see Imhoff to Magliabechi, September 12, 1695, GNM Letters, unpaginated. Sometime before November 1672, Imhoff also reported "ex Venetiis" to Christoph Wagenseil in Nuremberg; see Wagenseil to Magliabechi, November 28, 1672, in Targioni-Tozzetti, *Clarorurm Germanorum*, 315–17.

39. Imhoff to Fabricius, March 3, 1719, KB, Ms Thott 1227, unpaginated; Fabricius, "Epistola dedicatoria," in Saubert, ed., *Opera posthuma*, *2r–**v.

40. The work is cited hereafter as Imhoff, *Notitia Procerum* (1st ed., 1684, etc.). The first edition was flanked by the two volumes of [Imhoff], *Spicilegium Rittershusianum*.

41. In the late 1680s and early 1690s, he compiled the annual accounts of the Münzersche Kleiderstiftung. See StadtA N, B 35, nos. B797–99 (copies) and D 19, Br. 150–53 (drafts). On the foundation, see Rascher, "Kleiderstiftung." Imhoff also was involved in other foundations, for example, Imhoff to Fabricius, late 1702, KB, Ms Thott 1227, unpaginated.

42. In 1689, he cited administrative responsibilities as an excuse not to take on new projects. See Imhoff to Christian Franz Paullini, Jena, June 17, 1689, ThULB Jena, Ms Bud 347 f, fols. 448r–449v. Daniel Wilhelm Moller, Altdorf, to Paullini, October 7, 1689, ThULB Jena, Ms. Bud 348 f, fol. 136r considered this a pretext.

43. This is most obvious in his letters to Fabricius. The Helmstedt theologian wanted to withdraw his private capital from the city, but financial constraints made this a difficult task. Imhoff constantly informed Fabricius about the difficulties and compromises involved in the transaction.

44. StadtA N, A 1 no. 1697-09-08.

45. Imhoff to Fabricius, December 16, 1711, KB, Ms Thott 1227, unpaginated.

46. Fabricius, "Epistola dedicatoria," in Saubert, ed., *Opera posthuma*, *3r. Will 2:243: "er hat aber diese Ehre [of becoming a member of the city council] wegen seiner großen Liebe zu den Wissenschaften, und der Genealogie vornemlich, allezeit abgelehnet."

47. Followed by a fourteen-page addendum; see Imhoff, *Genealogiae familiarum Bellomaneriae*.

48. Again followed by a short appendix in 1691; see Imhoff, *Appendix*.

49. Imhoff had mentioned the "Europae procerum" already in an early letter. See Imhoff to Weise, March 25, 1682, CWB, Mscr.A 70–71 (32-1). Imhoff to Thulemeyer, September 11, 1690, UB F, Ms Thulemeyer, 741–44: "d. angefangenen cursu Historico-Genealogicum . . . vollenden."

50. Imhoff to Giovanni Behm, January 18, 1694, UB ER-N, Briefsammlung Trew: he suffers from a general "bücherkrankheit" (book disease).

51. Imhoff to Paullini, June 17, 1689, ThULB Jena, Ms. Bud 347 f, fols. 448r–449v. Like Leibniz or Pufendorf, Imhoff counted only among the "Patroni Fautores" of the Collegium, see ibid, fol. 3r. On the Collegium, see Wegele, *Geschichte der deutschen Historiographie*, 597–605.

52. Imhoff to Fabricius, December 16, 1711, KB, Ms Thott 1227, unpaginated, on the good news of a well-arranged marriage for one of his daughters. Concerning his son, he had previously thought about hiring a private tutor: Imhoff to Francke, May 3, 1701, AFSt/H C 825:1, unpaginated.

53. Hollenhagen, Nuremberg, to Francke, July 11 and August 5, 1702, AFSt/H C 72, nos. 5

and 6. Jakob Wilhelm Jr. was matriculated at Halle from 1702 until 1707, see AFSt/S A I 194, p. 146; 196e, p. 16; B I 1, p. 38; A I 118, p. 11, and appendix.

54. Sagittarius to Imhoff, July 25, 1691, GNM Letters, unpaginated. Imhoff to Fabricius, December 1, 1701, KB, Ms Thott 1227, unpaginated. Imhoff to Fabricius, July 9, 1705, ibid.: Imhoff had several positive conversations with Wilhelm Petersen. Imhoff to Fabricius, late 1702, and September 8, 1714, ibid.: positive about Gottfried Arnold. For his pietist contacts, see Imhoff to Francke, May 3, 1701, AFSt/H C 825:1, unpaginated; and Francke to Johann Daniel Herrnschmidt, July 15, 1701, AFSt/H D 43a, fols. 333–34.

55. Simon, "Der Prediger Tobias Winkler." Weigelt, *Geschichte des Pietismus*, 91–101. Other patricians, of course, opposed pietism; see Imhoff to Fabricius, July 9, 1705, KB, Ms Thott 1227, unpaginated; Ressel, *Protestantische Händlernetze*, 369–81.

56. He mentions Wirth, Werk, Stößner, and Krause as principal supporters of pietism.

57. Imhoff to Fabricius, May 16, 1704, KB, Ms Thott 1227, unpaginated: "Scheinchristen." Imhoff was also critical of orthodox religiosity; see Imhoff to Fabricius, September 9, 1701, ibid. Von Canstein to Francke, March 9, 1704, in Schicketanz, *Briefwechsel*, 256–57, wrote that "H. Imhoff sehnet sich von Nurnberg weg, allein Seine familie wird ihn aufhalten, über das Er vielen an häusern und güttern hatt, so Er nicht leicht verkaufen kan." Von Canstein had visited Imhoff in Nuremberg shortly beforehand and would have known these things firsthand (ibid., 255n54).

58. Imhoff to Fabricius, April 8, 1718, KB, Ms Thott 1227, unpaginated: "Der Meinung, daß beten, singen u. lesen der Schrift bessere Erbauung haben könne, als das alzuviele predigen, will ich nicht wiedersprechen."

59. Imhoff to Francke, May 3, 1701, AFSt/H C 825:1, unpaginated: "in victu und allen dingen simplicitatem liebe und keinen staat führe." Imhoff to Moritz Wilhelm, December 28, 1696, SächsHStA, loc. 8224, no. 7, fol. 41r: cultivation of "simplicitatem et frugalitatem" versus "luxuria." Imhoff to Fabricius, January 10, 1714, KB, Ms Thott 1227, unpaginated: ruminating about ways "um die unmössigkeit in der fröligkeit den Leuten zu verleiden."

60. The topic is understudied, cf. Bei der Wieden, "Pietismus und Genealogie," 11–21.

61. Imhoff to Thulemeyer, November 2, 1700, UB F, Ms Thulemeyer, 793–95, here 794: "Ich, der ich nicht aberglaubisch bin." This remark was made in the wake of a popular form of "superstition" that associated the unusually high infant mortality in that year to the turn of the century. Although his own six-month-old son, Wilhelm Christian, had died in October 1700, Imhoff refused to accept the widespread rumors. He did, however, ponder the astrological implications of age; see Imhoff, *Notitia Procerum* (5th ed., ed. Köhler), part 1, fol.)(2r.

62. See, e.g., Imhoff to Fabricius, August 5, 1720, KB, Ms Thott 1227, unpaginated.

63. Imhoff to Fabricius, July 6, 1708, KB, Ms Thott 1227, unpaginated: "christlich klugen vorsichtigkeit."

64. Imhoff to Fabricius, March 19, 1701, KB, Ms Thott 1227, unpaginated.

65. See Lohmeier to Imhoff, October 18, 1690, BSB, Cim. Lohmeier.

66. See anonymous, *Neu-erläuterte Zeit- und Jahr-Rechnung*. The author's acronym "A.N.P.I." on the title page is: "auctor Notitiae Procerum Imperii," i.e., Imhoff. The Lohmeier book reappeared in 1701 in two slightly different versions, both published by Johann Georg Lipper, but one in Lüneburg, the other in Frankfurt. The Frankfurt version contains more tables.

67. It did not contain any diagrams or visualizations, thus discontinuing Imhoff's standard layout. The *Recherches* was also the only book by Imhoff to be translated. See Imhoff, *Historische und Genealogische Nachrichten*. The translation appeared "ohne vorwißen des herrn authoris" (without the author's prior knowledge), BSB, Clm 28115, fol. 52r.

68. It seems that this book did not sell too well; see Imhoff to Fabricius, August 5, 1720, KB, Ms Thott 1227, unpaginated.

69. Imhoff to Leibniz, October 20, 1705, AA I, 25:210.

70. Imhoff to Fabricius, May 22, 1710, KB, Ms Thott 1227, unpaginated: "da ich niemand um mich habe, der mir etwas neues saget." See also, for a parallel case, Imhoff to Fabricius, January 10, 1714, ibid.

71. Imhoff to Fabricius, April 13, 1714, KB, Ms Thott 1227, unpaginated.

72. Imhoff to Moritz Wilhelm, March 11, 1712, SächsHStA, loc. 8224, no. 9, fols. 6r–v. See Imhoff to Moritz Wilhelm, February 22, 1715, ibid., fol. 23v.

73. Imhoff to Fabricius, May 13, 1712, KB, Ms Thott 1227, unpaginated: He senses her death and dreads the prospect of losing "der langen, liebreichen und getreuen beywohnung"; the loss would "sehr schwär fallen."

74. Imhoff to Moritz Wilhelm, September 22, 1712, SächsHStA, loc. 8224, no. 9, fol. 9r. Moritz Wilhelm of Sachsen Zeitz to Imhoff, November 8, 1712, Veste Coburg, Ms A, I, 349, (2), 3, unpaginated. Imhoff, *Notitia Procerum* (5th ed., 1732), fol.)(2r: "meditationi unici necessarii & praeparationi ad futurum eandemque meliorem . . . vitam." Imhoff to Fabricius, September 22, 1712, KB, Ms Thott 1227, unpaginated. Imhoff to Fabricius, September 22, 1712, ibid.: "weil ich meinen Laden damit [sc., with his last book] zumache." StA N, Handschriften 286, fol. 542r: "diesen [sc., genealogischen] Studiis gute Nacht gegeben." Biagio Majoli de Avitable to Imhoff, October 4, 1712, GNM Letters, unpaginated: "ritirarsi dal mondo." First mention of his intention "nichts mehr zu schreiben, sondern meiner von denen immer mehr und mehr anwachsenden Ambtsgeschäfften übrigen Stunden mit Lesung erbaulicher bücher zuzubringen" appears in Imhoff to Fabricius, no date (but second half of 1702), KB, Ms Thott 1227, unpaginated. Imhoff to Fabricius, May 1, 1711, ibid.: "mich die Lust ferner etwas zu edirn verlassen hat."

75. This was a printed version of a family tree provided by Heinrich XIII, Count of Reuß-Untergreitz, in preparation for a (ultimately abandoned) fifth edition of the *Notitia Procerum*. Imhoff to Moritz Wilhelm, September 25, 1715, SächsHStA, loc. 8224, no. 9, fols. 33r–v. See Czech, *Legitimation*, 57.

76. Imhoff to Fabricius, August 5 and December 19, 1720, KB, Ms Thott 1227, unpaginated; Imhoff, *Albanensis familiae*.

77. Imhoff, "ad lectorem," in *Notitia Procerum* (5th ed., 1732): "auspiciis & ductu ipsius a me susceptum esse."

Chapter 1 • Genealogy circa 1700

1. Knorn to Imhoff, January 18, 1713, GNM Letters, unpaginated: "Ich bin für einiger Zeit in einem wirthshauße mit einem mir unbekandten Passagier auff die Materie was für Auctores von Englischen und Schotländischen familien geschrieben zurehden kommen, da dieser viel rühmete von einem Kox von Kelton undt einem anderen Belcastre genand, ich habe aber bey keinem Buchführer diese Autores noch deren notitiam ausforschen können, möchte woll wißen, ob Sie in rerum natura, oder die nahmen nur von dem Passagier erdichtet waren." On Knorn and Leibniz, see AA I, 1:107, 310–11, 318, 321, 377, 386; I, 5:54.

2. Broadway, *"No historie,"* 167; Plumb, *Death of the Past*, 29 ("genealogical craze"); Woolf, *Social Circulation*, 105, 113.

3. Long-term histories of genealogy are rare; see Butaud and Piétri, *Les enjeux*; Weil, *Family Trees*; Soria Mesa, "Genealogía y poder," 21–56. See also Rouchon, *L'opération généalogique*; Jettot and Lezowski, *L'entreprise généalogique*; Piétri and Luciani, *L'incorporation des ancêtres*; Eickmeyer et al., *Genealogical Knowledge in the Making*.

4. Zerubavel, *Ancestors and Relatives*; Sahlins, *What Kinship Is*; Jussen et al., *Blood and Kinship*; Sabean and Teuscher, *Kinship in Europe*; Hummer, *Visions of Kinship*. For a lucid summary, see Crouch, *Birth of Nobility*, 99–123.

5. Genicot, *Les généalogies*, unpaginated ("Avant-propos"): "genre mineur," "peu nombreuse." For a more recent discussion, see Hummer, *Visions of Kinship*.

6. Wickham, *Framing the Middle Ages*, 552.

7. Duby, *Chivalrous Society*, 149–57 and passim.

8. See, e.g., Jettot and Lezowski, *Entreprise généalogique*, on the "occasion-based" nature of genealogy.

9. Genicot, *Généalogies*, highlights, and focuses on, the stand-alone works titled "Genealogia."

10. Genicot, *Généalogies*, 13–24.

11. Klapisch-Zuber and Herlihy, *Les Toscans*; Rouchon, "L'enquête généalogique"; Raines, *L'invention du mythe aristocratique*.

12. On bourgeois German genealogies, see, e.g., Bei der Wieden, "Pietismus und Genealogie." For Italian artisans, see Canepari, "Généalogie et transmissions." See also Burguière, "La mémoire familiale."

13. Klapisch-Zuber, *L'ombre*, 19–33; Genicot, *Généalogies*, 14.

14. For a succinct summary, see Crouch, *Birth of Nobility*, 99–123. See also Teuscher, "Bilatéralité"; van der Lugt, *L'hérédité*. Sabean and Teuscher, *Kinship in Europe*, and Geevers and Marini, "Introduction," 13, also give a discussion of the chronology.

15. For the distinction between dynasty as "social group" and dynasty as a "cultural construct," of which only the latter is dominated by agnatic thinking, see Geevers, "Danish Habsburgs," 283–85. See also Morsel, *Noblesse, parenté et reproduction*.

16. Haddad, "Qu'est-ce qu'une 'maison'?," 131. For the important role of "blood" in premodern European society, see, e.g., two recent special journal issues, curated by Doyon et al., "Les lois du sang"; and García González and Gasperoni, "La sangre y el parentesco." As these articles also make clear, the well-known phenomenon of *limpieza de sangre* must be seen as only one expression—albeit an excessive one—of a widely shared consensus.

17. For a classic statement, see Goody, *Development of the Family*.

18. Lawyers and other legal personnel involved in Italian matrimony courts, for instance, used rough sketches of simple family trees to better understand the genealogical relationships among those seeking dispensation for consanguinity; see Gasperoni and Hauck, "La représentation graphique."

19. Roumy, "La naissance de la notion canonique." Kellner, *Ursprung und Kontinuität*, 17–21.

20. See, e.g., Gottschalk, *Eigentum*, 42–44. For a case study of a three-hundred-year-long inheritance conflict, see Berghorn, *Verwandtschaft*.

21. For one example, see, e.g., Voets, *Tractatus de jure revolutionis*.

22. Fichtner, *Protestantism and Primogeniture*.

23. Some of the best work on these developments concerns Portugal; see, e.g., Rosa, *O morgadio em Portugal*.

24. See, e.g., Broadway, *"No historie."*

25. See, e.g., Trevisan, *Royal Genealogy*.

26. Anything else was unusual and called for explanation. See Imhoff to Moritz Wilhelm, June 23, 1701, SächsHStA, loc. 8224, no. 7, fol. 94v (on the Duchy of Swabia). For a general discussion, see Pečar, "Dynastien."

27. Crouch, *Birth of Nobility*, 148–55.

28. "Né in modo alcuno negar si può che qualunque di nobile nasce, nobile non si chiami: ma che egli poi sia vero nobile, non sì generalmente è vero," G. B. Nenna wrote in Naples in 1529. See Muto, "I trattati napoletani," 334; and Vitale, "Modelli culturali nobiliari," 64–83 ("Nobilitas = Gentilitas + virtutum fortunarumque copia").

29. Schalk, *From Valor to Pedigree*, 158. See ibid., 164 ("separation of virtue and nobility and the shift toward birth"), 202 ("The end of nobility as a profession").

30. Drévillon, *L'impôt du sang*. Although he acknowledges the abiding importance of military service to the French nobility, Shovlin, "Nobility," 116, nevertheless observes a broad shift in emphasis away from feats of arms.

31. See the printed booklet in Stadtarchiv Nürnberg, A 31, no. 332/2, fol. a3v–[a4]r: *sanguinis nobilitatem generisve celebritatem*.

32. Smith, *Culture of Merit*, 61–65 (quotation on 64).

33. Mori, *L'archivio Orsini*, 175. See also Sannino, "Le storie genealogiche," 111, 120. Vitale, "Modelli culturali nobiliari," 74, also links the publication of Ammirato, *Delle famiglie nobili fiorentine* in 1577–80 with a decisive uptick of blood-based concepts of nobility.

34. See, e.g., Czech, *Legitimation*, 68; Schuster, "Familien- und Geschlechtsbewusstsein."

35. See the table in Smith, *Culture of Merit*, 58–59.

36. "Genealogists" as a social group have received little attention, but see Butaud and Piétri, *Enjeux*, 151–86; Piétri, "Une offrande." See also Poncet, "Cercles savants"; Klocke, "Ein westfälischer Genealogenkreis."

37. Geevers and Marini, *Dynastic Identity*.

38. Holladay, *Genealogy and the Politics*, 1.

39. Flacchio, *Généalogie*, vol. 1, unpaginated ("Epitre dédicatoire"): "d'exposer aux yeux de VOTRE ALTESSE & du Public, l'Origine, l'étendu, & la grandeur d'une Maison aussi illustre que la sienne."

40. ". . . sans naissance non plus que la riviere du Nil": Montaigne, *Essais* III,5, ed. P. Villey and V.-L. Saulnier, online edition by P. Desan, https://artflsrv03.uchicago.edu/philologic4/montessaisvilley/navigate/1/5/6.

41. Belleguise, *Traité de la noblesse*, 51–52.

42. Sant'Anna, *Della storia genealogica*, 1.

43. Trevisan, *Mythical Ancestry*. On the Habsburgs, see Lhotsky, "Apis Colonna"; Silver, *Marketing Maximilian*. See also Kagerer, *Macht und Medien*.

44. Biffi, *Gloriosa Nobilitas*; Bizzocchi, *Généalogies fabuleuses*, 84–86; Giannettasio, "dedicatio," in *Bellica*; Heinemann, *Herkommen*.

45. Maťa, "False Orsini"; Paravicini, "Colonna und Orsini"; Czech, *Legitimation*, 33–35 (Henneberg-Römhild), 36 (Henneberg-Schleusingen).

46. Cardinal Boullion distanced himself from "comparaison sur l'antiquité et les preuves de la descente de nos rois avec l'antiquité et les preuves de notre généalogie," yet this was precisely how his activities were understood. For the quotation, see Fage, *Étienne Baluze*, 40–41.

47. Berger, "Essai sur la nature," 209–35.

48. Natale, "Falsari milanesi," 457–506. On Atto's writings, see Gavinelli, "Il testamento," 339–45. Gavinelli's work, however, is based entirely on Natale's in this regard.

49. Aldimari, *Carafa*, unpaginated ("lettera dedicatoria" to the Spanish king): "singolare per la moltitudine, e per la chiarezza de' grandi evenimenti." For a careful comparison of claims to ancientness and claims to moral excellence, see Solignat, "Perpétuer la maison."

50. Flacchio, *Généalogie*, vol. 1, unpaginated ("Epitre dédicatoire"): "marquer les traces que Vous & vos Freres suivent si dignement."

51. Solignat, "Perpétuer la maison," 184.

52. Flacchio, *Généalogie*, 1: unpaginated ("Epitre dédicatoire"): "ils [sc., les descendens] trouveront dans la leur [généalogie] tout ce qui pourra les conduire à la plus haute & veritable gloire."

53. Flacchio, *Généalogie*, 1: unpaginated ("Epitre dédicatoire"): "qu'aiant les veines remplies de leur sang, l'ame le soit aussi de leurs vertus."

54. Laborderie, *Histoire*; Helgerson, *Forms of Nationhood*, 133, 135; Broadway, *"No historie."* See also Bernstein, *Historical Communities*, 54–56, 234–76.

55. Sant'Anna, *Della storia genealogica*, fol. b^{r}: "miglior'uomini."

56. For all of this, see Sant'Anna, *Della storia genealogica*, fol. b^{v}.

57. Callard, "La fabrication," 402.

58. Wrede, "Zwischen Mythen"; Boislisle, "Appendice VIII," 533–50.

59. Flacchio, *Généalogie*, vol. 1, 1: "Les differens sentimens des Auteurs sur l'Origine de l'Illustre Maison de la Tour, ne peuvent rien diminuer de l'idée qu'on s'est formée jusqu'à present de sa grandeur, & s'ils ne conviennent pas du même principe; ils sont cependant tous d'accord de son Ancienneté & de sa Noblesse."

60. Other authors use almost the same words; see, e.g., Fouilleron, "Français par le livre," 618.

61. For the expression, see Ranum, *Artisans of Glory*. There has been considerable skepticism about Ranum's interpretation of early modern historiography; see, e.g., Fossier, "A propos du titre"; Dubois, review of *Artisans of Glory*.

62. "Con tutto ciò sarà non che bene, ma necessario che l'Ecc.za V. pigli tempo di leggerla di mano in mano"; quoted from Santi, *Precedenza*, 36.

63. Imhoff to Moritz Wilhelm, May 28, 1700, SächsHStA, loc. 8224, no. 10, fol. 76v.

64. Richter, *Briefwechsel*, 319–20: "Falsa, conficta, spuria sciens volensque scribam nuncquam: vera tamen ut quandoque celem, summa flagitat necessitas."

65. Aldimari, "al Lettore," in *Carafa*, unpaginated: "Il scrivere delle Famiglie, è altretanto difficile, quanto pericoloso, se si vuole scrivere la verità, rare volte si soddisfano gli huomini, sempre avidi, di apparire assai più di quello siano, e se si vogliono seguitare i loro desiderii, bisognerebbe componere più favole, che historie, con poca estimatione dell'Autore, e con detrimento gravissimo del Pubblico, e de' Letterati."

66. Quotation from Esteve, "Censorship, Censure, and Historical Thought," 144. See also Esteve, "Contenerse."

67. Important case studies include Pumfrey and Dawbarn, "Science and Patronage"; Biagioli, "Galileo's System"; Petto, *When France Was King*.

68. Richter, *Briefwechsel*, 319–20: "si quae fuerint Dominis ingrata omittam, donec aliud tempus et locus haec quoque in apricum proferendi dabuntur."

69. Beckler, *Chronicon Bohemiae*, 2 (§2).

70. Poncet, "L'usage des chartriers seigneuriaux," 252–53. Bernstein, *Historical Communities*, 250–55.

71. He owned, for instance, the four volumes of de la Roque, *Histoire généalogique*. Vols. 3 and 4 of that book contained only documentary "preuves." StadtB N, Hist 263.2°–266.2°.

72. Aldimari, "al Lettore," in *Carafa*, unpaginated: "haver'acquistato grandissime notitie delle scritture dell'Archivio della Zecca di Napoli, ch'è forse il più antico, fuor di alcuni pochi de' Monasteri della stessa Città, e di tutti gli altri Archivii, così della Regia Camera della Summaria, come della Real Cancelleria del Regno, & havuto per l'assidue diligenze usate, in mio potere, i manoscritti di quante Famiglie, e loro gesta, e Privilegii, siano registrate in essi Archivii."

73. Aldimari's pride in his genealogical collection is evident from his letters to Imhoff; see, e.g., Aldimari to Imhoff, June 10, 1690, GNM Letters, unpaginated. Cf. also Aldimari to Magliabechi, January 21, 1688, in Quondam and Rak, *Lettere*, 9–10, on Aldimari as bibliophile. On his collection and library, see Iasiello, *Il collezionismo*, 183–85, and Soria, *Memorie storico-critiche*, 1:15.

74. On Birken as genealogist, see Hausenstein, "Der Nürnberger Poet," 219–20; Rohmer, "Die Hirten"; Kagerer, *Macht und Medien*, 380–99.

75. See Herdegen, *Historische Nachricht*, 123–53.

76. Herdegen, *Historische Nachricht*, 124: "die ganze Chronick auf jetzige Art zu stilisiren." Cf. Jürgensen, *Utile cum dulci*, 42. See also Jöns, "Sigmund von Birken," 168: Birken was a "Berufsschriftsteller."

77. Birken, *Der Briefwechsel*, 138, 139, 198 (see also the commentary). Birken, *Chur- und fürstlicher Sächsischer Helden-Saal*.

78. There are no marginalia or other annotations, for instance, in StadtB N, Hist 264.2° and Hist 265.2° (*Preuves de l'histoire genealogique de la maison de Harcourt*, vols. 3–4, Paris 1662).

79. Schrenck von Notzing, "Die 'Genealogiae' des Nicolaus Rittershausen."

80. In the first edition of the work (1653), Rittershausen had (unintentionally) offended the

Duke of Bavaria by mentioning a "concubine." The duke complained, and Rittershausen had to devise an elaborate excuse. It is hard to fathom what lay behind the episode. The entire first edition was plagued by an "obvious confusion of names and lines," as the author admitted; see Strobel, *Conradi, Georgi et Nicolai Rittershusiorum*, 109–10. Perhaps the insult was indeed only an accident. The affair was closely monitored in German intellectual circles; see Enoch Glaeser, Tübingen, to Rittershausen, April 19, 1654, SUB GÖ, Ms. philos. 102, no. 30: the literati of Tübingen were uncertain whether the prince's intervention was adequate ("quid credendum nobis hac in parte sit, cognoscere"). The 1653 edition is rare; I consulted a digitized copy housed in the library of the University of Erlangen-Nürnberg (call number: H00/2 HIST 24 d).

81. Diephold, *Genealogiarum Imperatorum*. There has been some debate about the reasons behind Sannemann's decision to narrow the chronological framework of the work; cf. Schrenck von Notzing, "Die 'Genealogiae' des Nicolaus Rittershausen."

82. Bély, *Société des princes*.

83. Rittershausen, "ad lectorem," in *Genealogiae* (3rd ed., 1664): "Actum agerem, si Genealogias commendare vellem. Utilitas earum . . . ipsa se commendat, nec eget praeconio res, quae Regum ac Principum inter se negotia, jura, praetensiones, matrimonia, successiones, foedera & id genus alia comprehendit, & quoties difficultatis occasio emergit, luculenter ostendit: cujus ignorantia, rixas, altercationes, quandoque etiam gravia bella excitavit."

84. Cf. Lingnau, *Lektürekanon*, 174, 184–85, who shows that early modern literature on the "ideal politician" regularly presupposed significant genealogical reading.

85. Important earlier trailblazers included Hieronymus Henniges (?–1597) and Elias Reusner (1555–1612). Nevertheless, Rittershausen was among the first to situate (encyclopedic) genealogy in the field of political commentary. Henniges, a Lutheran pastor, by contrast, presented his four volumes of genealogical tables as a re-telling of Christian world history, while Reusner cast his own collection of family trees as but one element of a larger project to acquire a synoptic overview over universal history, from which he wanted to draw moral lessons. Neither author intended to use genealogical information for understanding contemporary politics, as Rittershausen suggested. Cf. Henniges, "ad lectorem," in *Theatrum*, vol. 1, unpaginated; Henniges, "ad lectorem," in *Theatrum*, vol. 4/3, unpaginated. Both prefaces use the biblical schema of the Four Monarchies but have nothing to say about genealogy, except that the reader was to find all the rulers and their offspring relevant to the biblical framework in the following pages. Reusner, "ad lectorem," in *Basilikon*. There is, again, nothing specific on genealogy.

86. See, e.g., the positive reply from Carl Ludwig Ernst, Count zu Sulz, Amberg, to Rittershausen, September 7, 1645, SUB GÖ, Ms. philos. 91, fol. 88r. Daniel Waldschmidt (on behalf of Georg Friedrich Count zu Waldeck-Pyrmont), Arolsen, to Rittershausen, August 24, 1657, SUB GÖ, Ms. philos. 91, fol. 118r–v. Rittershausen to ? (certainly a noble person), December 16, 1662, SUB GÖ, Ms. philos. 94, no. 60.

87. See, e.g., Christoph Sigismund Donauer Sr., Regensburg, to Rittershausen, no date (ca. 1645–48), SUB GÖ, Ms. philos. 93, no. 64; Ludolph, Gotha, to Rittershausen, September 1, 1662, SUB GÖ, Ms. Philos. 91, fol. 49r–v; Rittershausen to [Georg] Seger, no date, SUB GÖ, Ms. philos. 94, no. 59 (second letter). Rittershausen to Spener, no date (1664 or earlier), ibid., no. 68. Others were enlisted to review and correct his first drafts, see Rittershausen to ?, June 2, 1644, SUB GÖ, Ms. philos. 94, no. 49.

88. The term comes originally from François Hartog; I adopt it from Clark, *Time and Power*, 1–3.

89. Dooley, *Dissemination of News*. Landwehr, *Geburt der Gegenwart*.

90. Important studies include Raymond, *Invention of the Newspaper*; Scholz Williams, *Ways of Knowing*; Scholz Williams, *Mediating Culture*; Infelise, *Prima dei giornali*; Keller and Molino, *Die Fuggerzeitungen*.

91. Bauer, "Jetztherrschend," 271–300.

92. Slauter, "Periodicals."

93. See Lohsträter, "Zeitgeschichte," 418–22. For parallel observations on France, see Schuwey, "Les périodiques."

94. This group included, for instance, the Hamburg literary figure Peter Ambrosius Lehmann (1663–1729), whose journal *Hamburger Historische Remarques* Imhoff consulted (Juncker to Imhoff, April 8, 1700). Lehmann also personally forwarded some material on Portugal; his name appears several times in GNM Généalogies PAR–PZ (vol. 18), unpaginated (Pereyra).

95. On Beer, see Koschlig, *Das Ingenium Grimmelshausens*, 297–505, although Koschlig says next to nothing about the works under consideration here.

96. Beer, *Der Könige in Franckreich*; Beer, *Der Herzogen und Königen* (this was purely a translation of Emanuele Tesauro); Beer, *Der Durchleuchtigsten Erz-Herzogen.*

97. Koschlig, *Das Ingenium Grimmelshausens*, 318–19.

98. Wurster, "Die Regensburger Geschichtsschreibung," 119–21; Freytag, "Die Seifertschen Stammtafeln"; Korb, "Seifert und Hildebrand."

99. Seifert, *Genealogische Tabellen.*

100. Bourcy, "Nachweisungen und Inhalt," 310; Seifert, *Genealogie*, vol. 1, fol.)(2r–v. Gebhardt, "Ein Brief," 185–86.

101. Franz Anton von Imhoff to [Jakob Wilhelm Jr.], October 5, 1723, GNM Nachlass Imhoff VII (Collection Imhoff), no. 4, unpaginated.

102. Seifert, *Genealogie*, vol. 1, fol.)(2r–v: "auf meine eigenen Kosten." Many of his tables bear a remark that Seifert wished to "hand them over" (*übergeben*); see many examples in Seifert, *Ahnen-Taffeln.*

103. There were plans to counter Seifert's book by printing an official Imhoff genealogy; see Franz Anton von Imhoff to [Jakob Wilhelm Jr.], January 30, 1724, GNM Nachlass Imhoff VII (Collection Imhoff), no. 4, unpaginated. Nothing ever came of the idea.

104. On the family, see Grolée-Virville, *Les d'Hozier.*

105. On this, see Bluche, *Les pages.*

106. Defauconpret, *Les preuves de noblesse*; Harding and Hecht, *Die Ahnenprobe*; Smith, *Culture of Merit*, 59–62; Loiseau, "Much Ado about Nothing?". Wood, *Nobility of the Election*, remains foundational.

107. For this and the following, see, e.g., the compelling summaries in Dewald, *European Nobility*; Asch, *Europäischer Adel*; Figeac, *Les noblesses.*

108. Quotation from the 1614 debate in Grolée-Virville, *Les d'Hozier*, 16.

109. On the broader process, see, e.g., Gutiérrez-de-Armas, "Construir una identidad"; Gutiérrez-de-Armas, "Identidad nobiliaria."

110. Bluche, *Les pages*, unpaginated (at footnote 65).

111. Donati, *L'idea di nobiltà*, 252: "invasioni e guerre continue . . . non v'è stato regno, provintia, città, né lugo, che non habbi patito in manera tale."

112. Gersmann, *Adlige Lebenswelten*, 187–92.

113. See, e.g., Piétri, "Bonne renommée."

114. Müller, *Die Ritterschaft*; Pfeiffer, "Westfälische Aufschwörungstafeln."

115. This version of their genealogy connected Duke Ercole to the eighth-century Lombard king Desiderius. It was published as Biffi, *Gloriosa Nobilitas*. Based on Biffi's book, the new genealogy also appeared in Palazzi, *Aquila Saxonica*, 176–79, quotation on 179. Many thanks to Dr. Federico del Tredici (Rome) for helping me better understand this case.

116. See preceding note. Biffi's book was very rare. Leibniz asked Imhoff whether he owned it. Imhoff, however, had only been able to borrow a copy from Giuseppe Scipione, Count of Castelbarco, sometime before 1701; see Imhoff to Leibniz, August 4, 1705, in AA I, 25:10.

117. See the printed document, available in Imhoff's collection GNM Généalogies M (vol. 15), unpaginated (Milan/Visconti).

118. Speidel, *Speculum*, 474: "Et in Jure nostro Civili & Canonico multum refert scire statum & conditionem natalium, quibus quis ortus sit parentibus, unde jura legitimationum exsurgunt, aut progrediuntur: praecipue vero in materia Successionum & Matrimoniorum, contractibus[,] tutelis, bonorum possessionibus & testimoniis per vim suam & utilitatem Genealogiae exerunt."

119. Zschackwitz, "Vorbericht," in *Historisch-genealogischer Schauplatz*, unpaginated: "vornehmlich aber kan in grosser Herren ihren Rechts=Ansprüchen / ohne Beyhülffe der Genealogi, nichts mit Nachdruck gerichtet werden."

120. Schalk, *From Valor to Pedigree*, 153–55. Donati, *L'idea di nobiltà*, 254. Berghorn, *Verwandtschaft*, 25, 35, 167, 339.

121. Holzschuher, *Deductions-Bibliothek*, fol.)(r–v: *Deduktionen* are "Schriften, deren Absicht dahin geht, die Beschaffenheit einer streitigen Rechtssache näher zu entwickeln, dieselbe dem Publikum in ihrer vorteilhaftesten Gestalt darzustellen, und die Leser, so viel als möglich, von der in einem Fall vorhandenen Gerechtigkeit zu überzeugen, es mag dabey auf den Beweis der Tatsachen oder die Ausführung der hier enschlagenden Rechtssätze ankommen." See further Gebhardt, "Patrizische Leidenschaften."

122. For a typical title, one may quote the "Memoire de Mr. le Marquis d'Alegre, Prince d'Orange, sur la principauté de Neufchatel," mentioned in Lünig, *Bibliotheca Curiosa*, 3–4.

123. One famous German case, also relevant for Imhoff, concerned the succession in the northern German territory of Saxony-Lauenburg. Among many others, Leibniz also was involved in this protracted battle of books and legal briefs. Cf., e.g., Junge, *Leibniz*.

124. For a general survey, see Grafton, *What Was History?*, 1–62. Cf. also Dewald, *Status, Power*, 32.

125. For a broad survey, see Turner, *Philology*.

126. For "révolution dans les mentalités," cf. Menant, "La connaissance," 438. For a very optimistic version of this argument, see Williams, *First Scottish Enlightenment*, 215–43.

127. Reimmann, *Historiae Literariae*, 1:71: "maturitatem."

128. See, e.g., Reimmann, *Historiae Literariae*, 1:56 to the end. The author establishes a periodization for the history of genealogy: Antiquity (Tacitus to Carolingians), from Charlemagne to Maximilian I, and from Maximilian until 1700. This periodization, however, is somewhat ambiguous, since many "modern" genealogists considered many of their sixteenth-century forerunners uncritical. On the following, see the seminal works Bizzocchi, *Généalogies fabuleuses*, and Bizzocchi, "Unglaubliche Genealogien."

129. Benigno, "Il ritorno dei Vespri," reprinted nearly verbatim in Benigno, *Favoriti e ribelli*, 193–209. See also Benigno's entry in the *Dizionario Biografico degli Italiani*, http://www.treccani.it/enciclopedia/filadelfo-mugnos_(Dizionario-Biografico).

130. See, e.g., Gittio to Imhoff, September 14, 1688, GNM Letters, unpaginated: "piena di favole, et inventioni"; "non si dasse credito alle sue opere." Equally harsh: Gittio to Imhoff, February 19, 1697, ibid. A noncritical reference to Mugnos, *Teatro genealogico*, 3:474, appears in GNM Généalogies A (vol. 6), unpaginated (Altieri).

131. Pusterla to Imhoff, August 12, 1693, GNM Letters, unpaginated: Crescenzi was producing "fabulae" just as literary authors did for the theater. This allegation was particularly pointed for an author who had produced an "amfiteatro" of the nobility; see Crescenzi, *Anfiteatro romano*.

132. See the harsh quotation in Fouilleron, "Français par le livre," 623.

133. D'Hozier to Imhoff, November 13, 1711, GNM Letters, unpaginated: "ridicule vanité."

134. D'Hozier to Imhoff, November 25, 1710, GNM Letters, unpaginated: "aujourdhuy la folie des hommes est de vouloir etre jssus de souverains."

135. Imhoff to Moritz Wilhelm, August 17, 1695, SächsHStA, loc. 8224, no. 7, fol. 10r. Cf. also Wegleiter to Imhoff, September 11/21, [1685], BSB, Autogr. Wegleiter, unpaginated: "inveni in choro templi Coelestinorum Ducis Aerschottensis (cujus a missae sacris redeuntis fausto as-

pectu frui mihi licuit) familiam inde ab Adamo (et quidni a Praeadamitis) usque ad nostra fere tempora accuratis picturis eductam. Tunc ego mecum: oportet sane hanc gentem Austriacae nonnullisque alijs arcto consanguinitatis nexu implicitam esse, nam et harum natales e Paradiso vel arca Noae repetunt ineptuli." See also Speidel, *Speculum*, 473.

136. Pusterla to Imhoff, August 12, 1693, GNM Letters, unpaginated.

137. Haddad, "Question of the Imprescriptibility." Haddad, "Parenté et pratiques généalogiques."

138. Imhoff to Muratori, October 30, 1696, in Lieber, "Jacob Wilhelm von Imhofs Korrespondenz mit Muratori," 50–52. His comments hardly ever amount to more than generalizing statements, such as, e.g., Imhoff to Leibniz, July 24, 1696, AA I, 13:190–92: genealogists need to trace lines "ad claros limpidosque fontes."

139. Babin and van den Heuvel, *Gottfried Wilhelm Leibniz*, 86–93. See Antognazza, "Leibniz as Historian," 593–94, for the biographical context and a competent summary of Leibniz's findings.

140. Genealogies now had to be "substantiated," to borrow a term frequently used in Ostenfeld-Suske, "Official Historiography," 415.

141. Cyriacus Spangenberg, in 1571, for instance, had "no doubts" in connecting some "historical" Counts of Mansfeld from the tenth century to a (mythical) ancestor of the sixth century, although "no one [of this lineage] is mentioned by name in the Chronicles" (daß in Chronicken keines namhaftig gedacht wird), quoted in Czech, *Legitimation*, 42–43.

142. GWLB, Ms VIII 634, fol. 2r: "Genealogia accurata quae reperiat Originem familiae, quousque certo haberi potest."

143. All the following quotations appear in Bacchini's manuscript report on Damaideno as excerpted in Golinelli, *Benedetto Bacchini*, 1–11. See also Bacchini, *Dell'istoria del monasterio*, fol. **r ("prefazione"): "il Poeta crea il verosimile, e lo Storico lo produce, quegli se ne serve per elettione, questi per necessità."

144. Babin and van den Heuvel, *Gottfried Wilhelm Leibniz*, 93.

145. Babin and van den Heuvel, *Gottfried Wilhelm Leibniz*, 93: "Aujourdhuy on a bien d'autres lumieres dans l'histoire, apres tant de monumens tirés des tenebres, et l'on agit avec une exactitude entiere, c'est pourquoy celuy qui veut écrire maintenant avec approbation des personnes sçavantes en matiere de genealogies y doit apporter presque autant de soin, que s'il vouloit produire des preuves recevables dans un chapitre de quelque Eglise Cathedrale de l'Allemagne."

146. Imhoff to Moritz Wilhelm, February 26, 1718, SächsHStA, loc. 8224, no. 9, fols. 63v–64r.

147. Imhoff to Thulemeyer, November 2, 1700, UB F, Ms. Thulemeyer, 793–95, here 794: "nichts ex archivis oder communicatione privata veterum & ineditarum chartarum beybringet."

148. Juncker to Imhoff, March 7, 1704, GNM HA V, Juncker, unpaginated, about Cyriacus Spangenberg: "mirari satis non potui, tam negligentem deprehendi in consulendis Archivorum diplomatibus, ut nihil supra esse possit."

149. See Moritz Wilhelm to Imhoff, October 22, 1695, Veste Coburg, Ms A, I, 349, (2), 3, unpaginated: he had no access to an "ancient archive" (kein alt Archiv), so he could not comment on certain details of the princely house of Saxony.

150. This episode is from an anonymous report, written for Moritz Wilhelm, about Tobias Pfanner's text on Saxon genealogy, SächsHStA, loc. 8224, no. 8, fols. 23r–25v, here 23v–24r: "dieser in einem solchen seculo, als das 14te war, gelebet haben soll, binnen welchen verschiedene fleißige scribenten in- und auserhalb Sachsen floriret, welche gleichwie sie von mehrernanden Sigismundi Hhl Brüdern . . . uns die ausführlichste Nachrichten hinterlaßen, gewißlich von diesem Sigismundo, so gleich wohl das 28te Jahr erreichet haben soll, wo nicht so vollkommene, jedoch nur einige Anzeige gegeben haben würden." The report goes on to produce numerous further arguments about why Sigismund was an invented person.

151. Reimmann, *Historiae Literariae*, 250: "quis quaeso tam lippientibus oculis est praeditus, qui non perspiciat *Scepticismum etiam in studio Genealogico esse summopere necessarius.*"

152. On skepticism circa 1700, see Dooley, *Social History of Skepticism*; Völkel, *"Pyrrhonismus historicus."*

153. Imhoff to Moritz Wilhelm, October 28, 1697, SächsHStA, loc. 8224, no. 10, fol. 25v: "Wer der Marggraf Heinrich ohne Land eigentlich gewesen seye, wird wol ungewis bleiben, wann nicht noch alte richtige documenta sich hervorthun, die ein mehreres Licht geben."

154. See the pertinent notes "certa"/"probabilis" in the manuscript family tree of the Burgraves of Kirchberg (post 1721), in GNM Généalogies J–K (vol. 13), unpaginated (Kirchberg).

155. "ein Tempus Obscurum, das ist solche Zeiten, da man gantz keine Gewißheit von hat, nähst ein Tempus Mythico-Heroicum, da man zwar etwas absiehet, iedoch unter vielen ungewissen umbständen; Endlich aber ein Tempus Historicum, von solchen Dingen, welche mit mehrer gewißheit können gemeldet werden"; quoted from Hecht, "Production of Genealogical Knowledge," 158n32.

156. On the details of chronology, see Gittio to Imhoff, July 9, 1697, GNM Letters, unpaginated. D'Hozier to Imhoff, December 18, 1695, ibid., unpaginated.

157. Muratori to Imhoff, March 7, 1697, in Busino and Dufour, "Quattordici lettere," 422–23: "familiam suam a fabulis arcere." Cf. also Dewald, *Status, Power*, 25.

158. ". . . rem reor longe investigatu difficillimam, cum vel fabulis vel conjecturis aliorum niti nos oporteat," Franckenau to Imhoff, April 1, 1703, GNM Letters, unpaginated.

159. "recte conjecturasti," Franckenau to Imhoff, April 1, 1703, GNM Letters, unpaginated.

160. Lohmeier to Imhoff, March 19, 1691, BSB, Cim. Lohmeier, simply corrects an erroneous "conjectura" made by Imhoff.

161. Aldimari to Magliabechi, December 14, 1680, in Quondam and Rak, *Lettere*, 8–9. The lineage is presented in Aldimari, *Memorie historiche*, 512–85 (including a few edited documents on 566–85), as if written by Carlo de Lellis. Although Aldimari probably owned many of the late de Lellis's papers (Soria, *Memorie storico-critiche*, 1:15), this was in all likelihood a fraud. See Giustiniani, *Memorie istoriche*, 1:34: "finto nome di Carlo de Lellis."

162. See, for Florence, Klapisch-Zuber, *Das Haus*, 19–22; Fouilleron, "Français par le livre."

163. Ailes, "Can We Trust?"

164. "Doppia anima": Sannino, "Le storie genealogiche," 110.

165. For an example, see Liddy and Steer, "John Lord Lumley," 208.

166. Quotation from Broadway, *"No historie,"* 156. For a parallel assessment in a neighboring genre of historical literature, see Bernstein, *Historical Communities*, 142–69.

Chapter 2 • A Patrician Genealogist and His City

1. *Fränckische Acta*, 228–39.

2. *Fränckische Acta*, 231: "Nun könte man gar leicht nicht nur 16. sondern auch viel mehr Ahnen dessen Uhralten und Hoch=Adelichen Stamms, Vätterlicher und Mütterlicher Seiten, in unverruckter Reyhe erzehlen und anführen."

3. See the numerous publications of Christiane Klapisch-Zuber about Florentine genealogy. See also Tucker, *City of Remembering*.

4. Munck and Romano, "Introduction." A classic study is van Damme, *Le temple de la sagesse*.

5. Endres, "Nürnberg."

6. See Kenny, *Born to Write*. See also Rabinovitch, *The Perraults*.

7. Wüst, "Patrizier."

8. There is little literature on the artwork sponsored by the Imhoffs. On the parallel case of the Tucher, see Schwemmer, "Das Mäzenatentum."

9. GNM, Nachlass Imhoff I (Archiv der Gesamtfamilie), fasc. 3, no. 2, unpaginated: "Abschrift des in der alten kirchen zu St. Johannis bey beyreth gelegen an der Mauer gegen das

Bräuhaus zu sich eingemauert befundenen steinernen bey den Anno 1742 vorgenommenen Kyrchumbau aber in den Grund mit eingemauerten Imhofschen Monumentis, von Herrn Johann Georg Bart (?) Pfarrer zu St. Johann abcopirt." There follow several additional documents (drafts of letters in particular) from the 1740s documenting how eagerly the surveying of "outlying" monuments proceeded. See also GNM, Nachlass Imhoff I (Archiv der Gesamtfamilie), fasc. 3, no. 2: "Imhoffische Monumenta betreffend." GNM Nachlass Imhoff I (Archiv der Gesamtfamilie), fasc. 3, no. 2, unpaginated: "Von St. Johannis Altar in St. Laurentij kirch". See also the *Antiqua Imhofianae prosapiae monumenta* in GNM, Nachlass Imhoff II (Helmstädter Archivschrank) no. 9.

10. Meyer, *Die Stadt als Thema.* More broadly, see Wüst, "Patrizier."

11. Kagerer, *Macht und Medien.*

12. See, e.g., Landois, *Gelehrtentum.* See also Wuttke, "Humanismus"; Fuchs, *Medizin.*

13. Endres, "Sozial- und Bildungsstrukturen," 56–60: "[das] Analphabetentum [kann] nicht sehr weit verbreitet gewesen sein"; Landois, *Gelehrtentum*, 31–33.

14. Sauer, "Reformation der Bücher."

15. Mährle, *Academia Norica.*

16. Schmeisser and Birnstiel, "Gelehrtenkultur."

17. Brennecke, *Akademie.*

18. See the recent survey in Hamm, "Die Reformation."

19. Schröttel, *Johann Michael Dilherr*; Sommer, "Das Wirken Johann Michael Dilherrs"; Dülmen, "Orthodoxie und Kirchenreform." The name "Saubert" appears several times in Imhoff's correspondence, although generally only in passing. I have not been able to find references to Dilherr.

20. Cf. Steiger, *Ikonographie.*

21. For the end of *Meistersang*, see Paul, *Reichsstadt und Schauspiel*, 30–36.

22. Rohmer, "Literatur und Theologie." Niefanger and Schnabel, "Literarische Gruppenbildung." On the early history of the Blumenorden, see Jürgensen, *Utile cum dulci.*

23. Paul, *Reichsstadt und Schauspiel.*

24. Paul, *Reichsstadt und Schauspiel*, 91–110.

25. Paul, *Reichsstadt und Schauspiel*, 86–88. M. D. Omeis to von Birken, August 8, 1680, in Laufhütte and Schuster, *Sigmund von Birken*, 73: one (unidentified) "Monsieur Im-Hof" is closely associated with Omeis. On the (increasingly) religious implications of the Blumenorden's activities, see Rohmer, "Literatur und Theologie," and Jürgensen, *Utile cum dulci*, 59–60. Jürgensen (ibid., 198–99) is more skeptical regarding connections between the patricians and the Blumenorden.

26. Endres, "Nürnberg," 148–49.

27. Gaab, "Zur Geschichte der Emmert-Sternwarte."

28. Kreiner, "Nürnbergs und Altdorfs Anteil"; Seitz, "Johann Paul Wurffbain"; Uwe Müller, "Die Leopoldina."

29. Barnett, "Anspruch und Wirklichkeit"; Barnett, "Medical Authority," 255–342. See further Grulich, *Geschichte der Bibliothek.*

30. Arbeitskreis Orangerien in Deutschland, *Nürnbergische Hesperiden.* For a broader perspective, see Wüst, "Citronen."

31. Volckamer, *Nürnbergische Hesperides.*

32. Sporhan-Krempel, *Nürnberg als Nachrichtenzentrum.*

33. "Comp. Norib. 1692 per 4 krol IWH," in Imhoff's copy (StadtB N, Hist 1206.8°) of *Discorso della nobilta di Firenze e de fiorentini, di Paolo Mini Medico, filosofo e cittadino fiorentino* (Florence, 1593). Imhoff's copy of Ammirato, *Dell'istorie fiorentine* (StadtB N, Hist. 299.2°), for instance, was also "Comp. Norib. 1692 p[er] 1 fl. IWH."

34. Behringer, *Im Zeichen des Merkur.*

35. Several letters from "Dr. Wölcker" are copied/excerpted in GNM, Généalogies PAR–PZ (vol. 18), unpaginated (Pfalz). This was probably Christoph Carl Wölcker (1632–80) or perhaps his son Carl Georg (1660–1723). On both men, see Will 4:272–74, 274–79. Wölcker Jr. married Imhoff's sister Sibylla Dorothea in 1688: Seifert, *Genealogische Tabellen*, table 16. For Wölcker Sr. and his connections to the Blumenorden, see Stauffer, *Sigmund von Birken*, 1067–68.

36. Two individuals whom Leibniz recommended to Imhoff serve as a good example; see Leibniz to Imhoff, October 28, 1713, GWLB, LBr 449. They were the Count of Oropesa and the Marquis de Westerloo. Concerning the former, Imhoff replied to Leibniz (November 13, 1713, ibid.) that he knew him from a diplomatic mission that had taken Oropesa to Nuremberg a few months earlier. With respect to the latter, it is remarkable that two personal meetings became possible simply because Westerloo passed through Nuremberg. See Merode-Westerloo, *Mémoires*, 2:131, 289. On Westerloo, see also Imhoff to Fabricius, November 12, 1717, and April 8, 1718, KB, Ms Thott 1227, unpaginated; Imhoff to Moritz Wilhelm, December 27, 1695, and March 3, 1699, SächsHStA, loc. 8224, no. 7, fols. 27v, 69v.

37. See Hoppen, *Common Scientist.*

38. Blom, *Christoph and Andreas Arnold*, who seems to get the chronology wrong. Ashe did not visit Nuremberg from Vienna but the other way around: he reached Vienna via Nuremberg. Ashe also corresponded with Magliabechi; cf. Simunotti, "Prime osservazioni," 399–405.

39. Imhoff, "ad lectorem," in *Regum Pariumque.*

40. Schunka, *Ein neuer Blick.*

41. On this, see Blom, *Christoph and Andreas Arnold*, 4.

42. Stauffer, *Sigmund von Birken*, index. Jürgensen, *Utile cum dulci*, 88–90, 191.

43. Wegleiter to Imhoff, December 23, 1686, BSB, Autogr. Wegleiter: connections to Edward Bernard, the orientalist, and Elias Ashmole.

44. Jürgensen, *Utile cum dulci*, 23, 37–38. Stauffer, *Sigmund von Birken*, index. Blom, *Christoph and Andreas Arnold*; Jürgensen, *Bibliotheca Norica*, 533–800.

45. Targioni-Tozzetti, *Clarorum Germanorum*, 286: "ocellus."

46. Christoph Arnold's letters in Georg Richter, *Epistolae Selectiores*, 470–518.

47. Ashmole to Imhoff, October 3, 1687, and June 6, 1688, in Ashmole, *His Autobiographical and Historical Notes*, 4:1843–45, 1854–55. Christoph to Andreas Arnold, November 1684, quoted in Blom, *Christoph and Andreas Arnold*, 89: "Wann wird uns denn Haakius antworten von dem, was Mr. Imhof und ich an ihn und Mr. Ashmol gesandt?" A copy of a letter from Andreas to Christoph Arnold (July 14, 1680) about the English king Charles II, his mistress, and his natural son made it into Imhoff's collection, GNM, Généalogies S–T (vol. 21), unpaginated (Stuart). A "Relatione cavata d'una lettera del S.r Anton. Magliabecchi al S.r Christoforo Arnold, scritta 1675 17. Nov." concerning the death of Cardinal Leopold de Medici, appears in GNM, Généalogies M (vol. 15), unpaginated (Medici).

48. E.g., Andreas Arnold, Leiden, September 16, 1680, GNM, Généalogies N–O (vol. 16), unpaginated (Nassau). Blom, *Christoph and Andreas Arnold*, 29.

49. For personal encounters between Sigmund von Birken and Wilhelm Imhoff III, see Birken, *Tagebücher*, 1:86, 92, 163, 339, 344, and 2:225, 255, 428. On Wilhelm's death, the Blumenorden produced a commemorative pamphlet in praise of him, titled "Das Ebenbild des HochEdelgebohrnen / Fürsichtigen und Hochweisen HERRN Wilhelm Im Hof / nach seinem Tod entworfen von einem Mitglied des Peginesischen Blumen Ordens"; see Feuerlein, *Dreyfach-Göttlicher Liebes-schein*, 137–82.

50. For the details of Pfeiffer's life, see Gryphius, *Vitae Selectae*, 581–600. See Pfeiffer to Imhoff, August 23, 1681, GNM, Généalogies W (vol. 22), unpaginated (Wallenroth).

51. It seems as if Hardesheim's brother Justin served as intermediary: see Tobias Hardesheim to "his brother," March 18, 1679, a neat copy of which ended up in Imhoff's papers, GNM, Généalogies J–K (vol. 13), unpaginated (Isenburg). Hardesheim also reported on the Counts of

Erbach: see GNM, Généalogies E (vol. 9), unpaginated (Erbach): "commun. a Tob. Hardesiano d. 31. Jul. 1679."

52. Ziegra, *Nicolaus Wilckens*, 545–47. Fabricius dedicated his edition of Saubert's posthumous works to Imhoff; see Fabricius, "Epistola Dedicatoria ad Jacobum Wilhelmum Imhofium," in Saubert, *Opera Posthuma*, fols. *2r–**v. This is mentioned in Imhoff to Fabricius, November 11, 1693, KB, Ms. Thott 1227, unpaginated, and Imhoff, "praefatio ad lectorem," in *Notitia Procerum* (1st ed., 1684), fols. a5r–v. See further Schlüter to Imhoff, July 5, 1681, GNM, Letters, unpaginated, with material on the Danish royal family. Some of this material was later forwarded to Zeitz: see Imhoff to Moritz Wilhelm, January 16, 1703, SächsHStA, loc. 8224, no. 10, fols. 103r–v. In June 1689, Schlüter sent Imhoff a manuscript copy of Mithoff, *Historischer Kurtzer Bericht*. See the note on the copy, in GNM, Généalogies S (vol. 20), unpaginated (Sachsen): "Communic. a. Sev. Walth. Slutero Superintendente Saxo-Leoburgico m. Jun 1689." See also "Ex commun. Joh. Mothii Secretarii Regis Dan. intercedente Sev. W. Slutero pr. 3.Dec. 1683," on a fact-sheet in GNM, Généalogies R (vol. 19), unpaginated (Rantzau).

53. This group included the legal scholar Heinrich Linck (1642–96); cf. GNM, Généalogies S–T (vol. 21), unpaginated (Schoenberg), referring to "Nov. 1678"; and GNM, Généaogies R (vol. 19), unpaginated (Rutheni), referring to 1680, and the physicist Johann Christoph Sturm (1635–1703); cf. GNM, Généalogies G (vol. 11), unpaginated (Gravenegg), referring to 1692.

54. In 1694, Imhoff also turned to a local contact (one Jacob Reiger) to transfer money to Johann Christoph Pommer in Venice, who was to pass it on to Giovanni Behm: Imhoff to Böhm, January 18 and February 14, 1694, UB ER-N, Briefsammlung Trew. On Böhm and the Pommer family, see Ressel, *Protestantische Händlernetze*, index, and Minuzzi, "Sul filo dei segreti medicinali," 235–53. The German merchants of Venice also appear collectively, e.g., in Sant'Anna to Imhoff, January 16, May 28, and October 10, 1696, BSB, Cod. Ital. 506, fols. 7r–8r, 15r–v, 17r–18r, also mentioning one Paolo Baglioni in Naples as intermediary for the leg to Venice. See also a similar reference in Magliabechi to Imhoff, August 20, 1695, GNM, Letters, unpaginated.

55. Endres, "Verfassung und Verfassungswirklichkeit." Endres, "Grundzüge der Verfassung." Fleischmann, "Das Verhältnis." See the important recent reflections on the genesis of the patriciate as a social group in Morsel, "Sociogenèse d'un patriciat." There is no comprehensive study of the Nuremberg patriciate comparable to Hansert, *Geburtsaristokratie*. On patrician income, see Endres, "Sozial- und Bildungsstrukturen," 42–43.

56. Fleischmann, "Das Verhältnis," 18–19.

57. Hirschmann, "Das Nürnberger Patriziat." Hofmann, "Nobiles Norimbergenses." See also Stromer, "Reichtum und Ratswürde."

58. Bruscoli, "Der Handel."

59. Seibold, "Die Imhoffsche Handelsgesellschaft." See Bartelmess, "Die Patrizierfamilie Tucher," 226–29, for the parallel case of the Tucher firm, which existed until 1646 and was the last to close.

60. Endres, "Nürnberg," 133–42. On the new, nonpatrician economic elite, see also Stromer, "Reichtum und Ratswürde," 15–19.

61. Fleischmann, *Rat und Patriziat*, 601–31. See also C. F. Imhoff, "Die Imhoffs, Handelsherren und Kunstliebhaber"; Imhoff, s.v. "Imhoff" in *NDB*; and Jahnel, "Die Imhoffs."

62. The list of goods comes from C. F. Imhoff, "Imhoff," 146.

63. Stromer, "Reichtum und Ratswürde," 14.

64. Müller, "Die Geschäftsreisen"; Schultheiss, "Der Nürnberger Großkaufmann."

65. Weber, "Die Korrespondenzen"; Weber, "Der Briefwechsel"; Diefenbacher, *Das Nürnberger Buchgewerbe*, 449 no. 2840. On Johann Hieronymus's ongoing contacts with Nuremberg, see Will, *Commercium epistolicum*, 3:37–41.

66. Two different lists of Johann Hieronymus II's writings, both published and manuscript,

attest to the enduring appreciation of his work; see StA N, Handschriften 286, fols. 344r–347r, 348r–349r ("Herrn Johann Hieronymii Im Hofs MS. was Er in publicis pro Republica Norica als Er in der Obern Registratur Registrator war, verfertiget hatte").

67. Birken, *Tagebücher*, 2:200: "J[ohann] Jerem[ias] ImHof decollirt worden." Birken was informed about the "dieberey" on January 3, 1672, ibid., 2:83. An extensive account of this event, including a verbatim copy of the verdict, can be found in StadtA N, rep. F 1, no. 14/V, pp. 2540–42.

68. See Fleischmann, *Rat und Patriziat*, 622.

69. "demnach ich von erster jugent an aus angeporner art zw den medaylen und antiquiteten grosse naygung gehabt," quoted from Jante, *Willibald Imhoff*, 37. Budde, *Die Kunstsammlung*. On the Nuremberg collections, see Hampe, "Kunstfreunde."

70. Rosenthal, "Dürers Buchmalereien," 1–4, 41–54.

71. Jürgensen, *Bibliotheca Norica*, 1015–1309. Jürgensen, "Norimberga Literata," 473n321. Letters appear in Will, *Commercium epistolicum*, 2:57–68.

72. Christoph Jakob Imhoff, Nuremberg, to Johann Christoph Wolf, Hamburg, July 9, 1716, in UB L, Rep IX 5, no. 404. On Luther manuscripts in Nuremberg, see Will, *Bibliotheca Norica*, 3:143, citing Hirsch, preface to *Librorum*, vol. 1, unpaginated. He also had contact with Conrad Zacharius Uffenbach, the famous Frankfurt collector: see Christoph Jakob Imhoff, Nuremberg, to Christian Gottlieb Schwarz, Altdorf, September 2, 1722, BSB, Cgm 5458(1), fol. 161r; Schwarz to Uffenbach, December 23, 1722, ibid., fols. 165r–166v. A list of historical works available in Imhoff's library for Uffenbach's use follows ibid., fols. 167r–168v.

73. Koehler to Imhoff, n.d. (but either late May or June 1, as indicated by the date: "acc[eptum] d[ie] 2, Resp[onsum] d[ie] 5. Jun. 1722"), GNM, Letters, unpaginated.

74. The quotes ("Imhoffische Repositur," "zweiter Gang") come from Hirsching and Ernesti, *Historisch-literarisches Handbuch berühmter* [. . .] *Personen*, vol. 3, pt. 2, 58. Goldmann, *Geschichte der Stadtbibliothek*, 26, 69. Sauer, *642 Jahre Stadtbibliothek*, 65.

75. Koldau, *Frauen—Musik—Kultur*, 332–34.

76. See Davis, *Women on the Margins*, 145; Beuys, *Maria Sibylla Merian*, 100–103, 116–18, 204–8. Rücker and Stearn, *Maria Sibylla Merian*, 61–75. Ludwig, *Nürnberger naturgeschichtliche Malerei*. Friendship with patrician women granted Merian access to the city's elite.

77. Merian painted an entry in his *album amicorum*: Beuys, *Maria Sibylla Merian*, 120. Quote from StA N, Handschriften 286, fols. 605r–v (letter of van den Driesch to Count Nesselrode, June 1, 1716).

78. Hampe, "Kunstfreunde," 103–4.

79. For a list of Imhoffs living at a given date, see, e.g., "Namen aller anno 1733 lebenden Nürnberger Herrn Imhöfe," in GNM Nachlass Imhoff V (Mörlach-Höhenstein) no. 1, unpaginated (last page of a booklet detailing all the Imhoff "lines"). Errors in identifying the different Nuremberg family members easily occurred: see StA N, Handschriften 286, fols. 605r–v (van den Driesch to Leibniz, May 27 and June 13, 1716). For a comparable case, see Imhoff to Fabricius, July 6, 1708, KB, Ms. Thott 1227, unpaginated: Vincentius Placcius in Hamburg confused Jakob Wilhelm with Johann Hieronymus and made the genealogist a councilor of the bishop of Würzburg—a position held only by his relative.

80. Little is known about the Bari Imhoffs, yet the family seems to have played a part in the life of the Italian city. The founding of the local branch was reported in the early modern local historiography, often with references to "scritture autentiche della famiglia Incuria" that corroborated the strange-sounding details of a German ancestry; see, e.g., Beatillo, *Historia di Bari*, 207.

81. Lopes, *Portugal e a Europa*, in particular two papers by Jürgen Pohle (109–26, 243–62). See also Jakob, "Der Skandal"; Pohle, "Rivalidade e cooperação."

82. Werner, "Repräsentanten," 18–19, 27–41.

83. Koch, *Imhoff Indienfahrer.*

84. For the flow of information, see Werner, "Repräsentanten." Schultheiss, "Andreas Imhoff," 7–8. A few fragments of Imhoff's correspondence with family members—mostly unrelated to genealogy—survive in GNM, Nachlass Imhoff VII (Collection Imhoff), no. 3. There are, for instance, about forty letters from Rudolph Christian Freiherr von Imhoff to Jakob Wilhelm. The volume also contains correspondence unrelated to Jakob Wilhelm.

85. Andreas Lazarus to Jakob Wilhelm, August 7, 1687, GNM, Nachlass Imhoff VII (Collection Imhoff), no. 3, unpaginated: "le denombrement des Enfans de Mons.geur le Prince de Jean François de Nassaw."

86. Friedrich Wilhelm's grandfather was the uncle of Jakob Wilhelm's grandfather; he is mentioned several times in the Magliabechi correspondence, where the relevant file also contains a few of Friedrich Wilhelm's own letters to Magliabechi: BNC Firenze, Magl. VIII 259, nos. 17–19.

87. Carolus Volkamer, a military officer in the Bavarian army at that time (Imhoff to Volcamer, February 4, 1683, GNM, Généalogies R [vol. 19], unpaginated [Rechberg]), regularly transmitted genealogical information around 1680. See, e.g., "commun. Car. Volcamer. d. 10. Apr. 1681," GNM, Généalogies M (vol. 15), unpaginated (Maxlrain). On Peller, see, e.g., Imhoff to Muratori, September 10, 1696, in Lieber, "Jacob Wilhelm von Imhofs Korrespondenz," 48–50; and GNM, Généalogies S–T (vol. 21), unpaginated (Trivultio): "questo albore mi fù presentata dal S. Gio. Giac. Pellero, il quale l'haveva ricevuta à Milano dal Caval. Borromeo nel mese del Luglio 82." Imhoff received at least two pieces of information from Sigmund Ludwig Wurfbain concerning the counts of Berg in September 1680 and December 1682; see GNM, Généalogies B (vol. 7), unpaginated (Berg), as well as information about the Fuggers. See a remark on a Fugger-related fact sheet in GNM, Généalogies F (vol. 10), unpaginated (Fugger). See also Imhoff, "praefatio ad lectorem," *Notitia Procerum* (1st ed., 1684), fol. a5v. On Wurfbain, a convert to Catholicism and counselor to the counts of Waldburg-Zeil, see also Imhoff to Fabricius, January 10, 1714, KB, Ms. Thott 1227, unpaginated.

88. StA N, Handschriften 286, fol. 623r.

89. See Diefenbacher, "Stadt und Adel"; Pfeiffer, "Nürnberger Patriziat"; Wüst, "Patrizier," 26–29. For an impressive case study, see Bock, "Die Familiengeschichtsschreibung."

90. Haller von Hallerstein, *Dissertatio Juris Publici.*

91. Pfeiffer, "Nürnberger Patriziat," 53.

92. Hofmann, "Nobiles Norimbergenses," 83.

93. Eberle, *Das Reichsfürstentum*, 26–43, and passim.

94. See the quotation from an internal memorandum of the city council (1749) in Hirschmann, "Johann Gottfried Biedermann," 142.

95. Imhoff to Moritz Wilhelm, March 8, 1693, SächsHStA, loc. 8224, no. 7, fol. 2r ("ein schlechter Hofmann und in curialibus wenig erfahrenn").

96. For a very negative contemporary assessment, see Wurffbain, Nuremberg, to Paullini, April 3, 1688, ThULB Jena, Ms. Bud 348 f, fol. 632r: "Die Ursache, warum H. Arnold nicht in das Collegium historicum Imperiale mit ein tretten will ist eigentlich diese, weil man hier sehr ungern siehet, so man etwas de historia patriae schreibt und nichts als undanck verdienet, von außländischen sachen zu schreiben ist ebenfalls sehr schwehr, und wird leicht dieser und jener Potentat dadurch beleidiget, zu mahl wo nicht die Kay. May. zum Patron und Protectore erbetten wird, sintemaln ein oder anderer Potentat nicht gnugsamer gewallt haben würde, solche Protection in dem ganzen Reich zu extendiren." It is unclear, which member of the Wurfbain family wrote this letter.

97. See, e.g., Meyer, *Die Stadt als Thema.*

98. In the early seventeenth century, the city government ordered the city scribe ("Ratsschreiber") Johannes Müllner (1565–1634) to produce a history of Nuremberg. Müllner compiled

six volumes, one of the most important sources of the history of Nuremberg. Several patrician families, including the Imhoffs, were heavily involved in continuing Müllner's work after his death. Johann Hieronymus Imhoff II produced a continuation of the chronicle. See StA N Handschriften 44 (covering 1602–35, draft) and 45 (1633–66, clear copy), and StadtB N, Will I. 224.2° (covering 1650–1709, clear copy). See also StadtA N, Rep F1 no. 14/I–V, a chronicle of Nuremberg written by Christoph Andreas Imhoff.

99. Bock, "Die Familiengeschichtsschreibung." See also, for Frankfurt, Hansert, *Geburtsaristokratie*, 275–81.

100. Haller von Hallerstein, "Nürnberger Geschlechterbücher," 225. See also Hirschmann, "Das Geschlechterbuch." Kuhn, *Generation als Grundbegriff*.

101. Kirchhoff, *Gedächtnis*; Becker and Kirchhoff, "Neuedition und Übersetzung"; Becker and Kirchhoff, "Das Memorial."

102. Schmid, *Schreiben für Status*, 59–93, 107–16.

103. StA N, Handschriften 288, fols. 20v–21r (1679): "es ist auch hierbey nit unbillich in sonderheit anzumercken, dz demnach der Land Adel schon vor dießem sich nach und nach dießer landen in die Städt gezogen, unter andern auch die Hochadeligen Nürnbergische Familia der H. Imhof, wie bekant, sich dahin begeben, ohnangesehen dieselbe vorhero würckl. Freyherrn geweßen seyn, in welchem standt Ihnen dann auch die dermahlige Reichs Stadt Kaufbeyren zuständig war."

104. Cf. Wüst, "Patrizier," 14–17.

105. Imhoff'sches Geschlechterbuch, in GNM, Nachlass Imhoff II (Helmstädter Archivschrank), no. 1.

106. GNM, Nachlass Imhoff I (Archiv der Gesamtfamilie) fasc. 2, no. 1a, unpaginated ("Den Rathsgang der Familie betreffend").

107. Seifert, *Genealogische Tabellen*, table 14.

108. On Christoph Jakob's genealogical library, see his "Bibliotheca genealogica," in BSB, Clm 28115.

109. Examples in StA N, Handschriften 286, passim, where Jakob Wilhelm forwards questions that had arrived at his desk to Christoph Jakob. Family-related correspondence from Christoph Jakob to Johann Hieronymus (II), is copied in GNM, Nachlass Imhoff I (Archiv der Gesamtfamilie) fasc. 2, no. 1a, unpaginated. These pages are full of references to numerous archives in different locations that Christoph Jakob presumably had visited.

110. Wurffbain, *Drey unterschiedliche Summarische Bericht*. Biographical details in Diefenbacher and Endres, *Stadtlexikon*, 1205; Hirschmann, "600 Jahre Genealogie," 160. Wurffbain's mature genealogical work included publications on the Habsburg dynasty. He first made his name as a genealogical expert by researching the Nuremberg patriciate.

111. Ulmer, *Dissertatio Historica*.

112. Most of what follows comes from Sturm, *Memoriam Guilielmi Im-Hofii*. A copy of this rare text is held in StadtA N, A 31, no. 332/2. See also Feuerlein, *Drey-fach-Göttlicher Liebes-schein*, 137–82. StA N, Handschriften 286, fol. 620r.

113. Sturm, *Memoriam Guilielmi Im-Hofii*, fol. b2v: ". . . novumque quasi ingressus iter Historicum, Geographia praecipue ac Genealogia, Charitibus suis, comitibus, & ruminari quae viderat, audiverat, observaverat apud exteras gentes." On the breadth of his genealogical expertise, see Feuerlein, *Drey-fach-Göttlicher Liebes-schein*, 164: "darin [sc., in genealogy] er mit der Zeit so glücklich zunahm / daß er aller hohen Häubter in der Christen=Welt / aller Durchleuchtigsten Häuser Namen / Herkünffte / Verwandnissen und Alterthüme in den Tafeln seines Hirns / als in den ordentlichen Geschlechts=Registern aufgezeichnet hatte / und sie daraus wann er wolte / fast an einem Schnürlein oder Finger / wie wir Sprüchworts=weis reden / herzuzehlen wuste." On Imhoff's interest in historical documents, see Sturm, *Memoriam Guilielmi Im-Hofii*, fols. c3v–c4r. See also ibid., fol. [d4]r: "Viva bibliotheca & museum ambulans." Feuer-

lein, *Drey-fach-Göttlicher Liebes-schein*, 101, 163–64 ("daß er in die Cantzlei gienge / und alle in unserm Archiv aufgehäufte Urkunden / davon viele unter dem Staub und Schimmel begraben lagen / andere den Schaben und Motten zur Kost und Beute zu werden anfiengen / herfür stierte / durchstreunete").

114. On this, see Sturm, *Memoriam Guilielmi Im-Hofii*, fols. br–v.

115. Rittershausen, *Genealogiae imperatorum* (3rd ed., 1664), "preface." The other person explicitly mentioned is Spener. At least two short letters from Wilhelm Imhoff to Rittershausen, showing his enthusiasm for the professor's genealogical work (dated simply 1657 and 1658), survive in StadtA Ulm, V 154, unpaginated. Wilhelm Imhoff is also mentioned by Rittershausen to unknown, March 9, 1660, SUB GÖ, Ms. philos. 94, no. 68.

116. See the vita of Wilhelm Imhoff, published in Imhoff, *Regum Pariumque*, 232–33. A manuscript version is in StA N, Handschriften 286, fols. 354r–v.

117. "Unsere Patricii (sie wollen zwar diesen Namen durchaus nicht mehr leiden) halten nichts mehr von der Gelehrsamkeit. Olim non erat sic," wrote Georg Andreas Will, Altdorf, to Jakob Wilhelm Feuerlein, Göttingen, on February 8, 1759, WLB, Cod. hist. 4" 731, III, no. 683.

118. Kenny, *Born to Write*; Rabinovitch, *The Perraults*.

119. Pfeiffer, *Nürnberg*; Endres, "Sozial- und Bildungsstrukturen," 52–54; Bartelmess, "Die Patrizierfamilie Tucher," 226–28. For a broad survey, see Kraus, "Bürgerlicher Geist."

120. For an example of meritocratic sentiments, see a panegyric in honor of Wilhelm Imhoff, the genealogist's late father, StadtA N, A 31, no. 332/2, fols. a3v–[a4]r.

121. Imhoff, *Excellentium Familiarum in Gallia*, unpaginated (dedication to the Senate of Nuremberg): ". . . Ciceronis monitum, *ne quis in aliena gente ac republica nimis curiosus esse velit*." Imhoff is misquoting Cicero Cic. *De Off.* 1.125, in which Cicero states, "Peregrini autem atque incolae officium est . . . minime . . . esse in aliena re publica curiosum" (It is the duty of a foreigner and a resident alien . . . not to concern himself with another's state). He did not want to offend local sensibilities, though, and hence he almost apologized in his dedication for his curiosity about non-German families.

Chapter 3 · Genealogy and the Nobility

1. Schnabel, *Österreichische Exulanten*.

2. Jürgensen, *Utile cum dulci*, 199.

3. GNM, Généalogies S–T (vol. 21), unpaginated (Stubenberg). See also [Imhoff], *Spicilegium Rittershusianum*, unpaginated (preface). On Stubenberg's scholarly ambitions, see Schnabel, *Österreichische Exulanten*, 661. On his connections to the Blumenorden, see Stauffer, *Sigmund von Birken*, index. Andreas Lazarus to Jakob Wilhelm, June 5, 1687, mentions a personal exchange with Stubenberg (in Sulzbach?).

4. Imhoff to Leibniz, August 4, 1705, AA I, 25:10: "veteri fautore meo." Welz also came from an exiled family that had settled in nearby Ulm but had connections to Altdorf and Nuremberg. Welz acted as purveyor of information at least once, in late 1684; see "Ex lit. Mauritii Bar. a Freyberg ad Gotthardum Helfriedum L.B. a Welz, Ratisbona d. 15/25 Dec 1684," in GNM, Généalogies F (vol. 10), unpaginated (Freyberg). Welz is frequently mentioned in Imhoff's correspondence with Moritz Wilhelm; see, e.g., Imhoff to Moritz Wilhelm, January 31, 1699, and October 11, 1706, SächsHStA, loc. 8224, no. 10, fols. 57v, 53r.

5. I am expanding here on an argument about journals made in Schuwey, "Les périodiques."

6. Imhoff characterizes himself as a "homo umbraticus" in contrast to a knowledgeable courtier; Imhoff to Moritz Wilhelm, December 28, 1708, SächsHStA, loc. 8224, no. 8, fols. 90r–v.

7. Ministers and Counselors of Count Philipp Ernst von Hohenlohe-Waldenburg-Schillingsfürst, Wilhermsdorf, to their colleagues of Hohenlohe-Lauenstein, September 13, 1714, Hohenlohe, Zentralarchiv, Neuenstein, La 20, Bü 90, unpaginated. Friedrich Eberhard, Kirchberg, to

unknown (unspecified "Grafen"), November 13, 1724, Hohenlohe, Zentralarchiv, Neuenstein, Sf 25, Bü 220, unpaginated (no. 2). Another copy is in Hohenlohe, Zentralarchiv, Neuenstein, Wa 60 Bü 771. While many courts were well informed about Imhoff's plans, there were also false rumors: The counts of Nassau and the counts of Hohenlohe expected a fifth edition in 1714, but Imhoff firmly declined to take on this project; see, e.g., Imhoff to Moritz Wilhelm, March 30, 1715, SächsHStA, loc. 8224, no. 9, fols. 30v–31r. See also HHStA 130/I/10. Scholarly readers also waited for updates; see, e.g., Behrens to Leibniz, September 4, 1716, communicating details that should be changed "bey einer newen edition des Imhofii." StA BA, Markgrafentum Brandenburg-Bayreuth, Geheimes Archiv, 4676.

8. The Pappenheim family, for instance, informed Imhoff about their current dynastic situation; see GNM, Généalogies PAL–PAP (vol. 17), unpaginated (Pappenheim): "1. Herrn Graff Marquard von Pappenheim in d. Notitia angezeigte tochter ist noch im leben. seine andere Gemahlin mit welcher er sich zu anfang dieses jahrs vermählet, heißt: Maria Adelh. von Haßlang." This new piece of information was included in Imhoff, *Notitia Procerum* (3rd ed., 1693), 377. Imhoff received additional information on Marquard's children in two letters: September 22/ October 2, 1692; and October 6/16, 1692, GNM, Généalogies PAL–PAP (vol. 17), unpaginated (Pappenheim). This information also made it into the printed version. See also a series of "Notanda ad ea quae de familia Promniziana in appendice notitiae proceri Imperii dicuntur," that was "acc[eptum] in Com. Frid. Ern. d. Salms m Aug. 1698," in GNM, Généalogies PAR–PZ (vol. 18), unpaginated (Promniziana), referring to Imhoff, *Notitia Procerum* (3rd ed., 1693), 495–96. In the following edition (4th ed., 1699), however, this particular section, an appendix, no longer appears.

9. Compare different passages on Count Georg Friedrich (1625–88) in Imhoff, *Notitia Procerum* (2nd ed., 1687), 578, and Imhoff, *Notitia Procerum* (3rd ed., 1693), 341–42. Imhoff's revisions were based on a manuscript list of children provided by the family in 1692, in GNM, Généalogies S–T (vol. 21), unpaginated (Solms). Similarly, regarding Count Friedrich Sigismund, cf. Imhoff, *Notitia Procerum* (2nd ed., 1687), 579 and Imhoff, *Notitia Procerum* (3rd ed., 1693), 343. Imhoff also emended his description of the family crest.

10. This emerges from numerous, albeit unspecified, references to him in Hohenlohe, Zentralarchiv, Neuenstein, La 20 Bü 90, unpaginated. His special dedication to the imperial counts and their constitutional position in the Holy Roman Empire is mentioned explicitly in Solms-Laubach, *Geschichte*, 351.

11. Holtacker, *Abregé historique*, 11. I have been unable, however, to identify any pertinent passages in Imhoff, *Excellentium Familiarum in Gallia*. Holtacker's 1693 book was an updated and newly translated French version of an older genealogical work from 1680. Holtacker's short history appeared not only in German (1680) and French (1693), but also in Latin (1704). The family's most obvious genealogical problem lay in the fact that the family Lynden, of ancient Dutch nobility, did not de facto issue from the family Aspremont. It was the (fabled) Aspremont connection, however, that had propelled the family to the rank of imperial count in 1676.

12. The copy is StadtB N, Hist 337.2°, in which a handwritten note records: "Ex dono Com. Ferdinandj Goperti de Reckheim acc. an 1693," and that is partly hand-colored.

13. In 1684, he owned neither the earlier Latin nor the German version of the book; see Imhoff to Thulemeyer, January 28, 1684, UB F, Ms Thulemeyer, 723–25. He mentions it only vaguely in 1687: Imhoff, *Notitia Procerum* (2nd ed., 1687), 784.

14. Imhoff, *Notitia Procerum* (3rd ed., 1693), 463. Imhoff, *Notitia Procerum* (4th ed., 1699), 623.

15. Imhoff, *Notitia Procerum* (4th ed., 1699), 624. The doubts may have been prompted by a letter from d'Hozier to Imhoff, July 6, 1696, GNM, Letters, unpaginated, who criticized the chapter on the family in the third edition of the *Notitia Procerum*.

16. The Schwarzenbergs' involvement with Imhoff is but one part of the family's intense

genealogical activities in the later seventeenth century, many of which centered on Nuremberg. I tell the story surrounding the Imhoff connection more fully in Friedrich, "House of Schwarzenberg." On this important family, see Wüst, "Die Schwarzenberg"; Bérenger, "Les Schwarzenberg."

17. Count Ferdinand to his counselors, January 5, 1697, StA N, Herrschaft Schwarzenberg: Schwarzenberger Archiv 16/9, unpaginated. All the following details come from this file.

18. The intermediary was Dr. Christoph Peller von und zu Schoppershof (1630–1711). Cf. counselors to Peller, February 5, 1697; counselors to Count Schwarzenberg, February 6, 1697.

19. Peller to counselors, August 3/13, 1697.

20. GNM, Nachlass Imhoff VII (Collection Imhoff), no. 20, unfol. (ledger "Schwartzenberg"). StA N, Herrschaft Schwarzenberg: Schwarzenberger Archiv 16/7, unpaginated (October 5, 1701: the count finds further mistakes in Imhoff's version).

21. Aurich, Staatsarchiv, Rep 241 Msc, C 13. This refers to the first edition of the *Notitia* in 1684.

22. Marburg, Staatsarchiv, 81A, 14/6.

23. See, e.g,, an undated note from the von Galen family, in GNM, Généalogies G (vol. 11), unpaginated (Galen): "Linea vel schema Genealogicum Nostrorum DD Baronum de Galen, in libro, qui inscribitur Notitia Procerum Imperij etc. non poterit alio loco congruentius collocari, quam inter Proceres tractus Wetteravici, cum isthic hodie possideant bona Imperij immediata. Et quidem commodissime iuxta ordinem Alphabeticum, quem Author servat, inter Barones de Fleckenstein et Comites ab Hanau, id est, inter F et H. ac tunc ne quid in multis sit opus mutare, numerus paginae 295 296 297 et 298 bis repetatur, vel prouti id D. Author meliore modo fieri posse iudicaverit."

24. Metternich, Regensburg, to Imhoff, November 16/26, 1691, in GNM, Généalogies M (vol. 15), unpaginated (Metternich). The Metternichs were concerned about the relationship between different branches of the dynasty. Imhoff's informant specified in detail what the genealogist could and could not mention: should Imhoff choose to include the Protestant Chursdorf branch of the family residing in Brandenburg, he was forbidden to make any (explicit) use of the material he had received.

25. The following is taken from the small file WLA, A 274 Bü 63c.

26. It is unclear if the treatment in Imhoff, *Notitia Procerum* (5th ed., ed. Köhler), 311–12, represents a revised state. In any case, the passage is not sensational, although it presents the count's complicated private life in a matter-of-fact style and does not gloss over embarrassing details.

27. On Moser, see Schömbs, *Das Staatsrecht*. On Moser and Mömpelgard, see Stein, *Johann Jakob Mosers*.

28. On this, see Somsen, "Prussia's Franconian Undertaking."

29. The detail under consideration was Imhoff's account of the history of the burgraves of Nuremberg, a politically fraught question. See StA BA, Markgrafentum Brandenburg-Bayreuth, Geheimes Archiv 4676. The criticism was directed at Imhoff, *Notitia Procerum* (4th ed., 1699), 198–202. In the fifth edition of 1732, Imhoff (and Köhler) added material on the generation between 1699 and 1732, but he did not change the overall political thrust of the text. See Imhoff, *Notitia Procerum* (5th ed., 1732), 267–71.

30. Aretin, *Das Reich*, 93–94, 123, 269–75.

31. Giustiniani, *Memorie istoriche*, 2:114.

32. See, e.g., the prince's portrait and the dedicatory epistle in Giannettasio, *Bellica*.

33. Imhoff, *Genealogiae Viginti Illustrium in Italia Familiarum*, 3.

34. Sant'Anna to the prince of Cellamare, undated, BSB, Cod. Ital. 506, fols. 64r–67v, a copy attached to Sant'Anna to Imhoff, January 24, 1701, ibid., fols. 57r–59v. The quote is on fols. 66r–v: "non sola una tromba Sonora, ma una chiara, e vera attestazione . . . in commendazione del Principe."

35. Sant'Anna to Imhoff, September 14, 1700, BSB, Cod. Ital. 506, fols. 53r–54v, here 53r: "acciò possa V.S.Ill.ma colla sua eruditissima penna dargli maggiore splendore."

36. Further examples can be found: In Tuscany, Magliabechi made contact with the prince of Butera, Filippo Strozzi, and Count Giovanni Battista Diana Paleologo, who worked for the house of Cybo in Massa. These nobles then compiled information about their families or peers and had it sent to Nuremberg, often via Florence. See Magliabechi to Imhoff, March 26, May 14 and August 4, 1692, GNM, Letters, unpaginated. In France, d'Hozier engaged the Abbé Salviati; see d'Hozier to Imhoff, September 21, 1708, GNM, Letters, unpaginated.

37. Sant'Anna to Imhoff, May 26, 1696, BSB, Cod. Ital. 506, fols. 13r–v, here 13r: "nella stampa farvi particolar menzione della sua persona." Sant'Anna to Imhoff, September 21, 1700, ibid., fols. 51r–52r, here fol. 51r: "fare particolare menzione." Antonio was mentioned in Imhoff, *Genealogiae Viginti Illustrium in Italia Familiarum*, 73.

38. The passage on Michele de Giudice in Imhoff, *Genealogiae Viginti Illustrium in Italia Familiarum*, 73, reads like a summary of a longer, original passage in Sant'Anna to Imhoff, November 18, 1697. Sant'Anna expressly mentioned to Imhoff that Michele "merita . . . particolare menzione."

39. Sant'Anna to Imhoff, March 22, 1700, BSB, Cod. Ital. 506, fol. 34r: "farne menzione."

40. *Dizionario biografico degli italiani*, http://www.treccani.it/enciclopedia/carmine-nicola-caracciolo_(Dizionario-Biografico). Scarano, "Un principe napoletano vicerè di Perù"; Suppa, *L'accademia di Medinacoeli*, 122–30.

41. Bulifon, *Giornali di Napoli*, 189.

42. Imhoff, *Corpus Historiae Genealogicae*, 282.

43. On attempts to incorporate (fictitious) saints into genealogical visions, see Tutino, *Fake Saint*.

44. On Thomas's mother, see Torrell, *Saint Thomas Aquinas*, 1–3. Santobuono's memorial is, unfortunately, no longer extant, but see, e.g., the marginal annotation by Imhoff: "Il Pr. di S. Buono la dice essere stata di casa Caracciola nella sua lettera MSa p. 31," in Campanile, *Dell'armi overo insegne dei nobili*, 237 (StadtB N, Hist. 302.2°). In print: Imhof, *Corpus Historiae Genealogicae Italiae et Hispaniae*, 83. Imhoff also incorporated material from Sant'Anna about the saint's parents, in particular a reference to recent intra-Dominican debates about the topic; compare Imhoff's printed text with Sant'Anna to Magliabechi, March 12, 1696, in Quondam and Rak, *Lettere*, 619.

45. Sant'Anna to Imhoff, May 26, 1696, BSB, Cod. Ital. 506, fol. 13r: "li sono piacivute assai le notizie della Famiglia Giudice per le quali ella [sc., Imhoff] si professa anco obbligato."

46. Sant'Anna to Imhoff, November 18, 1697, BSB, Cod. Ital. 506, fols. 30r–31r, here 31r: "voglia far grazia avisarmi haverle ricevute, acciò possa mostrare la medesima sua lettera al d.o S.r Principe di Cellamare."

47. Sant'Anna to Imhoff, July 20, 1700, BSB, Cod. Ital. 506, fols. 42r–43r, here 43r: "Confesso la verità a V.S. Ill.ma, che il d.o S.r Principe merita ogni ossequio, et attenzione, cosi per la sua vari qualità, e talenti, come anco per la sua gran nascita, e parentato, essendo suoi congionti e parenti il S.r Duca di Modena, il S.r Duca di Parma et il S.r Duca della Mirandola la sorella del quale fu sua moglie." Tommaso d'Aquino was married to Fulvia Pico (1663–91), daughter of the prince of Mirandola. Cf. ibid.: "verità historica deve farsi cosi," "condescendere alli sentimenti del sud.o S.r Principe."

48. D'Hozier to Imhoff, March 17, 1695, GNM, Letters, unpaginated, where d'Hozier mentions several proactive transmissions of material from major French noble houses, just in case Imhoff would ever plan to produce a second work on France. No such work was planned at the time. D'Hozier to Imhoff, April 6, 1696, ibid.: French nobles point Imhoff to their families and insist that they were "non pas indigne de votre recherche."

49. Sant'Anna to Imhoff, November 26, 1707, BSB, Cod. Ital. 506, fols. 109r–110r, here 109v:

"Tengo pronte tutte le notizie della detta famiglia Saracena quale non solo è molto nobile, ma anco assai antica, e stimata, se VS.Illma vuole onorarla nella sua opera li manderò le sudette notizie e con ogni confidenza li dico, che questi Signori concorreranno anco alla spesa della stampa."

50. There were a few moderately critical voices heard in southern Italy: e.g., Sant'Anna to Imhoff, July 30, 1703, BSB, Cod. Ital. 506, fols. 95r–96v. Other families also felt that Imhoff had "mistreated" (*maltrattato*) their histories; see Sant'Anna to Imhoff, July 9, 1703, ibid., fols. 97r–98v, on Giulio Antonio Acquaviva.

51. On the prince d'Aquino, see Niccolini, "La fine del dominio spagnuolo," 144–80. On the prince and his mother, cf. also the remarks in Mancino, "I 'Panni ricamati.'" The idea of the Lombard descent of the d'Aquino family was popular at the time; cf. Giannettasio, *Bellica*, unpaginated (preface to Tommaso d'Aquino).

52. The family is first mentioned by Sant'Anna to Imhoff, October 25, 1695, BSB, Cod. Ital. 506, fols. 3r–4v. They do not seem to have pursued the subject, since Aquino does not reappear until 1700, in Sant'Anna to Imhoff, July 12, 1700, ibid., fols. 40r–41v. Bulifon had botched the dispatch of information to Nuremberg. When the Aquino dossier finally left Naples (Sant'Anna to Imhoff, July 20, 1700, ibid., fol. 42r–43r), it was too late for inclusion in Imhoff, *Historia Italiae et Hispaniae Genealogica*.

53. All of these come from one questionnaire, sent back from the counts of Nassau on June 30, 1692, in GNM, Généalogies N–O (vol. 16), unpaginated (Nassau).

54. See, e.g., several questionnaires for the counts of Nassau, in GNM, Généalogies N–O (vol. 16), unpaginated (Nassau).

55. The count, for instance, corrected the figure of his sister-in-law's heritage (Imhoff had "40m fl," but the correct figure was "80m fl"); cf. Imhoff, *Notitia Procerum* (3rd ed., 1693), 386, and Imhoff, *Notitia Procerum* (4th ed., 1699), 521.

56. See Councilors to Count Ferdinand Wilhelm Euseb, November 2, 1701, StA N, Herrschaft Schwarzenberg: Schwarzenberger Archiv 16/9, unpaginated, including a summary of a letter of [Johann Kaspar] von Mohr, Nuremberg, to the Councelors, October 28, 1701, ibid. Von Mohr had talked to Imhoff in Nuremberg and had conveyed the count's dissatisfaction.

57. Around 1530, the Alsatian noble Wilhelm von Rappoltstein (1468–1547) could still rudely rebuff a request for information from the genealogically inclined author Wilhelm Wernher von Zimmern: "[Rappoltstein] took it [sc., von Zimmern's request for information] so badly that in reply he only asked [Zimmern] why he [i.e., Zimmern] cared about his [i.e., Rappoltstein's] family, since [Rappoltstein] was not interested in learning about [Zimmern's] family." See Götz, *Wege und Irrwege*, 79–80.

58. D'Hozier to Imhoff, December 18, 1695, GNM, Letters, unpaginated: "Si vous voules une table généalogique de la Maison du Baufflers, et que vous voulies la faire Imprimer je vous l'enverrai aussi."

59. Imhoff to Moritz Wilhelm, February 4, 1697, SächsHStA, loc. 8224, no. 7, fol. 46v.

60. Imhoff to Thulemeyer, April 24, 1693, UB F, Ms Thulemeyer, 761–63, here 762: "habe ich für rathsam befunden, solches blat umdrucken zu lassen, u. zu trachten, wie es denen buchhändlern nicht allein hier, sondern auch anderer Orten in die hände komme, damit sie es denen exemplarien, vor weiteres distraction beylegen, u. das alte durchschneiden mögen."

61. Compare Imhoff, *Notitia Procerum* (4th ed., 1699), 398, with Marburg, Staatsarchiv, 81A, 14/6, unpaginated (final clear copy of the first part of the file pertaining to Imhoff). The most significant changes occurred at the beginning of the chapter about Hanau in nos. 1–6 of the new version. There, Imhoff largely follows the suggestions documented in the Marburg file, although not verbatim. He also shortens and rephrases the wording of the draft, omitting additional examples and details. In particular, he cuts an allusion to a Roman ancestor right at the beginning (which had been put forward only reluctantly in the counts' draft anyway). The ref-

erence to France is in Imhoff, *Notitia Procerum* (3rd ed., 1693), 296 (last sentence of no. 5). A marginal note in the manuscript states that "this could be left out" (dies könde gehen usgelassen werden).

62. Pieper, *Einheit im Konflikt*, 48–53; Arndt, *Das niederrheinisch-westfälische Reichsgrafenkollegium*.

63. Imhoff generally did not discuss the lowest rank of the imperial aristocracy, the 350–500 free imperial knights (*Reichsritter*); see, e.g., Godsey, *Nobles and Nation*. Since many families among the imperial counts had family ties to free imperial knights, a handful of the latter are mentioned at least briefly in the *Notitia Procerum*, e.g., in the chapter on the counts of Metternich, originally a "knightly" house. Imperial knights also occasionally appear when one is elected as archbishop of an imperial diocese.

64. Quotations from Metternich, Regensburg, to Imhoff, November 16/26, 1691, in GNM, Généalogies M (vol. 15), unpaginated (Metternich), and Luise Katharina von Rautter, Countess of Truchsess, Königsberg, to Imhoff, May 15, 1703, GNM, Généalogies S–T (vol. 21), unpaginated (Truchsess).

65. Friedrich Eberhard, Kirchberg, to unspecified "Grafen," November 13, 1724, Hohenlohe, Zentralarchiv, Neuenstein, Sf 25 Bü 220, unpaginated (no. 2): Cooperation would "zu dem Lustre des ReichsGrafen Stands unter anderm worauß mit gehöret, daß man dessen alte Häuser in solchen Genealogischen büchern nebst deren mit Königl. = Chur= und Fürstl. Häusern getroffenen Allianzen jedesmahlen exacte beschrieben finde, hingegen zu beklagen ist, daß bishero hin undt wieder in diesem Stuck zu sehr an sich gehalten undt denen jehnigen so dergleichen arbeit unternohmen auß gantz ohnbegreifflichen ursachen nicht nothdurfftig an hand gegangen worden."

66. At least occasionally, Imhoff was cited together with Schmid, *Die Durchläuchtige Welt*. See Luise Katharina von Rautter, Countess Truchsess, to Imhoff, May 18, 1703, GNM, Généalogies S–T (vol. 21), unpaginated (Truchsess). Another authority was Pistorius, *Historisch- und juridische Anmerkungen*.

67. Forwarded by a court official from Giessen, Ernst Christoph Köhler, to Imhoff, June 6 and July 2, 1698, in GNM, Généalogies J–K (vol. 13), unpaginated (Kirchberg). The new chapter was added at the very end of the volume. See Imhoff, *Notitia Procerum* (4th ed., 1699), 643–47.

68. Imhoff, *Notitia Procerum* (4th ed., 1699), 643. Ibid., 646, contains a brief discussion of the very complicated recent ownership situation in Sayn-Hachenburg; the paragraph was presumably based on a manuscript German summary provided to Imhoff (probably) by Köhler. See GNM, Généalogies J–K (vol. 13), unfol. (Kirchberg).

69. This scenario did not come about until 1715: Czech, *Legitimation*, 270–78 (chapter on "Der Wiederaufstieg der Burggrafen von Kirchberg") with Avemann, *Vollständige Beschreibung*. Avemann's name appears as an authoritative source in the documents cited above several times.

70. See the dossier in GNM, Généalogies G (vol. 11), unpaginated (Gahlen).

71. Imhoff to Thülemeyer, September 5, 1698, UB F, Ms. Thulemeyer, 777–79, here 778: "Mhherr hat mir vor etlichen Jahren ein scriptum von der familia der Herren von Gahlen communicirt, daraus ich gesehen, daß selbige Herren verlangen, es der Notitiae Proc. Imp. unter denen Wetterauischen Grafen einzuverleiben; ich habe nun bey der unter der Preß stehenden neuen edition dieses buches zwar reflexion darauf gemacht, aber nothwendig befunden mich in comitiis zu erkundigen, ob Sie auf solcher Grafenbanck sessionem & votum haben; weil ich nun zur Antwort bekommen, quoad non, u daß ihre angegebene immediate Guter wol in der Ritterschafftlichen matricul sein mögen, aber in der Gräflichen seyen sie nicht, werden mich hoffentlich dieselbe Herren nicht verdencken, wann ich illo loco von Ihnen abstrahire, u mir mit selben hochgel. Collegio keine affaire mache. Ich werde aber unter denen Herrn Hoffräthe, darunter ein Herr von Gahlen ist, Gelegenheit nehmen von dieser familia etwas u. so viel sich thun lässet zu melden."

72. In August 1684, Franz, Count of Haunsberg, provided information about recent developments in his family, which the genealogist had actively solicited—but no chapter on the house of Haunsberg ever appeared in Imhoff's printed works. See the remark, "Notas istas Franciscus C. de Haunsperg manu propria adscripsit Aug 1684 Monachii," on a neatly drawn genealogical table on which Imhoff had added specific questions about individual family members in red ink. See GNM, Généalogies H (vol. 12), unpaginated (Haunsperg).

73. Hoffmann, Wetzlar, to Imhoff, May 21, 1708, GNM, Généalogies S–T (vol. 21), unpaginated (Virmond), writing on behalf of Count Ambrosius Franciscus Fridericus Christianus Adelbertus von Virmond (1684–1744): he requested an "eigenes Capitel." The fifth edition of the *Notitia Procerum*, edited by Köhler in 1732, does not contain a chapter on the family.

74. Imhoff to Thulemeyer, March 11, 1692, UB F, Ms. Thulemeyer, 749–52, here 750: "Von der neuen Fürsten von Nassau sessione & voto in Collegio Principum weis man zu Regenspurg noch nichts. u. weil ich in ultima revisione operis mei noch nicht dahin kommen, habe ich noch nicht resolvirt, wie ich es mit ihme angreife." Arndt, *Das niederrheinisch-westfälische Reichsgrafenkollegium*, 105–6.

75. In 1687, the Nassau family was still split across two sections, V, 6 (princes) and VI, 11 (counts). In 1693, all the branches of the family were discussed in section V, 6 (princes), although Imhoff still alluded to some lingering problems; Imhoff, *Notitia Procerum* (3rd ed., 1693), 241.

76. Imhoff, *Historia Italiae et Hispaniae Genealogica*, 4: "fabulosis . . . initiis." For the medieval developments of this myth, see Busch, *Die Mailänder Geschichtsschreibung*, 225–33.

77. Imhoff, *Notitia Procerum* (4th ed., 1699), 644: "Haud tamen intercidit memoria Comitum Burggraviorumque Kirchbergensium, qui ante Albertum [sc., before ca. 1400] vixere, sed superest omnino, ut continuae tantum seriei filus desideretur."

78. Imhoff, *Excellentium Familiarum in Gallia*, 89–91. On the context, see Fouilleron, "Français par le livre," 623–25. Many other contemporaries also still believed the traditional story about Grimaldus and Otto.

79. Muratori to Imhoff, March 3, 1697, in Busino and Dufour, "Quattordici lettere," 422; Muratori, *Anecdota*, 1:243–48. Gittio also adhered to the Desiderius story; see GNM, Généalogies M (vol. 15), unpaginated (Milan/Visconti).

80. The section on the Visconti is in Imhoff, *Historia Italiae et Hispaniae Genealogica*, 158–88. He repeated the story in his 1710 book on Italy, printing two supporting letters from one Daniel Milan-Visconti of The Hague, who claimed dynastic ties to the Italian Visconti. In these letters, Milan-Visconti rehearsed and refined Imhoff's earlier position on the medieval origins of the families under consideration; cf. the comment in Imhoff, *Genealogiae Viginti Illustrium in Italia Familiarum*, fols. [*4]r–v: "sicque comiter me levat errore, in quem adductus fueram." On Imhoff's subsequent attitude toward Eckhart, see Imhoff to Fabricius, December 27, 1709, KB, Ms Thott 1227, unpaginated. D'Hozier (to Imhoff, July 25 and November 25, 1710, GNM, Letters, unpaginated) criticized this endorsement of Milan-Visconti.

81. Eckhart, *Genuinum Stemma*. Bizzocchi, *Généalogies fabuleuses*, 84–85. Erdner, "Plagiat an Leibniz?"

82. HHStA, 130/I/10, unpaginated (internal memorandum). Imhoff's version was, according to inner government circles, no longer up to date. The counts instead based their alternative reconstruction of these ancient times on a local author's *Opusculo de Originibus Domus Nassoviae*, directly refuting Imhoff; the counselors obviously hoped to avoid embarrassment on account of Imhoff's seemingly outdated origin story.

83. Eckhart, *Genuinum Stemma*, fol. [A4]r: "Interea tamen optandum esset etiam atque etiam, ut tandem aliquando de Genealogico studio promovendo cogitetur, &, quod deerat, *Systema integrum*, ad veterum monumentorum amussim, & rei veritatem, rejectis puerilibus fabulis, elaboraretur."

84. Cf. once again Duindam, *Dynasty*; Duindam, *Dynasties*; Pieper, *Einheit im Konflikt*.

85. GNM, Généalogies F (vol. 10), unpaginated (Falkenstein). Imhoff, *Notitia Procerum* (2nd ed., 1687), 498. The printed stemma in Imhoff, *Spicilegium Rittershusianum*, distinguishes the "linea Brucheana, linea Falckensteiniana, linea Obersteiniana" much less explicitly.

86. These were printed tables from the *Tübinger Tafeln* (Anonymous, *XIV Tabulae Genealogicae*), unpaginated. Compare with the annotated version in GNM, Généalogies PAR–PZ (vol. 18), unpaginated (Pfalz).

87. See, e.g., Imhoff, *Notitia Procerum* (3d ed., 1693), 233 (Dietrichstein), 427 (Nostiz), and passim.

88. Examples abound, e.g., Imhoff, *Genealogiae Viginti Illustrium in Italia Familiarum*, 36–40 (Birago family).

89. Imhoff, *Genealogiae Viginti Illustrium in Italia Familiarum*, 102–19.

90. Imhoff, *Corpus Historiae Genealogicae*, 304–5, 310–11, 338. For a bibliography of texts, see d'Afflitto, *Memorie degli scrittori*, 267–68. Giacinto Falletti, a nobleman himself, actually acquired the title of "vice principe dei Carafa della Spina principi di Roccella"; hence he was promoting his own agenda.

91. See the "Tabla genealogica de los Condes de Munzanto marqueses de Cascaes," in GNM, Généalogies PAR–PZ (vol. 18), unpaginated (Portugal). This sheet bears the remark in the upper right corner: "Acc Mense Martio an. 1698 Lutetia Parisionem à Legato Lusitanico extremo loco nominato opera Dom. Car. Hozierii." The family and the dispatch of the table are mentioned in Imhoff, *Historia Italiae et Hispaniae Genealogica*, 96, 100–101 ("narrandi deficit argumentum").

92. Imhoff, *Genealogiae Viginti Illustrium in Hispania Familiarum*, 198, 210. On the count, see Rodríguez Villa, *Don Diego Hurtado de Mendoza y Sandoval.*

93. Imhoff, *Genealogiae Viginti Illustrium in Italia Familiarum*, 137: "Perfuncti jam ennarratione stirpis regnatricis in stemmate Mediceo, ad reliquos ejus progredimur ramos, minus negocii stylo nostro exhibituros."

94. E.g., Imhoff, *Notitia Procerum* (3rd ed., 1693), 427.

95. Imhoff, *Genealogiae Viginti Illustrium in Hispania Familiarum*, 108–9, 124 ("narrandi argumentum deficit").

96. Luise Katharina von Rautter, Countess Truchsess, to Imhoff, May 18, 1703, in GNM, Généalogies S–T (vol. 21), unpaginated (Truchsess). The family used Imhoff, *Notitia Procerum* (3rd ed., 1693), 395.

97. Ott, *Onomatologia*, fol. B3v: "Mutant [nomina] hodie vel ob infamiam stirpis, vel propter haereditatem adeundam."

98. Aldimari, *Memorie istoriche*, 541, 553, 555.

99. Sant'Anna to Imhoff, April 17, 1702, BSB, Cod. Ital. 506, fols. 89r–90v: "vero ramo," "non sono della vera famiglia d'Aquino." Imhoff, *Corpus Historiae Genealogicae*, 93–94. Receipt of an important dossier providing (allegedly correct) information about the Cosenza-Aquino had not yet been acknowledged by Imhoff by the end of 1702, when Imhoff's book went to print; Sant'Anna to Imhoff, December 18, 1702, ibid., fols. 91r–92r. By mid-1703, the printed book was in Naples, Sant'Anna to Imhoff, July 9, 1703, ibid., fols. 97r–98v.

100. Sant'Anna to Imhoff, April 30, 1705; December (?) 14, 1705; March 30, 1706 (reprinted folios have arrived), BSB, Cod. Ital. 506, fols. 101r–102r, 103r–v, 105r–106r.

101. Benvenuti to Imhoff, March 15, 1692, GNM, Letters, unpaginated.

102. Ibid.

103. Balbín to Imhoff, ca. September 1684, GNM, Letters, unpaginated: "difficilis erit confectio Genealogica ob eorum [sc., individuorum] multitudinem."

104. Balbín to Imhoff, ca. September 1684, GNM, Letters, unpaginated: "Innumerabiles sunt hujusmodi faeminae in Genealogijs, quarum Parentes ignorantur." It is unclear from his letter what kind of woman Katharina Slavatin ("hujusmodi") was.

105. One Mr. Pommer, as cited by Fabricius to Imhoff, April 5, 1684, GNM, Letters, unpaginated: "2 Töchter, nach deren Namen mich nicht informiret, weiln vermuthet, nicht zu importieren."

106. GNM, Généalogies L (vol. 14), unpaginated (Lippe). The list concerns only offspring of the count's first marriage (with Anna Katharina of Nassau). The information on the count's offspring comes from Wikipedia, https://de.wikipedia.org/wiki/Simon_VII._%28Lippe%29 (accessed January 7, 2022).

107. Ibid.: "dise drey freulein sind auch ohne zweifel in zarter kindheit verblich[en], zumahlen ich mir nicht erinnern kan, daß ich sie unter den freulein (: so noch unter 15 undt 20 jahren abgemahlet uff dem residentz-schlosse Detmold vor 4 jahren besichtiget, auch genau betrachtet undt zur nachricht auffgeschrieben:) solte wahrgenommen haben. Solten sie aber ad pubertatem kommen seyn, woran doch zweifele, so berichte doch vor gewiß, daß sie niemahlen verheiratet gewesen."

108. "Man erinnert sich nicht, daß H. Graff Simon solte einen Johann Ludwig gehabt haben." Similarly, Friedrich Philipp (1619–29) and Simon (1620–24), also mentioned on the original sheet, had dropped out of memory, as Thulemeyer's commentary indicated: "Von diesen beiden weiß man auch nichts."

109. On parental attitudes toward children and early modern emotional regimes of family life, see, e.g., Ben-Amos, "Reciprocal Bonding." On other occasions Imhoff's dossiers include detailed information about children who had died at an early age; see, e.g., a fact sheet regarding the offspring of Count Johann Ernst of Nassau-Weilburg (1664–1719), in GNM, Généalogies N–O (vol. 16), unpaginated (Nassau).

110. Wurffbain, *Drey unterschiedliche Summarische Bericht*, fol. Niiij v. See also StA N, Handschriften 284, a genealogy of the Imhoff family with—at best—rudimentary information about (young) children.

111. Juncker to Imhoff, April 8, 1700, GNM, HA V. Historiker Juncker, unpaginated.

112. Wagner to Imhoff, January 23, 1688, BSB, Autogr. Wagner. At that time, the current English king, James II, still had only one surviving child—Mary II, wife of William III of Orange. The last two children of James II, both surviving, were born later: James Francis Edward (b. June 10, 1688) and Louisa Maria Theresa (b. 1692).

113. See Spener, *Sylloge* (2nd ed., 1677), fol. ¶¶2r.

114. GWLB, Ms VIII 634, fol. 2r: "Nec minus foeminae. quoniam et illae in multis [?] successionis capaces. exempla dantur plurima in Imperio, et in domo nostra. Et . . . affinitates etiam faciunt ad lucem familiarum."

115. See Imhoff, *Notitia Procerum* (3rd ed., 1693), 441.

116. The following discussion relies on Rabinovitch, *The Perraults*, 16–18; and Chappey, *Ordres et désordres biographiques*. Chappey provides numerous examples for the following discussion.

117. There is only one entry, postdating Imhoff, under "noblesse" (and no entry under "généalogie") in Durey de Noinville, *Table alphabétique des dictionnaires*, 78, and that is Aubert de La Chesnaye Des Bois, *Dictionnaire généalogique*. Chappey, *Ordres*, 101–5, highlights Moréri, *Le grand dictionaire* [*sic*] as a partial surrogate.

118. Imhoff, *Genealogiae Viginti in Italia Familiarum*, 31.

119. Imhoff, *Genealogiae Viginti in Italia Familiarum*, 172–73. His literary activities appear only in passing.

120. Imhoff, *Genealogiae Viginti in Italia Familiarum*, 237–40. On Bulifon printing Colonna, see Crivelli, "Print Tradition," 128–29.

121. Imhoff, *Corpus Historiae Genealogicae*, 57–58.

122. A servant from the Acquaviva family had raised his hand against someone from the Carafa camp after catching the latter hunting on Acquaviva land. The Duke of Nola had the

servant in question apprehended, mutilated by having his ears and nose cut off, and sent him back to the Acquaviva family. Giulio Antonio then took revenge by leading three hundred men to the castle of Nola, rousing the duke out of bed—where he was lying with his wife, of Acquaviva origin no less—and cutting of the duke's nose and ears in turn. At this point, only a duel could settle the affair. For further details, see Sforza, *Giovanni Carafa Duca die Noja*, 104–15. More on the duel in the conclusion to this book.

123. Peronnet, "L'invention de l'ancien régime"; Caiani, "Re-inventing the Ancien Régime."

124. See, e.g., Bizzocchi, "Unglaubliche Genealogien," passim.

125. The point is made regarding England by Jettot, "Intelligible to the Mind." In remains to be seen to what extent English developments are paralleled on the Continent.

Chapter 4 • The "Genealogical Brotherhood"

1. Imhoff, *Appendix ad Historiam Genealogicam*.

2. Tentzel, *Monatliche Unterredungen*, July 1691, 611–13.

3. The phrase "confraternité généalogique" is taken from d'Hozier to Imhoff, December 14, 1708, GNM, Letters, unpaginated. Yakel and Torres, "Genealogists as a 'Community of Records.'" The term, with a slightly different meaning, originally comes from Bastian, *Owning Memory*.

4. Several informants remain unidentified, e.g., one "Dr. Mezger" who, in January 1681, forwarded a "Bericht wie viel der Zeit der Herren ReichsErbmarschallen und Graffen zu Pappenheim im leben"; see GNM, Généalogies PAL–PAP (vol. 17), unpaginated (Pappenheim).

5. For two recent studies, see Egmond, *World of Carolus Clusius*; Gorman, *Scientific Counter-Revolution*.

6. Imhoff, "Ad lectorem," in *Appendix ad Historiam Genealogicam*, unpaginated: "Adese ergo Viri populares! vos imprimis, qui passim libellis epistolisque ac tabulis praefecti estis."

7. Susanne Friedrich, *Drehscheibe Regensburg*.

8. Schnabel, *Österreichische Exulanten*, 661.

9. "Com. von dem Hochf. Hohenl. Abgesandten bei dem Creistag zu Ulm, durch vermittlung des hochf. Oetingen Abges. Dr. Multzen, 26. Dec 78," GNM, Généalogies H (vol. 12), unpaginated (Hohenlohe). This "Dr. Multzius" was Jakob Multz von Oberschönfeld (1637–1713), a well-known specialist on archives and archival thinker; see, e.g., Vogtherr, "Archivtheorie und Archivpraxis," 403–9. "Schönfeldius" also appears in Imhoff to Thulemeyer, January 28, 1684, UB F, Ms. Thulemeyer, 723–25, here 723.

10. Material on the Schönberg family came forward from Ehrenfried Gejer, counselor at Waldburg, via Rudolph Friedrich Schulte, counselor at Glauchau; see GNM, Généalogies S–T (vol. 21), unpaginated (Schönberg).

11. Thulemeyer to Imhoff, no date, UB F, Ms. Thulemeyer, 709–12, here 712: "Scribam autem proxime ad parentem meum, ut ex Ducis Consiliario mihi noti[tiam] expiscetur." Neither name nor biographical data nor occupation of Thulemeyer's father are currently known. From that letter, it seems clear, however, that he and the Duke of Bevern were at least occasionally in direct and personal contact, perhaps in some relation to the duke's role in the cathedral chapter in Strasbourg.

12. Ortt to Imhoff, January 17, 1691, and August 30, 1692, in GNM, Généalogies H (vol. 12), unpaginated (Hohenlohe). On June 26, 1692, a government official in Schaumburg (one "Baurmeister"?) wrote a letter to an unnamed individual who was to forward certain pieces of information about a recent wedding of a princess of Anhalt to "Herr Im Hoff"; see GNM, Généalogies A (vol. 6), unpaginated (Anhalt). See also a set of replies by Georg Achatz Richter (dates unknown), secretary to the prince of Eastern Frisia, from January 1687, in GNM, Généalogies J–K (vol. 13), unpaginated (Jever und Knipshausen).

13. All in GNM, Généalogies H (vol. 12), unpaginated (Hohenlohe).

14. GNM, Généalogies H (vol. 12), unpaginated (Hohenlohe).

15. See Friedrich, *Birth of the Archive*, 39–42.

16. A fact sheet on the Limburg family, "extrahirt auss der Sommerhäuser Tauff-Matricul d. 5. Octobr. 1692" by Johann Müller, pastor of Limburg, in GNM, Généalogies L (vol. 14), unpaginated (Limburg). Müller correctly states that the registers only give the date of baptism. Friess's extracts are from the church books from Markt Einersheim, a seat of the family in Franconia, ibid.

17. See Antognazza, "Leibniz as Historian."

18. See Imhoff to Leibniz, January 21, 1692, in AA I, 7, pp. 546–47.

19. Their shared correspondents included Muratori, Magliabechi, Moritz Wilhelm of Sachsen-Zeitz, Juncker, Greiffencrantz, and many others.

20. Leibniz appears mostly only as a specialist on the Este and Guelphs. On the Este, see below. On the Guelphs, see Imhoff, *Notitia Procerum* (4th ed., 1699), 205.

21. "Acc. 22. febr. 1683 ex commun. D. Speneri," in GNM, Généalogies R (vol. 19), unpaginated.

22. Spener to Imhoff, July 16, 1690, BSB, Autogr. Spener, unpaginated: Spener had read Alonso Lopez de Haro's Spanish works, borrowed from Maximilian Zum Jungen in Frankfurt. Less than two months later, Imhoff also asked Zum Jungen for the same volume; see Imhoff to Thulemeyer, September 11, 1690, UB F, Ms. Thulemeyer, 741–44.

23. Spener's is acknowledged in [Imhoff], "Epistola ad lectorem," in *Spicilegium Rittershusianum*, unpaginated.

24. See, e.g., Carhart, *Leibniz Discovers Asia*, passim. Benz, "Historiker um Gottfried Wilhelm Leibniz."

25. See Thulemeyer to Imhoff, no dates, UB F, Ms. Thulemeyer, 709–12, highlighting the relevant passage of Weise, *Schediasma*, 10–12.

26. But see a stemma of the Schaffgotsch, "commun. Christ. Weise Rector Zitaviensis m. Febr. 1684," in GNM, Généalogies S (vol. 20), unpaginated (Schaffgotsch).

27. Hiller, *Die Geschichtswissenschaft*; Schilling, "Caspar Sagittarius."

28. Richter, *Der Briefwechsel*.

29. In addition to Sagittarius, Weise suggested contacting Peter Müller in Jena and Israel Clauder (1627–83) in Coburg, who published, for instance, Clauder, *Stemma Saxonicum*.

30. Wegleiter to Imhoff, September 24, 1682, BSB, Autogr. Wegleiter, unpaginated.

31. Imhoff to Fabricius, November 17, 1693, KB, Ms. Thott 1227, unpaginated. Frauendorf, who eventually became mayor of Naumburg (Schmitt-Maass, *Fénelons "Télémaque,"* 218–19, with n. 524), knew Imhoff quite well. The two men seem to have met personally, and Frauendorf dedicated one of his works (his anonymous translation of Cave, *Erstes Christenthum*) to Imhoff, praising his piety. See Kantzenbach, "Gottfried Arnolds Weg," 214.

32. Ménestrier, *Gründliche Wiederlegung*. This book was dedicated to Imhoff for the many books that Wagner had received from him.

33. Wagner mentioned, for instance, the work of Anton Weck (1623–80); see Wagner to Imhoff, February 23, 1688, BSB, Autogr. Wagner, unpaginated. On Weck, see Gautsch, "Lebensbeschreibung," including a list of manuscript works.

34. Juncker also connected Imhoff and Ernst Wilhelm Tentzel. On Juncker, see Voss, "Christian Juncker"; Berndt, *Leben und Wirken*; Germann, *M. Christian Juncker*; Engel, "Christian Junckers wissenschaftlicher Briefwechsel."

35. Juncker had bought several manuscripts of the late Christoph Daniel Findekeller (1634–94); see Winter, *Die europäischen Handschriften*, 62, 71. He used Findekeller's manuscripts to update Imhoff's works; see Juncker to Imhoff, January 1, 1699, GNM, HA V. Historiker, Juncker, unpaginated.

36. Thulemeyer had a working relationship with Johann Just Winckelmann (1620–99), a leading historian of Hesse; see Imhoff to Thulemeyer, March 24, 1693, UB F, Ms. Thulemeyer, 757–60. Imhoff does not appear to have had any contacts besides Winckelmann, Johann Her-

mann Schmincke (1684–1743), and Johann Wilhelm Schannat (1683–1739). Examples include Imhoff to Thulemeyer, August 5, 1684, August 25, 1686, and September 5, 1698, all ibid. The earliest datable contribution came in February 1682; see GNM, Généalogies F (vol. 10), unpaginated (Falckenstein). On Hesse, see Fuchs, *Traditionsstiftung*. On Schannat, also see Weis, *Johann Friedrich Schannat.*

37. Tölle, "Generation, Theft, and Dynasty."

38. Pregitzer Sr. to Imhoff, August 5, 1686, GNM, Letters, unpaginated, mentions Johann Ernst Imhoff (1614–82) from Kirchendellinsfurth, close to Tübingen, whose testament he claims to have signed as witness. Concerning the Imhoff "gens," he writes, "ex qua Materterae meae maritus Generosus dn. Carolus ab Egen Aviam habuit, et in eadem ego Patronum agnovi generosum Dn. Johann Ernestum im Hoff ante paucos annos inter nos Tubingae defunctum." Pregitzer Sr. to Imhoff, December 25, 1686, ibid., mentions further details about these individuals. See Seifert, *Genealogische Tabellen*, 7 (table 7), for this branch. For meetings, see Pregitzer Sr. to Imhoff, October 19, 1693, GNM, Letters, unpaginated. They may have also met in 1695, when Pregitzer Sr. spent time in Nuremberg on behalf of the dukes of Wurttemberg. See Pregitzer, *Suevia et Wirtenbergica Sacra*, fols.)(3r–v.

39. On the son, see Pregitzer Sr. to Imhoff, June 20, 1692, GNM, Letters, unpaginated. On the son's work as a proofreader, see Pregitzer Sr. to Imhoff, June 20, 1692, ibid. There is one letter from Imhoff to an (unnamed) *corrector*, but it dates from 1696 and hence may not be related. See Imhoff to [unknown], mid-December 1696, SUB HH, Sup.ex. 36, fols. 520r–v.

40. GNM, Généalogies W (vol. 22), unpaginated (Württemberg/Teck): "Commun. a Dn. Pregizero, Consiliario Wirtemberg, m Oct 1701" on a list of descendants of one twelfth-century count. Pregitzer Sr. to Imhoff, August 5, 1686, GNM, Letters, unpaginated: "In quo libro [sc., the *Notitia Procerum*] saepius tecum locutus sum." See also Pregitzer Sr. to Imhoff, October 19, 1693, ibid., where he describes a visit to Heilbronn and enthusiastically mentions the wealth of genealogically relevant inscriptions in that city.

41. Imhoff made contact with Balbín in the summer of 1681, arranged by Christian Weise. Christoph Arnold often served as an intermediary between Balbín and Imhoff (as well as Spener); see Balbín to Imhoff, e.g., September 6, 1681, or August 12, 1682, GNM, Letters, unpaginated. Excerpts from letters from Balbín to Arnold (January 1, 1680, and June 23, 1681) in Imhoff's papers are extant in GNM, Généalogies F (vol. 10), unpaginated (Fürstenberg).

42. Balbín to Weise, November 29, 1685, in Richter, *Briefwechsel*, 152: Balbín boasts of his excellent relationship with many "aularum mareschallis."

43. "Inde tot sibi amicos paravit Balbinus, quod multorum utilitatibus inserviret." Weise to Otto Mencke, December 3, 1687, in Richter, *Briefwechsel*, 317.

44. E.g., Imhoff, *Notitia Procerum* (4th ed., 1699), 649.

45. For one example, see Balbín to Imhoff, February 25, 1682, GNM, Letters, unpaginated, a vague allusion to anti-Jesuit prejudices. See also Imhoff to Fabricius, September 18, 1704, KB, Ms Thott 1227, unpaginated: "weil mir einer der vornehmsten Kayserl. ministris hat vorhalten lassen, daß da ich mich sonst ziemlich unpartehilich in besagtem scripto verwiesen hette, doch darinnen eine Partheiligkeit vermerken hette lassen, daß ich denen lebenden fürsten ex protestantium classe grose . . . elogia zugelegt, bey denen Röm. Cathol. aber es so beym gleichen bewenden lassen, ich den Schlus gefast habe, bey der neuen edition alle solche elogia viventium invidiosa auszulassen." Fabricius reported, that the duke had "laughed" when he learned of Imhoff's policy. Whether he laughed in irony or in agreement is impossible to tell; see Imhoff to Fabricius, November 24, 1704, ibid.

46. Balbín described the newcomers as "non ex nostra nobilitate"; Balbín to Imhoff, February 25, 1682, GNM, Letters, unpaginated.

47. Balbín to Imhoff, December 8, 1681, and February 25, 1682 (pragmatism), GNM, Letters, unpaginated.

48. See Richter, *Briefwechsel*, 16–18 (introduction). For a nuanced analysis of Balbín's (mis) representation of the Bohemian context, see Maťa, "Patres Patriae?." On Balbín's position in Bohemian patriotic culture, see also Pípalová, "Bohuslav Balbín."

49. Imhoff, *Excellentium Familiarum in Gallia*, preface (to the Senate of Nuremberg).

50. Imhoff to Thulemeyer, August 8, 1683, UB F, Ms. Thulemeyer, 715–18. See also the discussion below in chapter 6.

51. Imhoff, "ad lectorem," in *Excellentium Familiarum in Gallia*. In particular, Greiffencrantz sent Imhoff the publications of Père Anselm, by far the most important French genealogist of his generation. Their cooperation seems to have continued, yet no details are available. See "Genealogia familiae de Sousa in Portugallia ex Codice Msto, quem Carolus Josephus de Ligne Marchi de Arronches, viennensi Legatione nobilitatus, Greiffencrazio communicavit, excerpta an 1696," in GNM, Généalogies E (vol. 9), unpaginated (Espagne).

52. There is a single letter from Allard to Imhoff, June 1, 1682, GNM, Letters, unpaginated. Besides writing novels and verse, Allard collected books and was an expert on genealogy. See Coulomb, "La publicité des archives," 92–94. Virieux, "Guy Allard."

53. Delisle, *Le cabinet des manuscrits*, 1:358.

54. Figeac, *Les noblesses*, 147–50.

55. Imhoff used his good relationship with d'Hozier to send his son, Jakob Wilhelm Jr., to Paris when the latter went on his grand tour. D'Hozier, meanwhile, claimed to be less fortunate with his offspring: he eventually fell out with his nephew and successor as *juge d'armes*, Louis Pierre d'Hozier de Sérigny (1685–1767). In several letters, d'Hozier opened his heart to Imhoff about this unpleasant affair. Once they reached old age, the two men also occasionally talked about their spiritual and mental state.

56. All mentioned in d'Hozier to Imhoff, March 17, 1695, GNM, Letters, unpaginated. On Pomponne de Refuge, see d'Hozier to Imhoff, December 18, 1695, ibid. The Salviati appear, e.g., in d'Hozier to Imhoff, December 31, 1707, and May 25, 1708, ibid.

57. D'Hozier to Imhoff, April 21, 1702, GNM, Letters, unpaginated: he is showing Imhoff's letters to the "curieux" in Paris. He also alerted the royal librarian Bignon to Imhoff's works; see d'Hozier to Imhoff, March 7, and July 25, 1710, ibid. Introduction to Gagnières: d'Hozier to Imhoff, February 26, 1698, ibid. On the complex, deteriorating relationship between d'Hozier and Gagnières, see Delisle, *Le cabinet des manuscrits*, 1:347; Ritz-Guilbert, *La collection Gaignières*, 93–94.

58. D'Hozier to Imhoff, February 17, 1699, and July 27, 1700, GNM, Letters, unpaginated. Gaignières to Imhoff, January 3, 1700, quoted in Bouchot, *Inventaire des dessins*, 1:XIX.

59. Ritz-Guilbert, *La collection Gaignières*. See also Ritz-Guilbert, "La collection Gaignières"; Brown, *Oxford Collection*; Grandmaison, *Gaignières*; Nozza Bielli, "Il collezionismo."

60. D'Hozier to Imhoff, February 26, 1698, GNM, Letters, unpaginated. On Gaignières and genealogy and on his exchanges with Claude Le Laboureur and Ménestrier, see Grandmaison, *Gaignières*, 13–15, 29.

61. D'Hozier to Imhoff, February 17, 1699, GNM, Letters, unpaginated: this concerned one "Conte Gilbert."

62. See d'Hozier to Imhoff, August 21, 1712, GNM, Letters, unpaginated: first mention of Clairambault. D'Hozier to Imhoff, April 4, 1713, ibid.: mention of Clairambault and Buffier. Clairambault sent the constitutions of the Order of St. Esprit, d'Hozier to Imhoff, December 29, 1713, ibid.

63. D'Hozier to Imhoff, July 29, 1700, GNM, Letters, unpaginated. Imhoff, *Genealogiae Viginti Illustrium in Italia Familiarum*, fol. *3v ("praefatio").

64. On La Valette's political role, see Boucher, "Introduction," 21–25. He is considered the author of [La Valette], *Abrégé nouveau*. On this, see Bernard, *Cartulaire*, 1:VII, XV. On the family's genealogical activities, see also Béroujon, "La satire."

65. Imhoff, *Genealogiae Viginti Illustrium Familiarum in Italia*, fol. *3v ("praefatio").

66. The connection between Du Fourny and Anselm is mentioned in d'Hozier to Imhoff, March 17, and April 22, 1699, GNM, Letters, unpaginated. Du Fourny worked as auditeur de la Chambre de Comptes, a position with a scholarly tradition. He subsequently prepared an expanded edition of the Augustinian's *Histoire généalogique* (1st ed., 1674), which appeared in 1712. On Du Fourny and Père Anselm, see also d'Hozier to Imhoff, March 17, 1695, GNM, Letters, unpaginated.

67. Only two letters survive from before its publication. There is a significant gap in the correspondence after d'Hozier acknowledged receiving his copy of the 1687 edition; their correspondence does not resume until 1694.

68. For the phrase "living archive," see chapter 2, passim. For "bureau d'adresse," both as institution and metaphor, see, e.g., Mazauric, *Savoirs et philosophie*.

69. D'Hozier to Imhoff, December 18, 1695 (*Notitia Procerum*) and April 13, 1694 (England), all in GNM, Letters, unpaginated.

70. D'Hozier to Imhoff, November 29, 1701, GNM, Letters, unpaginated.

71. D'Hozier to Imhoff, November 20, 1702, GNM, Letters, unpaginated.

72. Wartis provided in 1682 information about the Altemps and Acquaviva families. At least seven letters survive, five in GNM, Letters, unpaginated, one in GNM, Généalogies H (vol. 12), unpaginated (Hohenems/Altemps), and one (from October 24, 1693) in GNM, Généalogies N–O (vol. 16), unpaginated (Orsini). He greeted members of both the Imhoff and the Scheurl families.

73. Mentioned in Imhoff, *Historia Italiae et Hispaniae Genealogica*, "praefatio." Two manuscript dedicatory poems by Holthusen for Imhoff are in UB TÜ, Mi VIII 1.

74. Six letters between Battier and Imhoff are in BSB, Autogr. Battier. See also Pusterla to Imhoff, August 12, 1693, GNM, Letters, unpaginated.

75. There is little information about Pusterla available. See Castiglioni, "I prefetti della Biblioteca Ambrosiana." It was Magliabechi who secured the new post for Pusterla; see Viola, "Vecchia e nuova erudizione," 99n8.

76. Imhoff, *Historia Italiae et Hispaniae Genealogica*, 249: thanking him for a genealogy of the marquis of Caravaggio.

77. Imhoff to Magliabechi, November 20, 1690, BNC Firenze, Magl. VIII 259, unpaginated: "Mà essendo questa impresa tanto più difficile, quanto meno posso trovar' autori che hanno trattati di questa materia con accortezza."

78. The earliest record of epistolary contacts between the two men is a letter from Imhoff to Magliabechi, November 28, 1689, mentioned in Magliabechi to Imhoff, March 20, 1690, GNM, Letters, unpaginated. This letter may have been connected to a visit of a cousin or nephew of Imhoff in Florence mentioned several times in the next letters, e.g., Magliabechi to Imhoff, April 24, 1690, ibid. It would seem that Magliabechi had already learned of Imhoff's genealogical activities from other friends in advance. In his letter to Imhoff from March 20 (quoted above), he mentions a letter (undated) from Christian Wagner announcing Imhoff's plan to write a book on Italian genealogies—and dedicate it to Magliabechi.

79. Imhoff to Magliabechi, November 20, 1690, BNC Firenze, Magl. VIII 259, unpaginated: ". . . di degnarsi à secondare mio proposito, ed à darmi notitia dei libri ed huomini, mediante li quali io posso riuscirvi."

80. Mirto, *Stampatori*, 94.

81. Magliabechi to Imhoff, June 12, 1696, GNM, Letters, unpaginated. Imhoff to Muratori, July 3, 1696 (via Magliabechi), in Lieber, "Jacob Wilhelm von Imhofs Korrespondenz," 47–48; Magliabechi to Muratori, July 17, 1696, in Muratori, *Carteggi*, 305–6. Magliabechi had only recently, in summer 1695, established correspondence with Muratori; see Muratori, *Carteggi*, 278. On the two great Italian scholars, see, e.g., Viola, "Vecchia e nuova erudizione," and Bruni, "Muratori epistolografo."

82. Imhoff to Muratori, October 30, 1696, in Lieber, "Jacob Wilhelm von Imhofs Korrespondenz," 49–52. Imhoff to Muratori, August 13, 1697, ibid., 56. Some of Muratori's material on the Homodei is in GNM, Généalogies H (vol. 12), unpaginated (Homodei). Muratori was in personal contact with Daniele Biragi, see, e.g., Muratori to Imhoff, March 20, 1698, in Busino and Dufour, *Lettere*, 427–28.

83. The first reference to Bacchini appears in Magliabechi to Imhoff, January 2, 1693, GNM, Letters, unpaginated. On Bacchini see Golinelli, *Benedetto Bacchini.*

84. A letter from Imhoff to Bacchini is mentioned in Imhoff to Magliabechi, January 17, 1694, BNC Firenze, Magl. VIII 259, unpaginated. In the same letter, however, Imhoff asks Magliabechi to forward additional questions to Bacchini—the direct contact, it seems, was not too extensive. One forwarded letter from Bacchini to Magliabechi is in GNM, Généalogies F (vol. 10), unpaginated (Ferrara).

85. Mandosio is mentioned several times in the letters to Magliabechi from Naples and Sicily, see Quondam and Rak, *Lettere*, ad indicem.

86. Orlandi, "Biografia e bibliografia"; Mandosio, *Bibliotheca Romana*; Mandosio, *Theatron.*

87. Orlandi, "Biografia e bibliografia," 82–83. See also Crescimbeni, *Le vite degli Arcadi illustri*, 144–47. The very first part of the *Bibliotheca Romana*, the *prima centuria*, was dedicated to Cartari, while the last (printed) *centuria* was dedicated to Magliabechi. On Cartari, see Filippini, *Memoria della Chiesa*, who, however, does not mention Mandosio.

88. E.g., Mandosio to Imhoff, January 12, 1697, GNM, Letters, unpaginated. There is, indeed, some material from Mandosio in GNM, Généalogies S–T (vol. 21), unpaginated (Sforza). The provenance is beyond doubt, given Mandosio's characteristic handwriting.

89. E.g., Mandosio to Imhoff, January 17, 1701, GNM, Letters, unpaginated.

90. Mandosio to Imhoff, April 29, 1702, GNM, Letters, unpaginated.

91. Mandosio to Imhoff, April 29, 1701, and December 16, 1702, GNM, Letters, unpaginated: Mandosio expresses gratitude for being honorably mentioned in Imhoff, *Historia Italiae et Hispaniae Genealogica*, 257, in the very last sentence of the book.

92. See, e.g., Magliabechi to Imhoff, October 25, 1693, GNM, Letters, unpaginated: "diversi Amici in quelle Corti." Another of these friends was Ottavio Vernicci from Bologna, who procured information about the family Picco della Mirandola; Magliabechi to Imhoff, October 20, 1691, GNM, Letters, unpaginated. Imhoff to Magliabechi, November (no day given), 1691, BNC Firenze, Magl. VIII 259, unpaginated.

93. On Benvenuti, see Mirto, "Lettere di Antonio Magliabechi"; Mirto, "Lettere di Giovanni Battista Bidelli."

94. Magliabechi to Imhoff, February 2, 1691, GNM, Letters, unpaginated: "dà esso solo VS.Illma la potra avere perfetta"; Magliabechi to Imhoff, January 2 and December 15, 1693, ibid.: praise for Benvenuti's work, which has finally arrived. Imhoff to Magliabechi, January 3/13, 1694, BNC Firenze, Magl. VIII 259, unpaginated, acknowledges the importance of Magliabechi's help in contacting Benvenuti.

95. Benvenuti to Imhoff, March 15, 1692, and June 29, 1694, GNM, Letters, unpaginated; Imhoff to Benvenuti, December 3, 1694, BNC Firenze, Magl. VIII 282, unpaginated. This last letter, however, is only a brief note of thanks.

96. Callard, *Le prince*, 186–90.

97. Imhoff to Muratori, July 12, 1700, in Lieber, "Jacob Wilhelm von Imhofs Korrespondenz," 60: "haud satis mihi esse videtur explorata."

98. This was the famous *Priorista fiorentino*, available in a beautiful reproduction at http://www.archiviodistato.firenze.it/archividigitali/riproduzione/?id=630343. On the Priorista, see Baggio and Marchi, "L'archivio della memoria," 864–65. Boutier, "'L'Accademia dei Nobili'" 216–19. Lafage, *Côme III de Médicis*, 95–101.

99. Imhoff to Magliabechi, November 20, 1690, BNC Firenze, Magl. VIII 259, unpaginated: In Naples, "hò benissimi autori."

100. Angelillo and Stendardo, "Il Seicento," 70–71.

101. The date 1685 comes from a handwritten note in his personal copies of de Lellis, *Discorsi delle famiglie nobili* (StadtB N, Hist 311–313.2°). The volumes contain several marginal annotations, though some of them dating from later years.

102. Rovito, *Respublica dei Togati*; Spagnoletti, "The Naples Elites." On the *togati* and their connection to the nobility, see also Tutino, *Fake Saint*.

103. On his contact with royal archivists, see the article s.v. in the *Dizionario biografico degli italiani*. On his manuscript excerpts ("Notamenti") of public documents, see Gentile, "I manoscritti di Carlo de Lellis"; Filangieri, "Notamenti e repertori"; Filangieri, *Gli atti perduti*. In January 1943, most of the de Lellis manuscripts were destroyed; see Palmieri, *Degli archivi napolitani*, 345n78.

104. In the preface to de Lellis, *Discorsi*, vol. 1, unpaginated ("A' Lettori"), the publisher highlights how de Lellis discussed numerous families from the higher and lower nobility as one group, without necessarily following the stratified hierarchies maintained by the families themselves.

105. Gittio to Imhoff, February 3, 1688, GNM, Letters, unpaginated.

106. Pironti, *Bulifon—Raillard—Gravier*, 31–38; Pedìo, *Storia della storiografia*, 101–4; Di Marco, "Librai, editori e tipografi," 42–43; Bulifon, *Giornali di Napoli*.

107. I have not found any surviving letters between Bulifon and Imhoff, but several are attested in Imhoff's correspondence with Sant'Anna and Magliabechi, among others. On Bulifon and contemporary information mongering, see Angelillo and Stendardo, "Il Seicento," 88–89; Magnanini, "Postulated Routes," 81–82. For the quotation, see Rosa, "Un médiateur." For an example, see Bulifon, *Lettere memorabilia*, a publication of miscellanea intended to provide material on contemporary history. See also Vivo, "Pharmacies as Centres."

108. Gittio to Imhoff, February 3, 1688, GNM, Letters, unpaginated.

109. Giustiniani, *Memorie istoriche*, 3:112–15; Trombetta, "La Libraria," 19–20.

110. Gittio to Imhoff, April 6, 1688, GNM, Letters, unpaginated. On Aldimari, cf. Rovito, *Respublica del Togati*, 440–49; Giustiniani, *Memorie istoriche*, 1:32–40; Quondam and Rak, *Lettere*, 5–6; Pedìo, *Storia della storiografia*, 214–16.

111. Sant'Anna to Magliabechi, March 6, 1694, in Quondam and Rak, *Lettere*, 588–90. Sant'Anna did not write directly to Imhoff until August 1695, when rumor had it that Magliabechi had died. See Sant'Anna to Imhoff, August 20, 1695, BSB, Cod. Ital. 506, fols. 1r–2r. On their communication, see also Imhoff, *Corpus Historiae Genealogicae*, 282.

112. As suggested by Gittio to Imhoff, January 24, 1696, GNM, Letters, unpaginated. The Carmelite published two books, Sant'Anna, *Della storia genealogica*, and Sant'Anna, *Discorso*. On this, see Magliabechi to Imhoff, December 20, 1694, GNM, Letters, unpaginated.

113. Quondam and Rak, *Lettere*, 558–59 (biographical note on Sant'Anna).

114. For one of the exceptions, the prince of Butera, whom Imhoff did not contact via Sant'Anna, see Aldimari to Imhoff, September 5, 1690, and September 8, 1691, GNM, Letters, unpaginated. Imhoff seems to have sent a package of his most recent works to Butera. Magliabechi to Imhoff, February 2, 1691, ibid., mentioned that the prince was also cooperating with Bulifon. The prince was in direct contact with Magliabechi as well; see Magliabechi to Imhoff, November 18, 1691, ibid. Butera promised to send a "family history" to Imhoff; see Magliabechi to Imhoff, October 20, 1691, ibid.

115. Giustiniani, *Memorie istoriche*, 2:114: "infantastichì tanto per la sua nobiltà, che null'altra applicazione stimò più degna, che trovar monumenti, onde fissare la sua genealogia, e quella degli altri."

116. Gittio to Imhoff, April 6, 1688, GNM, Letters, unpaginated: Aldimari "sia cosi secreto nelle sue actioni." Imhoff, *Genealogiae Viginti Illustrium Familiarum in Italia*, preface.

117. Van Vugt, "Geografia e storia," 262.

118. Spener, *Theatrum*, vol. 2, unpaginated (preface). Cf. Lopez de Haro, *Nobiliario genealogico*.

119. Imhoff to Thulemeyer, September 11, 1690, UB F, Ms. Thulemeyer, 741–44: He borrowed Lopez de Haro's important work from the library of Maximilian Zum Jungen in Frankfurt. Imhoff bought his copy of Lopez de Haro, *Nobiliario genealogico* (StadtB N, Hist. 241.2°) in Brussels in 1692 (note next to ex libris). In 1695, Imhoff acquired his copy of Salazar de Mendoza, *Origen de las dignidades* (manuscript note next to his ex libris, StadtB N, Hist. 235.2°). Riederer to Imhoff, April 28, 1697, GNM, HA XIV Varia, Riederer, unpaginated, on (impossible) purchases in Amsterdam.

120. Carlos José de Ligne (1661–1713), marquis de Arronches and ambassador in Vienna, once provided material on his wife's family, the Sousa, in 1696; see GNM, Généalogies E (vol. 9), unpaginated (Espagne). De Ligne was ambassador in Vienna only for a few months in late 1695 and early 1696, before he became implicated in a complicated murder trial; see Münch Miranda and Salvado, "Justice, Politics, and Diplomacy."

121. Imhoff to Magliabechi, May 8, 1695, BNC Firenze, Magl. VIII 259, unpaginated: the genealogies of Spain and Italy "vi devono esser mescolate per la congiunzione di sangue."

122. Sant'Anna to Imhoff, October 10 and 20, 1696, and July 3, 1697 (lack of children), all in BSB, Cod. Ital. 506, fols. 17r–18r, 19r–20r. For the duke's genealogy, I rely on Fundación Casa Ducal de Medinaceli, http://www.fundacionmedinaceli.org/casaducal/fichaindividuo.aspx?id=169 (accessed January 18, 2022). There were illegitimate children, none of whom was still alive by the time of Imhoff's writing.

123. Imhoff to Magliabechi, May 8, 1695, BNC Firenze, Magl. VIII 259, unpaginated. Magliabechi forwarded this request to Naples; see Sant'Anna to Imhoff, October 25, 1695, BSB, Cod. Ital 506, fols. 3r–4v, here 4r: "una persona prattica nelle genealogie, e case di Spagna."

124. Sant'Anna to Imhoff, October 25, 1695, BSB, Cod. Ital 506, fols. 3r–4v, here 4r: "egli s'è scusato, perche tiene molte cose per le mani, e stà occupatissimo per il suo officio." Vidania's name reappears on several occasions in the correspondence; winning him over, thus, seems to have been an ongoing project. On Vidania, see Benedikt, *Das Königreich Neapel*, 380–85.

125. D'Hozier to Imhoff, April 13, 1695, GNM, Letters, unpaginated.

126. Imhoff, *Historia Italiae et Hispaniae Genealogica*.

127. Imhoff, "ad lectorem," in *Corpus Historiae Genealogicae*, unpaginated: "Circa Hispanicas familias cum aqua mihi haereret, destituto a nonnemine in Australi plaga degente, in quo spem meam collocaveram, apparuit ex Boreali, tanquam ex machina, egregius ac strenuus adjutor, videlicet GERARDUS ERNESTUS DE FRANCKENAU."

128. See the letters of Franckenau in Csepregi, *Pietas Danubiana*, along with Mulsow, "Freethinking"; Benavent Montolio, "Gregorio Mayans," 33–34; and Horn, "Literaturgeschichtliche Aspekte."

129. Ehrencron owned a large library; see anonymous, *Viri Illustris Friderici*.

130. Ehrencron to Leibniz, April 25, 1703, AA I, 22, pp. 386–87, and Leibniz to Ehrencron, [June 1703?], ibid., p. 457. In Leibniz's case, however, this proved not to be a fruitful way to learn about Spain.

131. Franckenau had potentially learned about Imhoff's research from his *Historia Italiae et Hispaniae Genealogica* (1701); Johannes Moth, a long-standing acquaintance of Imhoff's in Copenhagen, may have established contact between them, perhaps at the suggestion of Thulemeyer, who knew the genealogical scene in the Palatinate.

132. Franckenau to Imhoff, February 11, 1702, GNM, Letters, unpaginated. Imhoff, *Corpus Historiae Genealogicae*, 366–67.

133. Imhoff, *Viginti Genealogiae Illustrium Familiarum in Hispaniae*, 118 (Guzmán).

134. See, e.g., the quote from Franckenau's papers in Imhoff, *Viginti Genealogiae Illustrium Familiarum in Hispaniae*, 57.

135. Franckenau to Imhoff, March 31, 1703, GNM, Letters, unpaginated.

136. Ehrencron, who lived the final years of his life in Glückstadt, took his collection with him. This separation occasionally prevented Franckenau from helping; see, e.g., Franckenau to Imhoff, April 3, 1706, GNM, Letters, unpaginated.

137. See Kagan, *Clio and the Crown*, 273–78, for a general assessment of Salazar.

138. Gittio, who called the Spaniard "molto mio amico," had encouraged Imhoff on November 30, 1694, GNM, Letters, unpaginated, to write to Salazar y Castro, but nothing came of this. The name does not reappear in later letters.

139. See Franckenau to Salazar y Castro, Copenhagen, January 14, 1702, in Soria Mesa, *La biblioteca genealógica*, 147–48. Here, Franckenau also details his unsuccessful attempts to meet Salazar y Castro in person in Madrid.

140. Soria Mesa, *Biblioteca genealógica*. A letter from Salazar y Castro appears in Franckenau to Imhoff, September 21, 1702, GNM, Letters, unpaginated.

141. D'Hozier to Imhoff, September 6, 1702, GNM, Letters, unpaginated: "j'avois envoyé un memoire en Espagne par le fils d'un de mes amis qui y a acompagné le Roi."

142. Gittio to Imhoff, November 7, 1690, GNM, Letters, unpaginated: "Del resto se V.S.I. mi ricercasse notizie di questo Regno di Napoli potrebbe passare, ma ella mi richiede di continuo notizie di Spagna, come se io fossi Spagnolo, ò dimorante in Madrid: nulla pero dimeno cercherò sodisfarla con la mia debolezza, e poca prattica che tengo in quello che mi sovviene."

143. Cf. Meibom's remark in GWLB, Ms VIII 634, fol. 4r: "Nomina nos fallunt quia Latine scripta." He added, as if to remind himself, "NB de alta flamma," i.e., the Hohenlohe family, whose Italian branch styled itself "Comes de Alta Flamma." On difficulties posed by names, see also Spener, "ad lectorem," in *Theatrum*, vol. 1, unpaginated.

144. Franckenau to Imhoff, November 25, 1702, GNM, Letters, unpaginated.

145. Gittio to Imhoff, January 7, 1698, GNM, Letters, unpaginated.

146. E.g., Gittio to Imhoff, June 20, 1690, GNM, Letters, unpaginated.

147. Gittio to Imhoff, June 28, 1695, GNM, Letters, unpaginated: "e non sono cosi facili à scriversi da forastieri non prattichi del Paese, li quali stanno alle relationi che se l'inviano e non possono conoscere gli errori." Gittio to Imhoff, January 24, 1696, ibid.: "essendo assai facile che i forestieri non prattichi del Paese restino ingannati." Gittio to Imhoff, February 19, 1697, ibid.: "gli Autori forastieri non possono conoscere queste inventioni, e falsità." D'Hozier in Paris seems to have agreed. In praise of Imhoff's French genealogies, he wrote (to Imhoff, September 9, 1687, ibid.): "Un François originaire n'auroit pas mieux exécuté cette entreprise, que vous l'avés fait."

148. See Cooper, *Inventing the Indigenous*.

149. Geevers, "The Danish Habsburgs," 273–76; Del Tredici, *Comunità*, 282–307; Delille, "Namen und Linien." See also Andenna and Melville, *Idoneität—Genealogie—Legitmität*, 365–449.

150. On Imhoff lineages, see, e.g., StA N, Handschriften 285, and GNM, Nachlass Imhoff I (Archiv der Gesamtfamilie) fasc. 2, no. 3.

151. "Discorso nel quale si prova, che i Signori Aquini di Cosenza sono un vero Ramo della famiglia di Aquino," BSB, Cod. Ital. 506, fols. 85r–88r (attached to Sant'Anna to Imhoff, April 17, 1702). The quotation (fol. 85r) reads: "È cosa assai certa dall'esperienza giornalmente autenticata, ritrovarsi molte famiglie, quali benche habbiano l'ystesso origine, & derivano dal medesimo stipate, in ogni modo si trovano haver la loro residenza, non solo in diverse Città dell'istesso Regno o Provincia, ma in luoghi assai lontani et in Regni o Provincie di gran tratto distanti."

152. Nicolaus Rittershausen to [unknown], no date (draft), SUB GÖ, Ms philos 94, unpag-

inated. (no. 85): "Touchant le XVI ayeuls des Comtes de Furstenberg, s'est à la verité chose aussy malaisée de les rassembler que les royaumes sont esloignez oùz ils sont à chercher, la race des uns en Espagne, l'yssue des autres en Boheme."

153. Rittershausen to Georg Seger, no date (1659?), SUB GÖ, Ms. phil. 94, unpaginated (no. 59, second letter): "Nimis credulus fui amicis." Rittershausen later crossed out this sentence in this draft.

154. Imhoff to Moritz Wilhelm, August 18, 1699, SächsHStA, loc. 8224, no. 10, fol. 63v: "ich werde es anfänglich einem andern, ohne fernere untersuchung auf gut Glauben nachgeschrieben haben, und weil keine Anthung darwieder geschehen, ist es von einer edition in die andere also forgeführet worden; dann obwol Ew Hochfürstl durchl. mich zu End vorigen Jahres . . . gnädigst berichtet haben, . . . sondern ich mus auch kenennen desselben nicht einmal mich erinnert zu haben." Moritz Wilhelm to Imhoff, September 16, 1699, Veste Coburg Ms A, I, 349, (2), 3, unpaginated, replied cordially.

155. Classic statement in Shapin, *Social History*. See also Lux and Cook, "Closed Circles or Open Networks?" On trust generally, see, e.g., Ziegler, *Trauen und Glauben*.

156. In one instance, he dismissed a manuscript table of the Italian dukes of Bracciano "that I received from Rome," perhaps from Mandosio, and retained Francesco Sansovino's older, alternative version: Imhoff, *Genealogiae Viginti Illustrium*, 331.

157. D'Hozier to Imhoff, March 22, 1685, GNM, Letters, unpaginated: "vous verres qu'il est dificile de tout avoir, et de tout sçavoir."

158. Imhoff to Leibniz, December 1, 1701, AA I, 20, p. 638.

159. Spener to Imhoff, September 9, 1684, BSB, Autogr. Spener, unpaginated: "lubens concedere possum illis qui circa Salmios et Solmenses, errores meos tibi indicarunt."

160. Fabricius to Imhoff, October 3, 1710, GNM, Letters, unpaginated: "die nomina gehören unter die res facti, die man annehmen mus wie mans ab hominibus fide dignis empfänget, u bleibt es bey der alten Trostsprüchlein: errare humanum est."

161. Friedrich, "Genealogy as Archive-Driven Research Enterprise."

162. Only two of Imhoff's letters to the Jesuit survive; for a relevant example, see Imhoff to Balbín, May 1, 1682, in Rejzek, *P. Bohuslav Balbin*, 292–94n**.

163. Imhoff to Magliabechi, January 17 and March 13, 1694, BNC Firenze, Magl. VIII 259, unpaginated.

164. See, e.g., the list of numbered replies, in Balbín to Imhoff, September 6, 1681, GNM, Letters, unpaginated.

165. GNM, Généalogies S–T (vol. 21), unpaginated (Sforza).

166. Mandosio replied to Imhoff, again via Magliabechi, on July 7, 1696, GNM, Letters, unpaginated.

167. GNM, Généalogies B (vol. 7), unpaginated (Borromeo). GNM, Généalogies A (vol. 6), unpaginated (Acquaviva).

168. Franckenau to Imhoff, February 17, 1703 ("dubia"); September 1, 1703 ("scrupulum"), GNM, Letters, unpaginated.

169. Pusterla to Battier, January 1, 1693, attached to Battier to Imhoff, January 12, 1693, BSB, Autogr. Battier, unpaginated: "cujus nonnullas tabulas ab eodem Cl. Imhofio summario ordine descriptas mihi demum in clariori lumine ponendas ex priori tua Epistola intellexeram."

170. In the 1750s, Johan Joseph Vöhlin wrote, "zudeme sich die wenigsten, bei denen ich einige Wissenschaft und Nachrichten von ihren Sippschaften und Ahnen einholen wollte, die Mühe gaben, mir hierin zu willfahren und einige Auskunft zu geben"; quoted in Schiele, *Johann Joseph Vöhlin*, 44.

171. Schiele, *Johann Joseph Vöhlin*, 45: "Abgang gnugsamer Informationen."

172. Imhoff recounted that, at one point during his travels in Italy, he met a knowledgeable Portuguese man in Venice who might have been an informant about the Spanish families. This

was Rodrigo Mendez Silva (1606–70), a recently converted Christian who ran into serious trouble with the Spanish Inquisition. He fled to Venice in 1663, where he remained until his death. Mendez Silva was "historian to the king" and a well-known expert in historical and genealogical matters. See Guillén Berrendero, "Valores nobiliarios." Cf. Imhoff to Magliabechi, September 12, 1695, BNC Firenze, Magl. VIII 259, unpaginated.

173. In addition to Gittio's above-quoted remarks, cf. Imhoff to Thulemeyer, May 19, 1693, UB F, Ms. Thulemeyer, 765–67, here 767: "nur dessen versichere ich Mhherren nochmals, daß es mir gar nicht in Sinn kommen, die von Bernardo C. Lipp & Marg[arita] a Reden erzeugte Kinder pro natis e concubitu u also pro naturalibus zu halten, oder auszugeben[.] bitte auch dieses zu glauben, daß ich amicam contradictionem, in qua nihil si acrimoniae, gar wol aufnehme, auch wol daran einige Proben in meinen Schriften gegeben habe, ja daß ich um bessern unterricht wo ich geirret sincere bitte."

174. See also two letters by Andreas Riccius, OSB, to Imhoff, April 13/23, 1692, and December 6/16, 1697, in GNM, HA Signatur IV. Klerus: Riccius, unpaginated. Riccius (1634–1700), who came from Zerbst in Saxony, was a convert from Lutheranism and had served as master of novices in Fulda, before he became a parish priest in Holtzkirchen, from whence he wrote to Imhoff. His letters mostly concern the abbots of the monastery at Fulda, about which Riccius would have known from his time there. On Riccius, see the series of necrologues of the Benedictine abbey of Werden (to which Riccius originally belonged), in Jacobs, *Werdener Annalen*, 187 (many thanks to Thomas Wallnig, Vienna, for pointing me to this publication). See also Justin Redn, OFM, to Imhoff, April 14 and July 9, 1720, GNM, Letters, unpaginated.

Chapter 5 • The Genealogist at Work

1. Magliabechi to Benvenuti, no date (1686/7), in Mirto, "Lettere di Antonio Magliabechi," 237: "il cercarne nel caos de' miei fogli tutta la mattina."

2. The dossiers must have been bound into codices after Imhoff's death, because they include a few pieces later than 1728. There is no way of telling when or by whom the papers were assembled into these codices. It seems likely that Imhoff's son, Jakob Wilhelm Imhoff Jr., was instrumental in organizing and preserving his late father's papers. A volume of family letters, which also contains some letters by Imhoff Sr. (GNM, Nachlass Imhoff VII [Collection Imhoff], no. 3) opens with a "specification" of the contents of the "volume" clearly written by Imhoff Jr. If this codex is representative, it seems that Imhoff Jr. indeed collected and arranged what today is available as the "Collection Imhoff."

3. Key contributions include Blair, *Too Much to Know*, and Smyth, *Material Texts*. See also the special issue Fleming, "Renaissance Collage."

4. Becker and Clark, *Little Tools*.

5. Klapisch-Zuber, "Genesis of the Family Tree"; Klapisch-Zuber, *L'ombre*; Hellström, *Trees of Knowledge*.

6. D'Hozier to Imhoff, December 18. 1695, GNM, Letters, unpaginated: "exactitude, avec laquelle vous avés rassemblé *tous les vivans* de chaque race que vous traité" (my emphasis).

7. Wagenseil, *Der Adriatische Löw*, fol. Cv: "Wappen deren Familien / so fürlängst abgangen / welche zu wissen unnöthig."

8. Juncker to Imhoff, April 24, 1706, GNM, HA V. Historiker: Juncker, unpaginated.

9. See Spener to Imhoff, November 28, 1682, in Spener, *Briefe aus der Frankfurter Zeit*, 6:381. Spener had withdrawn from active genealogical research and hence only had information he had gathered "communiter ex publicis relationibus." Imhoff to Moritz Wilhelm, December 22, 1699, SächsHStA, loc, 8224, no. 7, fol. 76r: "[habe] nichts vernommen, als was public worden und Ew. Hochf. Durchl. nicht unbekant sein kan."

10. In 1709, news media "eine Zeit hero die genealogica mehreres als zuvor attentiret"; Imhoff to Moritz Wilhelm, May 4, 1709, SächsHStA, loc. 8224, no. 8, fol. 95v.

11. Early modern news media focused predominantly on politics and the social elite, hence their relevance for genealogy. Cf. Schröder, *Die ersten Zeitungen*, 129–45.

12. D'Hozier to Imhoff, October 16, 1700, GNM, Letters, unpaginated.

13. Imhoff to Moritz Wilhelm, December 24, 1706, SächsHStA, loc. 8224, no. 10, fol. 124r.

14. Juncker to Imhoff, April 8, 1700, GNM, HA V. Historiker: Juncker, unpaginated. Imhoff to Moritz Wilhelm, May 26, 1699, SächsHStA, loc. 8224, no. 10, fol. 61r.

15. Imhoff to Thulemeyer, October 9, 1694, UB F, Ms. Thulemeyer, 769–71.

16. Millstone, "Designed for Collection"; Claydon, "Daily News," 74–77. Cf. Schuwey, "Les périodiques."

17. These neatly written fact sheets may go back to a previous round of hasty excerpting, perhaps on small slips of paper. One such preliminary excerpt is attached to a clean copy of a fact sheet about the Bournonville family; see GNM, Généalogies B (vol. 7), unpaginated (Bournonville).

18. Behrens to Leibniz, August 21, 1710, AA prepublication, available at https://www.gwlb.de/leibniz/digitale-ressourcen/repositorium-des-leibniz-archivs/laa-transkriptionen-1710: "Weil die nahmen der beyden in vorige woche zu Berlin gebohrnen Prinzen in den Zeitungen nicht gemeldet werden, bitte dienst. umb deren geneigte communication, dan nicht gerne eine gelegenheit des Imhofii Notitiam Procerum Imp. zu continuiren verabseumen mag."

19. I have compared the large number of excerpts for the Colonna family (GNM, Généalogies C (vol. 8), unpaginated (Colonna)) with Imhoff, *Genealogiae Viginti iIlustrium Familiarum in Italia*, 228–42. Concerning Sigismondo III d'Este (1647–1732), the manuscripts note that from "Nov. 1678" he was going to be "Grand Maitre" of the house of Savoy, while the printed text merely states, "SIGISMUNDUS III . . . Sabaudiae Duci inserviendo aetatem peregit" (spent his life in the service of the Duke of Savoy). See Imhoff, *Corpus Historiae Genealogicae*, 64. This is as specific as it gets in terms of taking up the news excerpts. Imhoff, *Genealogiae Viginti Illustrium Familiarum in Italia*, 241 (remarks about a dispute over precedence between Lorenzo Colonna and the Spanish grandees in 1680; a passage concerning the second marriage of Filippo Alexander Colonna to Olimpia Panfilio on November 25, 1697), for instance, goes back directly to GNM, Généalogies C (vol. 8), unpaginated (Colonna).

20. Nevitt, "Ben Jonson."

21. Mandosio to Imhoff, January 12, 1697, GNM, Letters, unpaginated.

22. Imhoff to Moritz Wilhelm, August 27, 1695, SächsHStA, loc. 8224, no. 7, fol. 15r.

23. Imhoff to Thulemeyer, October 9, 1694, UB F, Ms. Thulemeyer, 769–71.

24. Fabricius to Imhoff, February 24, 1709, GNM, Letters, unpaginated.

25. Juncker to Imhoff, June 12, 1704, GNM, HA V. Historiker: Juncker, unpaginated: Duke Bernhard of Sachsen-Meiningen had died on April 27, not 17.

26. There is no specialized literature on this subject, but see Stollberg-Rilinger, *Maria Theresia*, 10–11, and Tölle, *Heirs of Flesh and Paper*, passim. Patrician (or recently ennobled) families like the Imhoffs seem to have imitated this practice. See Imhoff to Fabricius, November 12, 1717, KB, Ms. Thott 1227, unpaginated: he had heard about the death of Rudolf Christian von Imhoff, "aber von niemandem Seiner Angehörigen notificirt worden."

27. Occasionally, these letters also fell somewhat short of their promises. Sometimes they lacked details (Moritz Wilhelm to Imhoff, January 5, 1697, Veste Coburg A, I, 349, [2], 3) or contained only "generalia" (Andreas Lazarus to Jakob Wilhelm Imhoff, [month unknown] 11, 1691, GNM, Nachlass Imhoff VII (Collection Imhoff), no. 3, "Teil 1," unpaginated).

28. Andreas Lazarus to Jakob Wilhelm Imhoff, December 29, 1682, GNM, Nachlass Imhoff VII (Collection Imhoff), no. 3, "Teil 1," unpaginated: "scachant que les informations qui se tirent des gazettes publiques, se trouvent bien souvent assés trompeuses touchant les jours de la nativité et de la mort de personnes de qualité je me suis imaginé de ne vous desobliger pas, si je vous envoyois l'extract que voyes, de nos lettres de notifications, comme elles s'appellent; dans

lesquelles vous trouverés marqué exactement le jour du mariage, de la nativité, ou de la mort, de la plus part des Princes qui sont alliés à Son Alt. mon Maistre. Il est vray que les noms des enfans nés ne s'y trouvent pas, la coustume n'estant que d'avertir generalemment, que c'avoit esté un Prince ou une Princesse, mais j'espere que vous pourrés supplier facilement à ce defaut de vos propres memoires, et si vous me voules croire vous ne feriés pas mal si vous tachiés de tirer aussi de semblables connoissances de toutes les autres cours. C'est qui rendroit vostre oeuvre plus parfait et plus illustre."

29. On archiving notifications, see also Tölle, *Heirs of Flesh and Paper*, 119–20.

30. One such case can be found in Andreas Lazarus to Jakob Wilhelm Imhoff, January 24, 1691, GNM, Nachlass Imhoff VII (Collection Imhoff), no. 3, "Teil 1," unpaginated.

31. Wagner to Imhoff, March 27, 1688, BSB, Autogr. Wagner, unpaginated: "ein großer Liebhaber deren Genealogien." Moritz Wilhelm also was, until 1705, the lord of Christian Juncker in Schleusingen. Imhoff was probably connected to Moritz Wilhelm by Christian Wagner through his friend Johann Avenarius, court preacher in Zeitz. Avenarius had, in turn, approached the duke's secretary for correspondence (*Briefsekretär*), Tobias Wilhelm von Gebler; cf. Wagner to Imhoff, March 27, 1688, BSB, Autogr. Wagner, unpaginated. There is little research on Moritz Wilhelm, a fascinating figure not least because he changed religion several times. Some context about him may be found in Czech, *Fürsten ohne Land*. See also Buder, *Merckwürdiges Leben*.

32. For instance, the duke had personally reviewed historiographical works by Christian Juncker and discussed his impressions at three meetings with the author. See Engel, "Christian Junckers wissenschaftlicher Briefwechsel." See also Juncker to Imhoff, October 30, 1702, GNM, HA V. Historiker: Juncker, unpaginated.

33. Imhoff to Moritz Wilhelm, June 27, 1695, SächsHStA, loc. 8824, no. 7, fol. 3r mentions a recent visit of Imhoff at Mortiz's residence. A return visit "in Imhoff's house" in January 1711 is mentioned in Imhoff to Moritz Wilhelm, May 20, 1711, SächsHStA, loc. 8224, no. 8, fol. 114r; and Moritz Wilhelm to Imhoff, June 6, 1711, Veste Coburg A, I, 349, (2), 3, unpaginated.

34. *Ex negativo*, see Moritz Wilhelm to Imhoff, January 11, 1703, Veste Coburg A, I, 349, (2), 3, unpaginated: "die Notifications Schreiben nicht bey der Hand."

35. E.g., Moritz Wilhelm to Imhoff, May 10, 1695, and May 29, 1697, Veste Coburg A, I, 349, (2), 3, unpaginated.

36. Moritz Wilhelm seems to have filed his letters in chronological order. This is what I infer from Moritz Wilhelm to Imhoff, January 5, 1697, Veste Coburg A, I, 349, (2), 3, unpaginated: "die verlangten Notificationes derer in 4 Monaten her eingelauffenen Geburthen vermählungen und Todesfällen anlangend."

37. Imhoff prepared several such lists and sent them to Zeitz. Cf. Imhoff to Moritz Wilhelm, February 5, 1706, SächsHStA, loc. 8224, no. 8, fol. 44r. Summaries for the years 1710 and 1714 in SächsHStA, loc. 8224, no. 8, fols. 112r–113r; no. 9, fols. 24r–v. Moritz Wilhelm probably recorded incoming contemporary information about recent dynastic events on folio pages like those that he used when writing to Imhoff—occasionally, these fact sheets were subsequently (mis)filed with letter drafts. See SächsHStA, loc. 8224, no. 8, fols. 27r, 28r, and the draft of his letter to Imhoff from November 15, 1704, ibid., fols. 26r–v. The letter mentions the four dynastic events noted on the fact sheet, as well as many more.

38. Cf. Frisch, *Invention of the Eyewitness*; McDowell, *Invention of the Oral.*

39. "Dici non potest . . . , quam sit difficile ejusmodi res minutas, et quae ex libris disci non possunt, sed ex memoria viventium et paucissimorum quidem petuntur, percontando rescio: infiniti sumus, qui Kolowratos qui Schlickios etc. noverimus, at vix ullus qui sciat, quot et qui fratres fuerunt, quae sorores, quo ordine geniti, quis mater patris, qui affines, soceri, qui generi, et nomina singulorum. haec ex ipsa domo eorum petenda sunt." Balbín to Imhoff, August 12, 1682, GNM, Letters, unpaginated.

40. See, e.g., Moritz Wilhelm to Imhoff, January 19, 1698, Veste Coburg A, I, 349, (2), 3,

unpaginated: "Die Nachricht wegen des Hauses Holstein Plön hat mir ein [. . . Plön?]nischer Hoff Rath Questel mündlich gegeben."

41. Sant'Anna to Imhoff, March 30, 1706, BSB, Cod. Ital. 506, fols. 105r–106r, here 105r: "e da lui mi sono informato come quello che mori nell'anno 1692 Borgia Governatore in Fiandra, havere per moglie una S.gra di Casa Enriquez colla quale non fece figli, e cosi il titolo del Ducato di Vallaermosa si è estinto. Il medesimo S.r Principe mi hà detto che il fà S.r Marchese ultimo d'Astorga hebbe una sola moglie sorella del morto Duca di Benavento Pimintello, e queste notizie sono certissime."

42. Sant'Anna to Imhoff, November 8, 1700, GNM, Généalogies G (vol. 11), unpaginated (Gaetani). The "principessa Nicandro" was Isabella Gaetani di Sermoneta; hence she was in a position to provide information about the Caetani family.

43. Would the count of Pfalz-Birkenfeld really convert to Catholicism? No, Imhoff learned from Count Helferich von Welz, who had heard the Pfalzgraf discuss this issue "at dinner" (*über der Tafel*); Imhoff to Moritz Wilhelm, May 26, 1699, SächsHStA, loc. 8224, no. 10, fol. 61r.

44. GNM, Généalogies B (vol. 7) unpaginated (Baden): "Mad. la Comtesse de Wied hat mir gesagt d. 27. Jul. 1680, daß Prinz Ludwig von Baden von Jugend auf eine affection zu einem Frl. von Löwenstein Nahmenß Amalia getragen, u sich mit selbiger in ein versprechen de futuro eingelassen, weiln er aber adulta aetate alß ihm die Neuburgische Prinzessin in Sinn kommen, ihrer vergessen, habe Sie sich in ein Kloster begeben. Seine cousins germains deß Marggraf Leopoldi Zwey Söhne wären übel conditionirt, der ältere stumm u[nd] der andern nicht viel beßer."

45. Sant'Anna to Imhoff, March 30, 1706, BSB, Cod. Ital 506, fols. 105r–106r, here 105v: "queste notizie sono certissime."

46. McDowell, *Invention of the Oral.*

47. Wagner to Imhoff, February 25, 1689, BSB, Autogr. Wagner, unpaginated: "Hesterno die audivi, dum apud virum rerum exterarum gravissimum pranderem, Henricum Principis Dessaviensis filiae unicae maritum nunc esse. Domum reversus evolvi tabulas, sed plures una principes Dessavienses reperi."

48. As I see it, there was really only one "Prince of Dessau" at the time, Johann Georg II (d. 1693), who, however, had at least four surviving daughters, *two* of whom indeed married Heinrichs. My hunch is that Wagner is referring to Elisabeth Albertine von Anhalt-Dessau (1665–1706), who was married to Herzog Heinrich (Sachsen-Weissenfels-Barby) (1657–1728), because the Wettin duke is someone Wagner would have known about.

49. Balbín to Imhoff, August 12,1682, GNM, Letters, unpaginated: "at si Praga absunt, et in Moravia vel Silesia, vel Viennae alibique degunt, unde scias?"

50. See d'Hozier to Imhoff, December 31, 1707, and March 9 and May 25, 1708, all in GNM, Letters, unpaginated: all letters refer to waiting for Salviati. The letter from September 21, 1708, implies that Salviati had finally arrived in Paris.

51. Lohsträter, "Zeitgeschichte."

52. For a brief survey, Eddy, "Diagrams."

53. Even-Ezra, *Lines of Thought.*

54. Mary Bouquet, "Family Trees"; Weigel, *Genea-Logik*; Bauer, *Wurzel, Stamm, Krone.*

55. "Plures liberi immature mortui videantur" (lower left quarter of the double folio).

56. GNM, Généalogies T (vol. 21), unpaginated (Tilly).

57. This amounted to an alphabetical rearrangement of Rittershausen's book. For a similar alphabetical rearrangement of Rittershausen, see the copy at Vienna, ÖNB 74.N.5 (available digitally).

58. For a pioneering statement, see Knight, *Bound to Read.*

59. Imhoff, *Corpus Historiae Genealogicae*, 44–64.

60. See, e.g., a statement from Pigna (1570), in Santi, *La precedenza*, 67–68, and Girolamo

Falletti's Este stemma in Reineckius, *Annales Helmoldi*, 225–32. See also Erdner, "Leibniz," 33, 123.

61. On all of this, see the monumental work by Erdner, "Leibniz." Azo had first married Kunigunde, heir of the "older Guelfs," and mother to Welf IV, ancestor of the modern (i.e., the "younger") Guelfs. In a second marriage, Azo II wedded Grisande, daughter of the Count of Maine. From this second marriage, Fulco, the progenitor of the modern-day Este, was born.

62. Imhoff excerpted material from Louis Moréri's *Le grand dictionaire* [*sic*] (1674) and used it (in contrast to Leibniz) to record the name of Azo II's second wife: Ermengard; cf. Imhoff, *Corpus Historiae Genealogicae*, 53. He owned Leibniz, *Lettera su la connessione*, a text he also excerpted; see GNM, Généalogies F (vol. 10), unpaginated (Ferrara).

63. Leibniz to Imhoff, December 7, 1696, AA I, 13: "Ultra [sc., prior to Alberto Azo II] autem non descendi quod communia nostris et Estensibus hoc limine finiuntur." On Leibniz's later and much more advanced work on Alberto Azo II and his pedigree, see Erdner, *Leibniz*. None of this, however, would have been available to Imhoff around 1700.

64. Imhoff, *Corpus Historiae Genealogicae*, 52, closely resembles Leibniz, *Lettera su la connessione*,)11(.

65. He owned a "Relation von der Lehenempfänguß . . . ," in GNM, Généalogies F (vol. 10), unpaginated (Ferrara).

66. Muratori to Imhoff, September 28, 1701, and November 24, 1701, in Busino and Dufour, *Lettere*, 433, 434.

67. Muratori to Imhoff, September 28 and November 24, 1701, in Busino and Dufour, *Lettere*, 433, 434. Cf. Imhoff to Muratori, September 4 and October 10, 1701, in Lieber, "Jacob Wilhelm von Imhofs Korrespondenz," 63–64.

68. Pigna, *Historia de principi di Este*. On this, see Santi, *La precedenza*.

69. Imhoff, *Corpus Historiae Genealogicae*, 52, 55 ("peperam traditum esse a Pigna"), 57.

70. On Falletti's "Estensium Marchionum Genealogia," see Santi, *La precendenza*, 20n, 82; Henniges, *Theatrum genealogicum*, 142.

71. Donati, "Una famiglia lombarda."

72. See the small fact sheet "Acc. Vienna 20. Mart 1682," GNM, Généalogies F (vol. 10), unpaginated (Ferrara). Someone in the ambassador's entourage continued to send material; see the memorandum, "Acc. Vienna ex aedibus Marchionis de Borgomanero . . . m. Jan. 1694," in GNM, Généalogies F (vol. 10), unpaginated (Ferrara).

73. See the stemma "Acc. a P. Bacchino, Mutinensi . . . d. 10. febr. 1694," GNM, Généalogies F (vol. 10), unpaginated (Ferrara).

Chapter 6 • *Publishing and Reading Genealogy*

1. StA N, Handschriften 286, fol. 498v. See also ibid., fols. 541r–542v.

2. Andreas Lazarus to Jakob Wilhelm Imhoff, April 22, 1686, GNM, Nachlass Imhoff VII (Collection Imhoff), no. 3, unpaginated.

3. Petrus, *Suevia Ecclesiastica*, 311, 529 ("virum, ut in coeteris apparet, solertem, veritatem amantem, & a parium studio alienum"), 736.

4. Anonymous, *Der neueste Staat*, 682–83.

5. Reimmann, *Historiae Literariae*, 1:75, 94–95.

6. Zschackwitz, *Historisch-genealogischer Schauplatz*, unpaginated (Vorbericht); Ludewig, *Germania Princeps*, 26; Johann Hübner, *Bibliotheca Genealogica*, 237, 307 ("nichts zu verbessern"), 308, etc.

7. *Deutsche Acta Eruditorum* 6 (1690): 268–72.

8. *Deutsche Acta Eruditorum* 1 (1703): 1–7 ("pulcerrimo libro"; "movere salivam lectoribus"). *Deutsche Acta Eruditorum* 9 (1710): 374–80 ("splendidum opus"). Cf. the positive general statement about Imhoff, *Deutsche Acta Eruditorum* 3 (1713): 97–98. Another positive review of

the 1710 book appears in *Neue Bibliothec oder Nachricht und Urtheile von neuen Büchern und allerhand zur Gelehrsamkeit dienenden Sachen* (1710): 190–97. A very positive statement about the 1710 book on Italy, printed by Châtelain, appears in Sant'Anna to Imhoff, July 15, 1710, BSB, Cod. Ital. 506, fols. 125r–126r, here 125v.

9. See, e.g., Chartier, *The Order of Books.*

10. Bauer's overall perspective is brilliantly summarized in his monograph: Bauer, *Wurzel, Stamm, Krone.*

11. For the idea of a "genealogical brand," see also Tölle, "Generation, Theft, and Dynasty," 308. Numerous authors and works outside Germany also became standard references in the field, such as André Duchesne (1584–1640) in France and Wiliam Dugdale (1605–86) in England.

12. Ann Blair (Harvard) is currently working on a monograph on unnamed helpers in early modern scholarship. For a brief preliminary statement, see, e.g., Bevilacqua and Clark, *Thinking in the Past Tense*, 35–36.

13. Anonymous, *Tredecem Tabulae Genealogicae.*

14. Christoph Molitor, Tübingen, to Johannes Molitor, Nuremberg, February 25, 1656, SUB GÖ, Ms. philos. 102, no. 45, recounts the origins of the work: "Illustrissimorum enim Comitum Ostfrisiae ut & Ottingensium generosi & nobiles Ephori Dn a Krott & Dn. a Borck ut & Malecomesius a Schencken Ephorus Magno Conringio notissimus non tantum facie sed & per literas, tabellas quasdam conscripserunt Principum Germaniae Genealogicas, in quibus antiquiorum incuriosi, saltem ut diligentius hoc saeculum & novissima examinarent, desudarunt, usibusque sibi privatis, ut decitetur scribendi onus typis expresserunt." None of these men are mentioned in Reimmann, *Historia Literariae*, 77–78, who also seems very uncertain about the authors. Molitor, Tübingen, to Rittershausius, May 3, 1657, SUB GÖ, Ms. philos. 102, no. 48.

15. In his preface to the third edition of 1670 (anonymous, *XIV Tabulae Genealogicae*, fols.)(2r–v), Johann Georg Cotta, the publisher, situates the *Tafeln* alongside Rittershausen's own works. Other publishers and authors also tried to capitalize on Rittershausen's name shortly after the author's death in 1670; cf. Diefenbacher, *Das Nürnberger Buchgewerbe*, 381: The Nuremberg firm of Johann Hoffmann published a poorly produced book on Franconia under the name of the late Nicolaus Rittershausen in 1673.

16. Only a few manuscripts in Imhoff's dossiers can be traced back to Rittershausen with certainty, see, e.g., two genealogical tables of the counts of Tilly, GNM, Généalogies T (vol. 21), unpaginated (Tilly). This clear copy of a genealogical table almost perfectly matches the version printed in Rittershausen, *Genealogiae* (3rd ed., 1664), unpaginated.

17. There remain, among Jakob Wilhelm's papers, a few letters from genealogical informants written to his father. One Tobias Francke, pastor in Michelbach, for instance, sent Wilhelm Imhoff a few pieces of information on the counts of Hohenlohe; see GNM, Généalogies H (vol. 12), unpaginated (Hohenlohe), two letters dated March 15, 1678, and May 26, 1680.

18. Schlüter to Imhoff, June 27, 1676, GNM, Letters, unpaginated: "Feceris erudito orbi rem gratissimam, si per modum appendicis augeas Rittershusianas tabulas. potes enim prae aliis, quia Te aeque ac nobilissimum Dn. Parentem tuum harum rerum curiosum scrutatorem per annos non paucos fuisse scio."

19. Imhoff considered the *Spicilegium* a "prodromus" to the *Notitia Procerum*; see Imhoff to Balbín, April 16, 1683, Prague, Knihovna Národního Muzea v Praze, XIX.A.3, no. 343.

20. Juncker to Imhoff, April 8, 1700, GNM, HA V. Historiker: Juncker, unpaginated: "Cum vero utrumque librum, Notitiam videlicet et Tabellas ad mutuos usus commode destinaverim; inde contigit, ut modo in has aliqua retulerim, modo in illam, alia etiam ad familias Gallicas abs TE editas, et neutrum adeo opus perfectum sit quidem, simul tamen sumtis nihil deesse videri queat."

21. The print run of Lohmeier had not yet concluded by the end of 1701; see Imhoff to Thulemeyer, December 2, 1701, UB F, Ms. Thulemeyer, 799–801. A new, revised edition is al-

ready mentioned in Imhoff to Thulemeyer, May 7, 1700, ibid., and Juncker to Imhoff, April 8, 1700, and January 18, 1701, GNM, HA V. Historiker: Juncker, unpaginated. The tables of part 1 of the work (on the royal houses of Europe) are the same as those in the 1695 edition. The layout of many of the tables in the second part, however, was significantly improved. Only minor changes in content were made, sometimes affecting the tone and thrust of the genealogical information. Moreover, on several occasions, Imhoff added new material and noted recent dynastical events. For later discussion about yet another edition, see Juncker to Imhoff, January 12, 1706, and May 6, 1710, ibid. Juncker announced to Imhoff (October 8, 1710, ibid), that the publisher Thomas Fritsch in Leipzig might be interested in such a project.

22. See Heinrich Linck's happy remark in a dossier on the counts of Plauen, GNM, Généalogies R (vol. 19), unpaginated (Rheingrafen): "Et hactenus in Rittershusio nihil corrigendum, sed saltem ea, quae rubro colore notata, adhuc addenda sunt." This did not exclude the correction of obvious errors; see, e.g., a "Bericht auff verschiedene Fragen, das Hochg. Wild- und Rheingr. Haus betreffend," ibid., here "nr 1." See also Heinrich Linck's notes on "Notabiliae quaedam Circa Genealogiam Illustrissimorum Comitum Ruthenorum Dominorum de Plauen, Secundum quae Tabulae Genealogicae Beati D.N. Rittershusii partim corrigendae, partim supplendae sunt," ibid., appearing in print in Imhoff, *Notitia Procerum* (1st ed., 1684), 1047–60.

23. Rittershausen, "praefatio ad lectorem," in *Genealogiae* (3rd ed., 1664), unpaginated: "Si vita suppetet, & otium continget, personis breves at convenientes rerum gestarum descriptiones attexam." Cf. Rittershausen, *Brevis Exegesis.*

24. Spener, *Theatrum*, consists mostly of tables, whereas Spener, *Sylloge* (1st ed., 1668), is mostly text.

25. Imhoff to Weise, June 23, 1682, CWB, Mscr.A. 70–71 (32-2), unpaginated. Imhoff to Balbín, May 1, 1682, in Rejzek, *Bohuslav Balbín*, 292–94 note **: "sed diversam ab illa, quam Rittershusius tenuit methodum elegi neque nudas tabulas aut propagationum series tantum exhibere, sed alia quoque ad historiam et jus publicum pertinentia tradere constitui."

26. Balbín to Imhoff, October (no day mentioned), 1687, GNM, Letters, unpaginated: "Aliam ego, et diversam a Te rationem in Genealogicis secutus sum: Tabulas nudas, et sine Notis in hoc Tabulario attuli, origines Familiarum postea in Stemmatographia, quam paro, allaturus."

27. Pregitzer to Imhoff, December 25, 1686, GNM, Letters, unpaginated: "Gaudeo etiam notas historicas, quibus Galli maxime amant suas Genealogias illustrare & probare, illis Tabulis addi."

28. *Deutsche Acta Eruditorum* 9 (1710): 376: "Singulae stirpes in suos ramos diffusae per tabulas oculis curiosis subjiciuntur; ne vero ex sola nominum allegatione taedium Lectori subnasci queat, sitim explent abunde additae statim *Exegeses historicae*."

29. The copy under consideration is Staatsbibliothek Bamberg, Gen f 6. On a page dedicated to Duke Friedrich Wilhelm of Altenburg (1562–1643) and his four sons, the anonymous reader got lost in the maze of text and lines. Just to make sure the visual challenges did not lead him astray, he added handwritten commentaries to the names of the sons, highlighting their filial and brotherly relationships: he added a "son of Friedrich Wilhelm" or a "brother" to their names, thus making even clearer what was already there. He even drew an additional line by hand connecting one son to Friedrich Wilhelm, to make clear at first sight the son's descendance from his father. Strictly speaking, none of this was neccessary, as Imhoff's printed page provided all the information in unambigous terms and signs—but, perhaps, not with enough immediate visual clarity, at least not with enough clarity for this particular reader.

30. Albizzi, *Principium Christianorum Stemmata.*

31. Balbín to Imhoff, July 16, 1688, GNM, Letters, unpaginated: "Tua clariora et apertiora sunt et prorsus analytica et methodica."

32. Several contemporaries, for instance, preferred Imhoff to the equally popular Johann Hübner, *Drey hundert*; see Juncker to Imhoff, May 6 and October 8, 1710, GNM, HA V. His-

toriker: Juncker, unpaginated, and Zschackwitz, "Vorbericht," in *Historisch-genealogischer Schauplatz*, unpaginated. The book by Christian Hübner, *Aller Durchläuchtigen Hohen Häuser*, which was also praised by Zschackwitz, gave much shorter narratives than Imhoff.

33. Lohrer, *Cotta*.

34. Imhoff's relationship with Châtelain remained positive even after their cooperation ended, as the printer helped relay numerous shipments of Imhoff's books. See, e.g., d'Hozier to Imhoff, July 25, 1710, and August 21, 1711, GNM, Letters, unpaginated.

35. Juncker to Imhoff, March 13 and October 29, 1706 (cooperation with Rüdiger in Berlin falls flat because of the firm's bankruptcy), GNM, HA V. Historiker: Juncker, unpaginated. For ideas about publishing in Italy, see Imhoff to Fabricius, October 25 and November 19, 1706, KB, Ms. Thott 1227, unpaginated. See also Imhoff to Muratori, December 21, 1698, in Lieber, "Jacob Wilhelm von Imhofs Korrespondenz," 58–59: "sed defuit qui sumptus impressionis susciperet." Difficulties with printing appear in Imhoff to Moritz Wilhelm, May 6, 1701, SächsHStA, loc. 8224, no. 7, fol. 90v, and Imhoff to Moritz Wilhelm, September 30, 1698, SächsHStA, loc. 8224, no. 10, fol. 51v.

36. Imhoff to Moritz Wilhelm, December 28, 1696, SächsHStA, loc. 8224, no. 7, fol. 42r. Imhoff to Leibniz, January 8, 1697, AA I, 13:477: "enixissime flagitante bibliopola."

37. D'Hozier to Imhoff, April 22, 1699, GNM, Letters, unpaginated. The letter includes Anisson's formal offer that d'Hozier forwarded to Imhoff (no date, ca. end of April 1699, arriving in Nuremberg on June 5). Willemetz, *Jean Anisson*, does not mention Imhoff.

38. Cf. Anisson's offer, in d'Hozier to Imhoff, April 22, 1699, GNM, Letters, unpaginated (see preceding note).

39. D'Hozier to Imhoff, July 18, 1699, GNM, Letters, unpaginated: "Marchands libraires, ils veulent gratuitement l'oeuvre, la peine, . . . et ils ne trouvent pas d'jnjustice dans cette conduite."

40. For "histoire du temps," see Imhoff to Moritz Wilhelm, May 6, 1701, SächsHStA, loc. 8224, no. 7, fol. 90v. See also Imhoff, *Historia Italiae et Hispaniae Genealogica*, preface.

41. D'Hozier to Imhoff, July 8, 1701, GNM, Letters, unpaginated.

42. The introduction relied on an exposé by Franckenau; see Franckenau to Imhoff, April 3, 1706, GNM, Letters, unpaginated.

43. D'Hozier to Imhoff, March 17, 1695, GNM, Letters, unpaginated.

44. Juncker to Imhoff, March 7, 1704, GNM, HA V. Historiker: Juncker, unpaginated.

45. There is virtually no scholarly literature on the *Gotha*; cf. Fritsch, *Die Gothaischen Taschenbücher*. Cf. also Mix, "Genealogische Kalender."

46. D'Hozier to Imhoff, November 6, 1699, and October 16, 1700, GNM, Letters, unpaginated. At issue was a sixteenth-century grandmaster of the order, François Salviati. See Tiersonnier, "Compte-rendu"; Poli, "Introduction." Guénegaud authored an "Histoire de l'Ordre hospitalier régulier et militaire de Saint-Lazare de Jérusalem et de l'ordre militaire et régulier de Notre-Dame du Mont-Carmel, par M. de Guénégaud (1714)," Bibliothèque national de France, ms. fr. 24.967, about which contemporaries had few positive things to say. See Poli's, "Introduction," 546.

47. Gittio to Imhoff, April 6, 1688, GNM, Letters, unpaginated: Aldimari "sia cosi secreto nelle sue actioni, che non fa penetrare ad altri le cose sue, e percio non saprei dirli tempo per quando uscira detta Opera della Casa Carafa." Aldimari's letters to Imhoff contain little actual genealogical information, and his tone is often self-congratulatory. See also Aldimari to Magliabechi, December 14, 1680, in Quondam and Rak, *Lettere*, 8–9.

48. Bulifon to Aldimari, June 25, 1690, in Bulifon, *Lettere memorabilia*, 2:302–11. On their cooperation, see Ajello, *Arcana juris*, 31–32n6; and Bulifon, *Journal du voyage*, 71–72.

49. Bulifon to Magliabechi, September 1, 1691, in Quondam and Rak, *Lettere*, 155–56.

50. Weise, *Schediasma*, 10. Once Imhoff became aware of this announcement in early 1682,

he immediately wrote to Weise and introduced his own, already ongoing research. Imhoff to Weise, March 25, 1682, CWB, Mscr.A. 70–71 (32-1). Imhoff may have been alerted to Weise's piece by Thulemeyer; see Thulemeyer to Imhoff, no date (but, presumably, late 1681 or very early 1682), UB F, Ms. Thulemeyer, 709–12. Imhoff to Thulemeyer, August 8, 1683, ibid. Schlüter to Imhoff, July 5 and November 25, 1681, GNM, Letters, unpaginated, trying to calm the genealogist.

51. Spener to Imhoff, July 12, 1683, in Spener, *Briefe aus der Frankfurter Zeit*, 6:675–79. It is currently unclear whether the sale ever happened and, if so, what happened to the material—did it end up on Imhoff's desk?

52. Spener to Imhoff, November 15, 1686, in Spener, *Briefe aus der Dresdener Zeit*, 1:151–53. Spener contemplated entrusting the final editing to the Leipzig historian Franckenstein. In 1683, Cotta seems to have considered editing Spener's French genealogies to be too risky; see Pregitzer to Imhoff, December 25, 1686, GNM, Letters, unpaginated.

53. Spener to Rechenberg, October 26, 1686, in Spener, *Briefwechsel mit Adam Rechenberg*, 1:28: "ne alter alterius obsit laboribus." See also the deferential remark about Imhoff in Spener to Rechenberg, November 21, 1686, in Spener, *Briefwechsel mit Adam Rechenberg*, 1:49. Further letters of Spener to Rechenberg (ibid.) confirm the overall picture: Spener did everything he could to avoid interfering with Imhoff's forthcoming publication.

54. Spener to Imhoff, December 6, 1686, and May 17, 1687, in Spener, *Briefe aus der Dresdener Zeit*, 1:178–80, 350–52. Wagner to Imhoff, February 23, 1688, BSB, Autogr. Wagner, unpaginated.

55. Spener to Paul Anton, May 30, 1688, in Spener, *Briefe aus der Dresdener Zeit*, 2:244–45. See also Kekulé von Stradonitz, *Ausgewählte Aufsätze*, 191.

56. Spener, *Illustriores Galliae Stirpes*, unpaginated ("praefatio"). See Spener to Rechenberg, December 14, 1686, in Spener, *Briefwechsel mit Adam Rechenberg*, 1:63: "Et quid interest, qui genealogias Franciacas universas habere velit, duorum autorum an unius nomine editas conquirat?"

57. Cf. Imhoff's commentary on [Bressler und Aschenburg], *Die Heutigen Christlichen Souverainen*. This book, he assured himself, was of low quality; see Imhoff to Thulemeyer, May 29, 1702, UB F, Ms. Thulemeyer, 813–15.

58. Genicot, *Les généalogies*, 26–27. See also Trevisan, *Royal Genealogy*, 24–25.

59. Knight, *Bound to Read*, 6–7.

60. University Library Ghent, BIB.HIST.003466 (viewed electronically).

61. Bibliothèque Municipale de la Ville de Lyon, call number 24345 (viewed electronically).

62. Imhoff, *Notitia Procerum* (5th ed., ed. Köhler, 1732), unpaginated (preface by Köhler).

63. See, e.g., Crick and Walsham, *Uses of Script*.

64. Juncker to Imhoff, October 8, 1710, GNM HA. V. Historiker: Juncker, unpaginated: "Dabey aber wünschete er [the publischer Thomas Fritsch] ohnmaßgeblich, daß niht allein die vitae Principum, dem kurzen Inhalt nach, fortgesetzet, sondern auch, zumahl der abgestorbenen Fürsten Töchter, vornehmlich diejenigen, so verheyrathet worden, beygezeichnet würden, zum behuf derer, so umb die Ahnen, bey Vorfertigung einer Stammtafel oder Erklärung der Ahnen-Wapen, besorget seyn müssen."

65. Czech, *Legitimation*, 57 (Heinrich XIII), 301 (Albrecht Christian Ernst). See also Institut für Sächsische Geschichte und Volkskunde, Sächsische Biografie, http://saebi.isgv.de/biografie/Albert_Christian_Ernst,_Graf_von_Sch%C3%B6nburg_(1720-1799).

66. Imhoff, *Genealogia Ruthenorum*, Sächsische Landes- und Universitätsbibliothek, Dresden, H.Sax.F. 26 (viewed electronically).

67. E.g., the manuscript notes on Tabula IV.

68. Lohmeier, *Der Europäischen Reiche*, Staatsbibliothek Regensburg, 999/2Hist.pol.483 (viewed electronically).

69. Most of the annotations are more or less adjacent to the printed text that prompted them. For an exception, see a long note about Charles d'Anjou (Tabula XXVII) that appears only in the blank pages after Tabula LIV.

70. Nothing is known about the provenance of this copy (e-mail from Bernhard Lübbers, head librarian of the Staatsbibliothek Regensburg, June 11, 2018).

71. Many of the annotations were taken from the following two works: Weller, *Deutscher Adler*, and Ziegler und Kliphausen, *Historisches Labyrinth*.

72. See Mareš, "Augustiniánský historiograf," 407–9.

73. Prague, Knihovna Národního Muzea v Praze, Mss. VI F 14, fols. 97r–256r (Notata ex tertia Jac. W. Imhoffii editione de notitia historico-heraldico-genealogica [1693] eiusdemque Mantissa [1699], independent pagination).

74. Notata ex tertia Jac. W. Imhoffii editione, p. 21.

75. On one occasion, he adds "acatholicos"; on another, he adds that Cardinal Borromeo was related to Pope Pius IV: Notata ex tertia Jac. W. Imhoffii editione, pp. 9, 26.

76. Notata ex tertia Jac. W. Imhoffii editione, p. 59.

77. Numerous references to Imhoff appear in Lingnau, *Lektürekanon*, 306–7.

78. Staatsarchiv, Wolfenbüttel 112 Alt, no. 305.

79. Römer, "Die Grafen von Regenstein-Blankenburg," 83–87. For further details, see Köcher, "Der preußisch-welfische Hoheitsstreit."

80. Rassler, *Vindicatio*, 34. Rassler quotes Imhoff, *Corpus Historiae Genealogicae*, 83, where the latter quotes the Dominican General Chapter's description of St. Thomas Aquinas as "the pride of Christendom." Rassler, however, omits mention of the Dominicans and (mis)attributes the judgment directly to Imhoff, thus citing a Protestant voice in support of a Catholic saint.

81. Anton Ulrich to Jakob Wilhelm, October 12, 1712, GNM, Nachlass Imhoff VII (Collection Imhoff), no. 3, unpaginated: "Unterricht wegen unserer ahnen."

82. Anton Ulrich to Jakob Wilhelm, October 12, 1712, GNM, Nachlass Imhoff VII (Collection Imhoff), no. 3, unpaginated (concerning his presentation at the Stift in Naumburg).

83. Johann Joseph to Jakob Wilhelm, November 19, 1719, GNM, Nachlass Imhoff VII (Collection Imhoff), no. 3, unpaginated (no. 104). Johann Joseph asks for the production of a booklet in quarto (*libell*) containing the "origo familiae Im Hoffianae." He says he had himself contemplated this project, but now is calling on his expert cousin to produce such a family history.

84. GNM, Nachlass Imhoff VII (Collection Imhoff), no. 3, unpaginated (no. 117), April 28, 1721.

85. Andreas Lazarus to Jakob Wilhelm, October 6, 1686, GNM, Nachlass Imhoff VII (Collection Imhoff), no. 3, unpaginated: "J'ay trouvé qu'il y avoit encore beaucoup de choses a dire, en faveur de S. A. et pour faciliter la restitution qu'on brigue, qui ne dependant que de l'histoire, n'estoit pas conneu a ces autres Mrs les Advocats du Duc qui ne traitterent cette affaire qu'en bons Juristes." See also Andreas Lazarus's letter of November 14, 1686, in which he gives further details about the positive reception given to Imhoff's information by the ducal emissaries in Vienna; ibid. This second letter also shows that the newly arrived genealogical information about Saxe-Lauenburg was carefully vetted against other information in Sulzbach.

86. Andreas Lazarus to Jakob Wilhelm, November 14, 1686, GNM, Nachlass Imhoff VII (Collection Imhoff), no. 3, unpaginated.

87. Andreas Lazarus to Jakob Wilhelm, January 20, 1687. Jakob Wilhelm dutifully obliged. Andreas Lazarus to Jakob Wilhelm, June 5, 1687, GNM, Nachlass Imhoff VII (Collection Imhoff), no. 3, unpaginated.

88. Andreas Lazarus to Jakob Wilhelm, December 4, 1686, GNM, Nachlass Imhoff VII (Collection Imhoff), no. 3, unpaginated, concerning a dispute between the Hanseatic cities of

Hamburg and Lübeck about lordship over the town of Mölln. Local conflicts are also mentioned, albeit vaguely, in Andreas Lazarus to Jakob Wilhelm, April 20, 1702, GNM, Nachlass Imhoff VII (Collection Imhoff), no. 3, unpaginated.

89. Imhoff nonetheless discussed the many succession crises that plagued the imperial nobility at the time of Duke Moritz Wilhelm; e.g., letter to Imhoff, April 7, 1696, Veste Coburg, Ms A, I, 349, (2), 3, unpaginated (on the Veldenz case). Occasionally, Imhoff called on his Sulzbach cousin Andreas Lazarus to obtain information specifically for Duke Moritz Wilhelm—a triangular exchange of information whereby the scholar, the prince, and the politician were connected by a shared interest in genealogy. See, e.g., Imhoff to Moritz Wilhelm, August 17, 1695, SächsHStA, loc. 8224, no. 7, fols. 8r–10v.

90. Genealogy was not always involved; see, e.g., eight letters from Johann Heinrich Acker to Imhoff, written from 1713 to 1718, GNM, Letters, unpaginated. See also Christoph Sonntag, Altdorf, to Imhoff, September 1, 1700, BSB, Autogr. Sonntag.

91. Ludewig to Imhoff, September 4, 1694, BSB, Autogr. Cim. Ludewig. Since the letter was addressed only to "Monsieur Im Hoff," it is unclear which family member was the recipient. Besides Jakob Wilhelm, Christoph Jakob might also have been the addressee.

92. Franck von Frankenstein to Imhoff, June 28, 1714, GNM, Letters, unpaginated.

93. Knorn to Imhoff, January 18, 1713, GNM, Letters, unpaginated. Frederick I died on February 25, 1713.

94. Redn to Imhoff, April 4 and July 9, 1720, GNM, Letters, unpaginated.

95. E.g., d'Hozier to Imhoff, July 6, 1696, GNM, Letters, unpaginated.

96. Testa, *Italian Academies*, 130–33.

97. See Stadtgeschichtliches Museum Leipzig, A/2014/2669/1, available online at https://www.stadtmuseum.leipzig.de/ete?action=queryDetails/1&index=xdbdtdn&desc=%22objekt+Z0112867%22 (accessed January 20, 2022).

98. Sant'Anna to Imhoff, July 9, 1703, BSB, Cod. Ital. 506, fols. 97r–98v.

99. D'Hozier to Imhoff, November 8, 1709, GNM, Letters, unpaginated. The new cache of prints arrived in late 1710. Obviously, this was a newly engraved version, one that pleased d'Hozier even more than the first one. See d'Hozier to Imhoff, November 25, 1710, and November 13, 1711, ibid. Available at Digitaler Portraitindex, http://www.portraitindex.de/documents/obj/33802136 (accessed January 20, 2022).

100. It appeared as the frontispiece of part 6, after page 456.

Conclusion

1. In Naples and elsewhere in Italy, public authorities had taken a firm stance against dueling since the early sixteenth century; see Muto, "I trattati napoletani," 335–36; Croce, "I duelli," 546.

2. Imhoff, *Corpus Historiae Genealogicae*, 106.

3. Bulifon, *Giornali di Napoli*, 196: "[1673] 16 novembre. Seguí in Norimberga, città franca del Palatinato superiore, nelle fortificazioni esteriori in una piazza detta il Lazzaretto, tra D. Giulio Acquaviva fratello del conte di Conversano e D. Francesco Carafa fratello del duca di Nola duello pubblico alla presenza di moltissimi popoli che vi erano andati di posta, quali ammiravano il valore e bravura de' cavalieri napoletani. Con que si pacificarono le loro case. Il Carafa fu ferito leggiermente al braccio."

4. Anonymous, *Fürtrefflicher Ritter-Kampf.*

5. Acquaviva's letter to the city government of Nuremberg, a copy of which survives in Imhoff's papers in GNM, Généalogies A (vol. 6), unpaginated (Acquaviva).

6. Bulifon, *Giornali di Napoli*, 196.

7. StA N, Reichsstadt Nürnberg, Ratsverlässe 2682, fol. 78r: the senate granted Acquaviva the right to take temporary residence in the city, while he waited for the affair to unfold.

8. Cf. Dewald, *Status, Power*, 13, 32–33.

9. Burguière, "L'état monarchique."

10. "Perceptual grid": from Spiegel, *Past as Text*, 105; "patterns of thought": from Heck and Jahn, *Genealogie als Denkform.*

11. Berghorn, *Verwandtschaft*, 277–81, understands this as an intentional transition from "communicative" to "collective" memory. Cf. the 258 relatives remembered by Pierre Bourdeille seigneur de Brantôme, *Oeuvres complètes*, 10:84–101, famously analyzed by Flandrin, *Families in Former Times*, 26–32.

BIBLIOGRAPHY

Archival Sources

Aurich

Staatsarchiv
Rep 241 Msc, C 13

Bamberg

Staatsarchiv (StA BA)
Markgrafentum Brandenburg-Bayreuth, Geheimes Archiv, 4676

Coburg

Veste Coburg
Kunstsammlung: Ms A, I, 349, (2), 3

Copenhagen

Royal Library (Det Kongelige Bibliotek: KB)
Ms Thott: 1227

Dresden

Sächsisches Hauptstaatsarchiv (SächsHStA)
Geheimer Rat 8224, 7–10

Erlangen

Universitätsbibliothek Erlangen-Nürnberg (UB ER-N)
Briefsammlung Trew

Florence

Bibliotheca Nazionale Centrale (BNC Firenze)
Magl. VIII 259, 282

Frankfurt

Universitätsbibliothek (UB F)
MS. FF. H.G. Thulemeyer

Göttingen

Staats- und Universitätsbibliothek (SUB GÖ)
Ms. philos. 91, 93, 94, 102

Halle

Archiv der Frankeschen Stiftungen
AFSt/H: C 72 Nr 5 and Nr 6; 825:1; D 43a
AFSt/S: A I 118; 194; 196e; B I 1

Hamburg

Staats- und Universitätsbibliothek (SUB HH)
Sup.ex. 36

Hanover

Gottfried Wilhelm Leibniz Bibliothek (GWLB)
Ms VIII, 634
LBr 449

Jena

Universitätsbibliothek (ThULB)
Ms. Bud: 347, 348

Leipzig

Universitätsbibliothek (UB L)
Rep IX, 5

Marburg

Staatsarchiv
81A, 14/6

Munich

Bayerische Staatsbibliothek (BSB)
Cim. Lohmeier
Cim. Ludewig
Autogr. Battier
Autogr. Sonntag
Autogr. Spener
Autogr. Wagner
Autogr. Wegleiter
Cgm 5458(1)
Clm 28115
Cod. Ital. 506

Neuenstein

Zentralarchiv Hohenlohe
La 20 Bü 90
Sf 25 Bü 220

Nuremberg

Germanisches Nationalmuseum (GNM)
Nachlass Imhoff:
I (Archiv der Gesamtfamilie): 2, 3
II (Helmstädter Archivschrank): 9
V (Mörlach-Hohenstein): 1

VII (Collection Imhoff): 1–22
Historisches Archiv:
V. Historiker: Juncker
IV. Klerus: Riccius
XIV. Varia: Riederer
Stadtbibliothek am Bildungscampus (StadtB N)
Will I. 224.2°
Staatsarchiv (StA N)
Handschriften: 44, 45, 286, 288
Herrschaft Schwarzenberg: Schwarzenberger Archiv 16/7, 16/9
Reichsstadt Nürnberg, Ratsverlässe 2682
Stadtarchiv (StadtA N)
A 1, A 31, B 35, D 19, F 1 Nr 14/I–V

Prague

Knihovna Národního Muzea v Praze
VI F 14
XIX.A.3

Stuttgart

Württembergische Landesbibliothek (WLB)
Cod. hist. 4" 731, III
Württembergisches Landesarchiv (WLA)
A 274 Bü 63c

Tübingen

Universitätsbibliothek (UB TÜ)
Mi VIII 1

Ulm

Stadtarchiv (StadtA Ulm)
V.154

Wiesbaden

Hessisches Hauptstaatsarchiv (HHStA)
130/I/10

Wolfenbüttel

Staatsarchiv
112 Alt, Nr. 305

Zittau

Christian Weise Bibliothek (CWB)
Mscr.A. 70–71

Sources Published before 1800

Albizzi, Antonio. *Principium Christianorum Stemmata*. Augsburg: n.p., 1608.

Aldimari, Biagio. *Historia genealogica della famiglia Carafa*. Naples: Antonio Bulifon, 1691.

———. *Memorie historiche di diverse famiglie nobili, così Neapolitane come Forastiere*. Naples: Raillard, 1691.

Ammirato, Scipione. *Delle famiglie nobili fiorentine.* 2 vols. Florence: Donato e Giunti, 1577–80.
Anonymous. *Fürtrefflicher Ritter-Kampf, der zweyen Neapolitanischen Fürsten D. Francesco Carafa de Noja und D. Guilio Antonio Aquaviva de Aragona: am 5./15. Novemb. 1673.* Nuremberg: n.p., 1673.
Anonymous. *Neu-erläuterte Zeit- und Jahr-Rechnung, darinnen so wohl der alten und vor Christi Geburt berühmten Haupt-monarchien [. . .] in XXXI. Tafeln verfasset und um mehrers Nutzens willen denen Stamm-Tafeln in teutscher Sprache hin zugefügt des A.N.P.I.* Lüneburg: Johann Georg Lipper, 1699.
Anonymous. *Der neueste Staat des Königreichs Portugall, und der darzu gehörigen Länder inn- und ausserhalb Europa: Anderer Theil.* Halle: Renger, 1714.
Anonymous. *XIV Tabulae Genealogicae Quibus Exhibentur Praecipuae Familiae Principum Imperii: Editio Tertia Auctior.* Tübingen: Cotta, 1670.
Anonymous. *Tredecem Tabulae Genealogicae: Quibus Exhibentur Praecipuae Familiae Hodiernorum Principum Imperii,* Tübingen: Brunn, 1656.
Anonymous. *Viri Illustris Friderici Adolphi Hansen ab Ehrencron dum in Vivis Erat S. Regiae Majestatis Daniae [. . .] Consiliarius Status [. . .] nec non ad Regem Catholicum Quondam Legati Extraordinarii Bibliotheca [. . .] , quod Publice Distrahet Abrah. de Hondt ad Diem 5. Septemb. Seqq. 1718.* Copenhagen: Abrahamus de Hondt, 1718.
Anselm de Sainte-Marie. *Histoire généalogique de la maison royale de la France et des grands officiers de la couronne.* 2 vols. Paris: Loyson, 1674.
———. *Histoire généalogique et chronologique de la maison royale de France, des grands officiers de la couronne et de la maison du roy, avec les qualitez, l'origine, et le progrès de leurs familles.* Edited by Honoré Caille Du Fourny. Paris: Guignard and Robustel, 1712.
Aubert de La Chesnaye Des Bois, François-Alexandre. *Dictionnaire généalogique, héraldique, chronologique et historique: Contenant l'origine & l'état actuel des premières maisons de France, des maisons souveraines & principales de l'Europe.* 3 vols. Paris: Duchesne, 1757.
Avemann, Heinrich Friedrich. *Vollständige Beschreibung des uralten und weitberühmten Hochgräflichen Geschlechts der Herren Reichsgraf- und Burggrafen von Kirchberg in Thüringen.* 4 vols. Frankfurt: Stock und Schilling, 1747.
Bacchini, Benedetto. *Dell'istoria del monasterio di S. Benedetto de Polirone nello stato di Mantoua libri 5.* Modena: n.p., 1696.
Beatillo, Antonio. *Historia di Bari, principal città della Publia.* Naples: n.p., 1637.
Beckler, Peter. *Chronicon Bohemiae: Das ist; Historische und Genealogische Beschreibung Der uralten Geschlechter im Könichreich Böhmen [. . .] insonderheit von dem mächtigen Hause Hovora [. . .].* Hof: Förster, 1694.
Beer, Johann Christoph. *Der Durchleuchtigsten Erz-Herzogen zu Oesterreich Leben, Regierung und Groß-Thaten [. . .].* Nuremberg: Endter, 1695.
———. *Der Herzogen und Königen in Böhmen: Leben, Regierung und Absterben [. . .].* Nuremberg: Johann Hoffmann, 1685.
———. *Der Könige in Franckreich Leben, Regierung und Absterben [. . .].* Nuremberg: Johann Hoffmann, 1668.
———. *Der Könige in Hispanien Leben, Regirung und Absterben [. . .].* Nuremberg: Johann Hoffmann, 1684.
———. *Der Könige in Italien Leben, Regierung, Thaten und Absterben [. . .].* Nuremberg: Johann Hoffmann, 1686.
Belleguise, Alexandre de. *Traité de la noblesse suivant les préjugez rendus par les commissaires deputez pour la verification des titres de noblesse en Provence.* N.p.: J. Morel, 1669.
Biffi, Girolamo. *Gloriosa Nobilitas Illustrissimae Familiae Vicecomitum.* Milan: Montia, 1672.
Birken, Sigmund von. *Chur- und fürstlicher Sächsischer Helden-Saal: Oder Kurze, jedoch ausführliche Beschreibung der Ankunft, Aufnahme, Fortpflanzung und vornemster Geschichten*

Dieses höchstlöblichen Hauses samt Dessen Genealogie, Wappen und KupferBildnisen. Nuremberg: Hoffmann, 1677.

Bressler und Aschenburg, Ferdinand Ludwig von. *Die Heutigen Christlichen Souverainen von Europa: Das ist; Ein kurtzer Genealogischer und Politischer Abriß.* Breslau: Speck, 1698.

Buder, Christian Gottlieb. *Merckwürdiges Leben des Durchlauchtigsten Fürsten u. Herrn, Herrn Moritz Wilhelms, Hertzogs zu Sachsen* [. . .]. 2nd ed. Frankfurt: n.p., 1720.

Bulifon, Antonio. *Journal du voyage d'Italie de l'invincible & glorieux monarque Philippe V.* Naples: Bulifon, 1703.

———. *Lettere memorabilia, istoriche, politiche, ed erudite.* 4 vols. Naples: n.p., 1685–98.

Cave, William. *Erstes Christenthum, oder Gottesdienst der alten Christen in den ersten zeiten des evangelii.* Translated by Johann Christoph Frauendorf (anonymously). Leipzig: Fritsch, 1694.

Clauder, Israel. *Stemma Saxonicum: Des* [. . .] *Haußes Sachsen Stamm-Baum; Wie solcher nach genawer Erforschung* [. . .] *über Tausend Jahr her aufwachsend befunden worden; Mit Genealogischer Feder entworffen.* Coburg: n.p., 1683.

Crescenzi, Giovanni Pietro. *Anfiteatro romano, nel quale con le memorie de' grandi si rappilogano in parte l'origine, & la grandezze de' primi potentati di Europa* [. . .] *parte prima* [the only one to appear]. Milan: Malatesta, 1648.

Crescimbeni, Giovan Maria, ed. *Le vite degli Arcadi illustri scritte da diversi autori: Parte seconda.* Rome: de'Rossi, 1710.

d'Afflitto, Eustachio. *Memorie degli scrittori del regno di Napoli.* Vol. 1. Naples: Simoniana, 1782.

de la Roque, Gilles-André. *Histoire généalogique de la maison de Harcourt.* Paris: Sebastien Cramoisy, 1662.

de Lellis, Carlo. *Discorsi delle famiglie nobili del regno di Napoli.* 3 vols. Naples: Paci, 1654–71.

Diephold, Rudolph. *Genealogiarum Imperatorum, Regum, Ducum, Marchionum, Comitum, &c: Europae, Asiae, et Africae* [. . .]. Halle: Oelschlägel Erben, 1628.

Durey de Noinville, Jacques Bernard. *Table alphabétique des dictionnaires, en toutes sortes de langues & sur toutes sortes de sciences & d'arts.* Paris: Chaubert, 1758.

Eckhart, Johann Georg. *Genuinum Stemma Desiderianum.* Helmstedt: Hammius, 1708.

Fabricius, Johann. "Epistola Dedicatoria ad Jacobum Wilhelmum Imhofium." In *Opera Posthuma, Edita a Joanne Fabricio,* edited by Johann Saubert, unpaginated. Altdorf: Meyer, 1694.

Feuerlein, Conrad. *Drey-fach-Göttlicher Liebes-schein: oder, heller Liebes-strahl und klares Merkmal der liebreichen Meinung Gottes; Aus den schönen Worten des Apostels Pauli, in der I. Thess. V. v. 9. 10. bey Hochansehnlicher doch Leidmütigster Beerdigung des* [. . .] *Wilhelm ImHof* [. . .]. Nuremberg: Endter, 1690.

Flacchio, Engelbert. *Généalogie de la maison Latour-Tassis.* 3 vols. Brussels: n.p., 1709.

Fränckische Acta erudita et curiosa: Die Geschichte d. Gelehrten in Francken, auch andere in diesem Crayß vorgefallene Curiosa u. Merckwürdigkeiten in sich haltend. Vol. 1. Nuremberg: Endter, 1726.

Giannettasio, Niccolò. *Bellica.* Naples: n.p., 1699.

Giustiniani, Lorenzo. *Memorie istoriche degli scrittori legali del regno di Napoli.* 3 vols. Naples: Simoniana, 1787.

Gryphius, Christian. *Vitae Selectae Qvorvndam Ervditissimorvm ac Illvstrivm Virorvm: Ut et Helenae Cornarae et Cassandrae Fidelis, a Clarissimis Viris Scriptae et Antehac Separatim Editae, Denuo ob Summam Raritatem et Praestantiam Recusae, ac in Unum Volumen Redactae.* Breslau [modern Wrocław]: Bauchius, 1711.

Haller von Hallerstein, Philippus Jacobus. *Dissertatio Juris Publici de Patriciis vulgo denen Adelichen Geschlechtern.* Kaliningrad: Reusner, 1684.

Henniges, Hieronymus. *Theatrum Genealogicum* [. . .] *in Quarta Monarchia.* 4 vols. Magdeburg: n.p., 1587.

Herdegen, Johann. *Historische Nachricht von dess Löblichen Hirten- und Blumenordens an der Pegnitz*. Nuremberg: Riegel, 1744.

Hirsch, Carl Christian. *Librorum ab Anno I. usque ad Annum L sec. XVI Typis Exscriptorum ex Libraria Quadam Suppellectile Norimbergae Collecta* [. . .]. Nuremberg: Felsecker, 1746–59.

Hirsching, Friedrich Carl Gottlob, and Johann Heinrich Martin Ernesti. *Historisch-literarisches Handbuch berühmter und denkwürdiger Personen, welche in dem 18. Jahrhunderte gestorben sind: Oder kurzgefaßte biographische und historische Nachrichten von berühmten Kaisern, Königen, Fürsten, großen Feldherren* [. . .] *und andern merkwürdigen Personen beyderley Geschlechts*. 17 vols. Leipzig: Schwickert, 1794–1815.

Holtacker, Johannes. *Abregé historique et généalogique de la tres-illustre, et tres-ancienne maison d'Aspremont, de laquelle est issue la maison des comtes de Lynden et Reckheim dedié a son altesse Madame la Princesse Charlotte de Nassau, Comtesse d'Aspremont & Reckheim, par Jean Holtacker, Doyen de nostre Dame a Worms l'an MDCLXXX: Traduit d'allemand en françois, par Robert le Voyer de S. Malo, Capitaine & Ingenieur de S.M. Imperiale*. Vienna: n.p., 1693.

Holzschuher, Christoph Siegmund von. *Deductions-Bibliothek von Deutschland: Nebst dazu gehörigen Nachrichten*. Frankfurt and Leipzig: n.p., 1788.

Hübner, Christian. *Aller durchläuchtigen hohen Häuser in Europa, wie auch der Grafen des heiligen Römischen Reichs neueste Genealogien*. Hamburg: Schiller, 1707.

Hübner, Johann. *Bibliotheca Genealogica*. Hamburg: Brandt, 1729.

———. *Drey hundert und drey und dreyßig Genealogische Tabellen*. Leipzig: Gleditsch, 1708.

Imhoff, Jakob Wilhelm. *Albanensis Familiae Hodierno Romano Pontifice Insignis Arbor Genealogica Illustrata Historica Relatione de Eiusdem Origine et Statu Praesenti*. N.p.: n.p., n.d.

———. *Appendix ad Historiam Genealogicam Regum Pariumque Magnae Britanniae Anno Superiore Editam*. Nuremberg: Endter, 1691.

———. *Corpus Historiae Genealogicae Italiae et Hispaniae: In Quo Stirpium Utriusque Regni Primariarum Genealogiae Exegesi Historica Perpetua Illustratae Ordine Alphabetico Exhibentur cum Insignium Iconibus & Indicibus Necessariis*. Nuremberg, Endter, 1702.

———. *Excellentium Familiarum in Gallia Genealogiae a Prima Earundem Origine usque ad Praesens Aevum Deductae et Notis Historicis, Quibus Memorabilia Regni Galliae Regumque et Clarorum Toga et Sago Virorum Facta Moresque ac Dignitates Recensentur, Illustratae, cum Iconibus Insignium & Indicibus Necessariis*. Nuremberg: Endter, 1687.

———. *Genealogiae Familiarum Bellomaneriae Claromontanae de Gallerande et Memmiae: Ex Quibus Illustris Oratorum Gallicorum Triga Nempe Marchio de Lavardin Nec Non Comites de Chiverny et d'Avaux Oriundi Sunt, Historiae Praesentis Aevi et Curiosorum Gratia Concinnatae ac Editae*. Nuremberg: Endter, 1688.

———. *Genealogiae Viginti Illustrium in Hispania Familiarum Ordine Alphabetico Exhibitae, Exegesi Historica Perpetua Illustratae Iconibusque Insignium Exornatae*. Leipzig: Gleditsch, 1712.

———. *Genealogiae Viginti Illustrium in Italia Familiarum in Tres Classes* [. . .]. Amsterdam: Châtelain, 1710.

———. *Genealogia Ruthenorum Comitum et Dominorum in Plauen: Prioribus, Quae Prodierunt, Emendatior, et ad Medium Annum MDCCXV Perducta*. Nuremberg: Endter, 1715.

———. *Historia Italiae et Hispaniae Genealogica: Exhibens Instar Prodromi Stemma Desiderianum ab Ima Radice*. Nuremberg: Endter, 1701.

———. *Historische und Genealogische Nachrichten von den Grandes oder Grossen in Spanien nebst dem Zustande derjenigen so heute darinnen leben* [. . .]. *Aus dem Frantzösischen in das Deutsche übersetzt und Biß auf die itzige Zeit continuiret*. Hamburg, Leipzig: Conrad and Wirnung, 1712.

———. *Notitia Sancti Romani Germanici Imperii Procerum tam Ecclesiasticarum quam Saecularum*. Tübingen: Cotta, 1684.

———. *Notitia Sancti Romani Germanici Imperii Procerum tam Ecclesiasticarum quam Saecularum*. 2nd ed. Tübingen: Cotta, 1687.

———. *Notitia Sancti Romani Germanici Imperii Procerum tam Ecclesiasticarum quam Saecularum*. 3rd ed. Tübingen: Cotta, 1693.

———. *Notitia Sancti Romani Germanici Imperii Procerum tam Ecclesiasticarum quam Saecularum*. 4th ed. Stuttgart: Cotta, 1699.

———. *Notitia Romani Germanici Imperii Procerum tam Ecclesiasticarum quam Saecularum*. 5th ed. Edited by Johann David Köhler. 2 vols. Tübingen: Cotta, 1732.

———. *Recherches historiques et genealogiques des grands d'Espagne: Avec un etat de ceux qui vivent aujourd'hui, contenant leur extraction, leurs noms, leurs qualitez, leurs alliances, leur posterité, leurs armes & blazons*. Amsterdam: Châtelain, 1707.

———. *Regum Pariumque Magnae Britanniae Historia Genealogica: Qua Veterum Juxta ac Recentium in Illa Familiarum Origines, Stemmata, & Res Memorabiliores, Ordine ad Novissimum Angliae Statum Aptato, Recensentur Atque Explicantur*. Nuremberg: Endter, 1690.

[———]. *Spicilegium Rittershusianum: Sive Tabulae Genealogicae Quibus Stemmata Aliquot Illustrium in Germania Familiarum quas Vir Cl. Nicolaus Rittershusius in Opere Suo Vel Omnino Praeteriit, Vel a Posteriore Saltem Stipite Enumeravit, Exhibentur; Decades Tres*. 2 vols. Tübingen: Cotta, 1683, 1685.

———. *Stemma Regium Lusitanicum, sive Historia Genealogica Familiæ Regiæ Portugallicæ*. Amsterdam: Châtelain, 1708.

Köhler, Johann David. *Der historischen Münz-Belustigung Anderer Theil Darinnen allerhand merkwürdige und rare Thaler, Ducaten, Schaustücken und andere sonderbahre Gold und Silber Münzen von manchem Alter, zusammen LIX Stücke, Accurat in Kupffer gestochen, beschrieben und aus der Historie umständlich erklärt werden*. Nuremberg: Bieling, 1730.

[La Valette, Laurent Pianelli de.] *Abrégé nouveau et méthodique du Blason*. Lyon: Amaulry, 1705.

Leibniz, Gottfried Wilhelm. *Lettera su la connessione delle serenissime case di Brunsvic e d'Este*. Hanover: Ammon, 1695.

Lohmeier, Georg. *Der Europäischen Reiche und Fürstenthümer Historische und Genealogische Erläuterung: In Stamm-Tafeln verfasset, und ans Liecht gegeben, Erstlich Durch Georg Lohmeiern, des Adelichen Collegii zu Lüneburg Inspect. Und Profess. Nunmehro aber nach seinem Tode wieder aufgelegt, biß auf gegenwärtige Zeit erstreckt, und durchgehends, wie auch mit den Stämmen der Teutschen Kayser, und verschiedener Könige und Fürsten, reichlich vermehret, durch den A. N. P. I*. Lüneburg: Lippers, 1695.

———. *Der Europäischen Reiche und Fürstenthümer Historische und Genealogische Erläuterung: In Stamm-Tafeln verfasset, und ans Liecht gegeben, Erstlich Durch Georg Lohmeiern, des Adelichen Collegii zu Lüneburg Inspect. Und Profess. Nunmehro aber nach seinem Tode wieder aufgelegt, biß auf gegenwärtige Zeit erstreckt, und durchgehends, wie auch mit den Stämmen der Teutschen Kayser, und verschiedener Könige und Fürsten, reichlich vermehret, durch den A. N. P. I*. Lüneburg: Lippers, 1701.

Lopez de Haro, Alonso. *Nobiliario genealogico de los reyes y títulos de España*. Madrid: Sanchez, 1622.

Ludewig, Johann Peter von. *Germania Princeps*. Halle: Zeidler, 1711.

Lünig, Johann C. *Bibliotheca Curiosa Deductionum*. Leipzig: Lanckisch, 1717.

Mandosio, Prospero. *Bibliotheca Romana, seu Romanorum Scriptorum Centuriae*. Rome: n.p. 1682.

———. *Theatron in Quo Maximorum Christiani Orbis Pontificorum Archiatros*. Rome: n.p., 1696.

Ménestrier, Claude. *Gründliche Wiederlegung der von Arnoldo Wion für des Irrländischen Bischoffs Malachiae Arbeit ausgegebenen und fast von jederman dafür angenommenen Prophezeyung, betreffend alle Römische Päbste, so von Anno Christi 1143 bis ans Ende der Welt regieren sollen*. Translated by Christian Wagner. Leipzig: Lanckisch, 1691.

Mithoff, Daniel. *Historischer Kurtzer Bericht wie die Vorfahren [. . .] jetziger noch lebender Hertzogen zu Sachsen [. . .] umb die Sächsische Churgerechtigkeit gekommen, und Sie selber bis dato noch davon abgehalten warden.* N.p.: n.p., 1629.

Moréri, Louis. *Le grand dictionaire [sic] historique; ou, Le mélange curieux de l'histoire sainte et profane* [. . .]. Lyon: Girin et Rivière, 1764.

Mugnos, Filadelfo. *Teatro genologico delle famiglie nobili titolate fevdatarie ed antiche nobili del fidelissimo regno di Sicilia, viventi ed estinte.* Palermo: Coppola, 1647.

Muratori, Antonio. *Anecdota, Quae ex Ambrosianae Bibliothecae Codicibus Nunc Primum Eruit.* Vol. 1. Milan: Malatesta, 1697.

Ott, Johann Heinrich. *Onomatologia seu de Nominibus Hominum Propriis.* Zurich: Gesner, 1671.

Palazzi, Giovanni. *Aquila Saxonica, sub Qua Imperatores Saxones [. . .] Elogiis, Hieroglyphicis, Numismatibus, Insignibus, Symbolis, Imaginibus Antiquis ad Vivum Exhibentur.* Venice: Herz, 1673.

Petrus, Franciscus. *Suevia Ecclesiastica.* Augsburg: Bencard, 1699.

Pigna, Giovanni Battista. *Historia de principi di Este.* Venice: Valgrisi, 1572.

Pistorius, Wilhelm Friedrich. *Historisch- und juridische Anmerkungen über allerhand den Ursprung Historie und Vorrechte derer Reichs-Grafen betreffende materien.* 4 vols. Frankfurt: Von Sand, 1726–28.

Pregitzer, Johann Ulrich. *Suevia et Wirtenbergica Sacra: Opus Posthumum et Plane Novum, Singulari Cura Revisum et cum Variis Supplementis Editum a B. Auctoris Filio, M. Gregorio Conrado Pregitzero.* Tübingen: Cotta, 1717.

Rassler, Maximilian. *Vindicatio contra Vindicias sive ad Vindicias Historicas Wilhelmi Ernesti Tenzelii Serenissimo Electori Saxoniae à Consiliis nuper & Historiis pro Hermanni Conringii Censura* [. . .]. Kempten: Mayr, 1711.

Reimmann, Jakob Friedrich. *Historiae Literariae Exotericae & Acroamaticae Particula sive de Libris Genealogicis Vulgatioribus & Rarioribus Commentatio.* Quedlinburg: Strunz, 1710.

Reineckius, Reiner, ed. *Annales Helmoldi, Presbyteri Buzoviensis in agro Lubecensi.* Frankfurt: Wechel, 1581.

Reusner, Elias. *Basilikon: Opus Genealogicum Catholicum de Praecipuis Familiis Imperatorum, Regum, Principum, Aliorumque Procerum Orbis Christiani.* Frankfurt: Basse 1592.

Richter, Georg. *Epistolae Selectiores, ad Viros Nobilissimos Clarissimosque Datae, ac Redditae.* Nuremberg: Endter, 1663.

Rittershausen, Nicolaus. *Brevis Exegesis Historica Genealogicarum Praecipuorum Orbis Christiani Procerum Imperatorum, Regum, Ducum, et Quorundam Comitum, Antehac Eodem Authorem Editarum: Opus Posthumum.* Tübingen: Brunn, 1674.

———. *Genealogiae Imperatorum, Regum, Ducum, Comitum.* 3rd ed. Tübingen: Cotta, 1664.

———. *Genealogiarum Praecipuorum Orbis Christiani Procerum.* Altdorf: n.p., 1653.

———. *Genealogiarum Praecipuorum Orbis Christiani Procerum.* 2nd ed. Tübingen: Cotta, 1658.

[———.] *Schema Genealogicum Gentis de Seinsheim.* Nuremberg: Felsecker, 1694.

Salazar de Mendoza, Pedro. *Origen de las dignidades seglares de Castilia, y Leon: Con relacion sumaria de los reyes de estos reynos, de sus acciones, casamientos, hijos, muertes, sepulturas, para el principe de Espana, Don Filipe nuestro señor, por el Doctor Salazar de Mendoza.* Madrid: n.p., 1657.

Sant'Anna, Girolamo Maria di. *Della storia genealogica della famiglia Del Ponte.* Naples: Mosca, 1708.

———. *Discorso nel quale si ragiona, ed insieme si detesta l'abuso di quei scrittori, ed istorici, quali mossi dall'adulazione, o spinti da altro fine particolare, scriuono l'istorie, e genealogie delle famiglie piene di bugie, e menzogne: Si commendano l'istorici sinceri, e veritieri, molti de' quali se ne rapportano.* Venice: Brigonci, 1693.

Schmid, Samuel Heinrich. *Die Durchläuchtige Welt, Oder Kurtzgefaßte Genealogische, Historische und Politische Beschreibung*. Hamburg: Schiller, 1701.
Seifert, Johann. *Ahnen-Taffeln Erster Theil Nach Ordnung des Alphabets*. 2nd ed. Regensburg: Peetz, 1716.
———. *Genealogie Hoch-Adelicher Eltern und Kinder*. Vol. 1. Regensburg: Hoffmann, 1724.
———. *Genealogische Tabellen des Im-Hoff'schen Geschlechts*. Regensburg: n.p., 1723.
Soria, Francescantonio. *Memorie storico-critiche degli storici napolitani*. 2 vols. Naples: Stamperia Simoniana, 1782.
Speidel, Johann Jakob. *Speculum Iuridico-Politico-Philologico-Historicarum Observationum et Notabilium: Verborum, Rerum et Antiquitatum, Germanicarum* [. . .]. Nuremberg: Endter, 1657.
Spener, Philipp Jakob. *Illustriores Galliae Stirpes*. Frankfurt: Zunner, 1689.
———. *Sylloge Genealogico-Historica*. Frankfurt: Zunner, 1668.
———. *Sylloge Genealogico-Historica*. 2nd ed. Frankfurt: Zunner, 1677.
———. *Theatrum Nobilitatis Europeae*. 4 vols. Frankfurt: Vogel, 1669–78.
Strobel, Georg Theodor, ed. *Conradi, Georgi et Nicolai Rittershusiorum, Patris et Filiorum, et Variorum ad Eos Datae Epistolae*. Nuremberg: Ammermuller, 1769.
Sturm, Johann Christoph. *Memoriam Guilielmi Im-Hofii, Viri Prosapia Generosi* [. . .]. Nuremberg: Endter, 1690.
Targioni-Tozzetti, Giovanni, ed. *Clarorum Germanorum ad A. Magliabechium Nonnullosque Alios Epistolae*. Florence: Typographia ad Insigne Apollinis in Platea Magni Ducis, 1746.
Tentzel, Wilhelm Ernst, ed. *Monatliche Unterredungen einiger guten Freunde von allerhand Büchern und andern annehmlichen Geschichten*. Leipzig: Geditsch, 1691.
Ulmer, Johann Georg. *Dissertatio Historica de Antiqua et Nobilissima Familia Im-Hof*. Altdorf: Hagen, 1651.
Voets, Melchior. *Tractatus de Jure Revolutionis: Ad Lucem Ordinationis Judiciariae Cap. Beschluß von Succession 88* [. . .]. 2nd ed. Düsseldorf: Schleuter, 1694.
Volckamer, Johann Christoph. *Nürnbergische Hesperides, oder Gründliche Beschreibung der Edlen Citronat, Citronen, und Pomerantzen-Früchte* [. . .]. Nuremberg: Endter, 1708.
Wagenseil, Johann Christoph. *Der Adriatische Löw, Das ist, Kurtze Anzeigung, von des Durchleuchtigen Venetianischen Adels gesammter Geschlechte Ursprung, Aufnahm, Wie auch Deren angebornen Stamm-Wappen Aus der Roem: Kays; Majestät und Eines Löbl. Magistrats der Stadt Leipzig Bibliothecken, alten Italianischen Manuscriptis, wie auch einigen neuen Hülffs-Mitteln, entnommen*. Altdorf: Kohles, 1704.
Weise, Christian. *Schediasma de Usu Novellarum*. Weissenfels: n.p., 1676.
Weller, Johann Tobias. *Deutscher Adler, das ist, sämtlicher Deutschen Kaiser Leben in allerhand jetziger Zeit üblichen Versen beschrieben*. Frankfurt: Wust, 1666.
Will, Georg Andreas. *Bibliotheca Norica Williana oder Kritisches Verzeichniß aller Schriften, welche die Stadt Nürnberg angehen*. Altdorf: Meyer, 1772.
———. *Commercium Epistolicum Norimbergense*. 3 vols. Altdorf: Schüpfel, 1755–59.
Wurffbain, Leonhard. *Drey unterschiedliche Summarische Bericht: I. Von den Nürnbergischen, Augspurgischen und andern fürnehmen Orten Rathsfähigen Geschlechtern deren Herrn Im Hof in gemein*. Nuremberg: Endter, 1637.
Ziegler und Kliphausen, Heinrich Anselm von. *Historisches Labyrinth der Zeit*. Leipzig: Gleditsch, 1701.
Ziegra, Christian. *Nicolaus Wilckens, weiland J.U.D. und Archivarius der Stadt Hamburg, Hamburgischer Ehrentempel*. Hamburg: Schröder, 1770.
Zschackwitz, Johann Ehrenfried. *Historisch-genealogischer Schauplatz auf welchem sowohl des durch lauchtigsten Ertzhertzogl: Hauses Oesterreichs, als auch der weltlichen Chur- & alten Fürstlichen Häuser Genealogien* [. . .] *vorgestellt* [. . .] *werden*. Lemgo: Meyer, 1724.

Secondary Literature

Ailes, Adrian. "Can We Trust the Genealogical Records of the Heralds' Visitations? A Case Study (the 1665–66 Visitation of Berkshire)." In *L'entreprise généalogique: Pratiques sociales et imaginaires en Europe (XVᵉ–XIXᵉ siècles)*, edited by Stéphane Jettot and Marie Lezowski, 143–62. Brussels: Peter Lang, 2016.

Ajello, Raffaele. *Arcana juris: Diritto e politica nel Settecento italiano.* Storia e diritto 1. Naples: Jovene, 1976.

Andenna, Cristina, and Gert Melville. *Idoneität—Genealogie—Legitimation: Begründung und Akzeptanz von dynastischer Herrschaft im Mittelalter.* Norm und Struktur 43. Cologne: Böhlau, 2014.

Angelillo, Fiorella, and Enrica Stendardo. "Il Seicento." In *Libri per vedere: Le guide storico-artistiche della città di Napoli; Fonti testimonianze del gusto immagini di una città*, edited by Francesca Amirante, 43–90. Studi di storia dell'arte. Naples: Scientifiche Italiane, 1995.

Antognazza, Maria Rosa. "Leibniz as Historian." In *The Oxford Handbook of Leibniz*, edited by Maria Rosa Antognazza, 591–608. New York: Oxford University Press, 2018.

Arbeitskreis Orangerien in Deutschland e.V., ed. *Nürnbergische Hesperiden und Orangeriekultur in Franken.* Orangeriekultur, 7. Petersberg: Imhof, 2011.

Aretin, Karl Otmar von. *Das Reich: Friedensgarantie und europäisches Gleichgewicht 1648–1806.* Stuttgart: Klett-Cotta, 1986.

Arndt, Johannes. *Das niederrheinisch-westfälische Reichsgrafenkollegium und seine Mitglieder (1653–1806).* Veröffentlichungen des Instituts für Europäische Geschichte, 133. Mainz: von Zabern, 1991.

Asch, Ronald G. *Europäischer Adel in der Frühen Neuzeit: Eine Einführung.* UTB Geschichte, 3086. Cologne: Böhlau, 2008.

Ashmole, Elias. *Elias Ashmole (1617–1692): His Autobiographical and Historical Notes, His Correspondence, and Other Contemporary Sources Relating to His Life and Work.* Edited by C. H. Josten. 5 vols. Oxford: Oxford University Press, 1967.

Babin, Malte-Ludolf, and Gerd van den Heuvel, eds. *Gottfried Wilhelm Leibniz: Schriften und Briefe zur Geschichte.* Veröffentlichungen der Historischen Kommission für Niedersachsen und Bremen, 218. Hanover: Hahn, 2004.

Baggio, Silvia, and Piero Marchi. "L'archivio della memoria delle famiglie fiorentine." In *Istituzioni e società in Toscana nell'età moderna*, edited by Giuseppe Pansini, 862–77. Rome: Ministero per i Beni Culturali e Ambientali Ufficio Centrale per i Beni Archivistici, 1994.

Baker, John. "Tudor Pedigree Rolls and Their Uses." In *Heralds and Heraldry in Shakespeare's England*, edited by Nigel Ramsay, 125–65. Donington, UK: Shaun Tyas, 2014.

Bann, Stephen, Michael Broers, and Ambrogio A. Caiani. *A History of the European Restorations.* Vol. 2, *Culture, Society and Religion.* International Library of Historical Studies. London: Bloomsbury Academic, 2020.

Barnett, Mason. "Anspruch und Wirklichkeit: Reformen in der frühen Academia naturae curiosorum." In *Gelehrte Gesellschaften im mitteldeutschen Raum (1650–1820)*, edited by Detlef Döring, 47–72. Abhandlungen der Sächsischen Akademie der Wissenschaften zu Leipzig, Philologisch-Historische Klasse 76,5. Stuttgart: Hirzel, 2002.

———. "Medical Authority and Princely Patronage: The Academia Naturae Curiosorum, 1652–1693." PhD diss., University of North Carolina at Chapel Hill, 1995.

Barry, Laurent. *La parenté.* Folio essais 498. Paris: Gallimard, 2008.

Bartelmess, Albert. "Die Patrizierfamilie Tucher im 17. und 18. Jahrhundert." *Mitteilungen des Vereins für Geschichte der Stadt Nürnberg* 77 (1990): 223–43.

Bastian, Jeannette A. *Owning Memory: How a Caribbean Community Lost Its Archives and Found Its History.* Contributions in Librarianship and Information Science 99. Westport, CT: Libraries Unlimited, 2003.

Bauer, Volker. "Jetztherrschend, jetztregierend, jetztlebend: Genealogie und Zeitungswesen im Alten Reich des ausgehenden 17. Jahrhunderts." *Daphnis* 37 (2008): 271–300.

———. "Monographizitaet, Serialitaet, Periodizitaet in der Universalgenealogie: Beschleunigung, Aktualisierung und der Einfluss des Zeitungswesens (1580er bis 1730er Jahre)." In *Die Entstehung des Zeitungswesens im 17. Jahrhundert: Ein neues Medium und seine Folgen für das Kommunikationssystem der Frühen Neuzeit*, edited by Volker Bauer and Holger Böning, 393–410. Presse und Geschichte—neue Beiträge 54. Bremen: Lumière, 2011.

———. "The Scope, Readership and Economy of Printed Genealogies in Early Modern Germany: 'Special Genealogien' vs. 'Universal Genealogien.'" In *L'entreprise généalogique: Pratiques sociales et imaginaires en Europe (XVe–XIXe siècles)*, edited by Stéphane Jettot and Marie Lezowski, 287–301. Histoire des mondes modernes 2. Brussels: Peter Lang, 2016.

———. *Wurzel, Stamm, Krone: Fürstliche Genealogie in frühneuzeitlichen Druckwerken*. Ausstellungskataloge der Herzog-August-Bibliothek 97. Wiesbaden: Harrassowitz 2013.

Becker, Hans, and Matthias Kirchhoff. "Das Memorial Endres I. Tuchers: Neuedition, Kommentierung und Übersetzung des ältesten Tucherschen Gedenkbuchs (1421–1440)." *Mitteilungen des Vereins für Geschichte der Stadt Nürnberg* 101 (2014): 1–37.

———. "Neuedition und Übersetzung des Memorials Berthold III. Tuchers (1386–1454)." *Mitteilungen des Vereins für Geschichte der Stadt Nürnberg* 100 (2013): 71–97.

Becker, Peter, and William Clark, eds. *Little Tools of Knowledge: Historical Essays on Academic and Bureaucratic Practices*. Ann Arbor: University of Michigan Press, 2001.

Behringer, Wolfgang. *Im Zeichen des Merkur: Reichspost und Kommunikationsrevolution in der Frühen Neuzeit*. Veröffentlichungen des Max-Planck-Instituts für Geschichte 189. Göttingen: Vandenhoeck & Ruprecht, 2003.

Bei der Wieden, Brage. "Pietismus und Genealogie: Überlegungen zu Joachim Justus Breithaupts Dransfelder Ahnenforschung." In *Aus dem Süden des Nordens: Studien zur niedersächsischen Landesgeschichte für Peter Aufgebauer zum 65. Geburtstag*, edited by Arnd Reitemeier and Uwe Ohainski, 11–21. Veröffentlichungen des Instituts für Historische Landesforschung der Universität Göttingen 58. Bielefeld: Verlag für Regionalgeschichte, 2013.

Bély, Lucien. *La société des princes: XVIe–XVIIIe siècle*. Paris: Fayard, 1999.

Ben-Amos, Ilana. "Reciprocal Bonding: Parents and Their Offspring in Early Modern England." *Journal of Family History* 25 (2000): 291–312.

Benavent Montolio, Jorge F. "Gregorio Mayans y el Hispanismo alemán en el siglo XVIII." *Saítabi* 48 (1998): 27–50.

Benedikt, Heinrich. *Das Königreich Neapel unter Kaiser Karl VI: Eine Darstellung auf Grund bisher unbekannter Dokumente aus den Österreichischen Archiven*. Berlin: Manz, 1927.

Benigno, Francesco. *Favoriti e ribelli: Stili della politica barocca*. Biblioteca del Cinquecento 152. Rome: Bulzoni, 2011.

———. "Il ritorno dei Vespri: Storia e politica nell'opera de Filadelfo Mugnos." *Pedralbes* 27 (2007): 131–50.

Benz, Stefan. "Historiker um Gottfried Wilhelm Leibniz." In *Leibniz und Niedersachsen*, edited by Herbert Breger, 148–72. Stuttgart: Steiner, 1999.

Bérenger, Jean. "Les Schwarzenberg à l'époque moderne." *Histoire, Économie et Société* 26 (2007): 29–46.

Berger, Jean. "Essai sur la nature du Grand cartulaire de l'Eglise Saint-Julien de Brioude." In *Brioude aux temps Carolingiens*, edited by Alain Dubreucq, Christian Lauranson-Rosaz, and Bernard Sanial, 209–35. Le Puy-en-Velay: n.p., 2010.

Berghorn, Detlef. *Verwandtschaft als Streitzusammenhang: Eine Fall-Geschichte in Beziehungen im hohen Adel des Alten Reiches, 16. bis 19. Jahrhundert*. Quellen und Forschungen zur höchsten Gerichtsbarkeit im Alten Reich 76. Cologne: Böhlau, 2021.

Bernard, Auguste, ed. *Cartulaire de l'Abbaye de Savigny: Suivi du petit cartulaire de l'Abbaye d'Ainay*. Vol. 1, *Cartulaire de Savigny*. Cambridge: Cambridge University Press, 2010. First published in 1853.

Berndt, Carsten. *Leben und Wirken von Christian Juncker (1668–1714): Pädagoge, Historiograph der Ernestiner, Luther-Biograph, Numismatiker, Übersetzer und Bibliothekar, sowie dessen Abhandlung zur Geschichte von Ilmenau, Schwarza und dem Rennsteig*. Bad Langensalza: Rockstuhl, 2017.

Bernstein, Hilary J. *Historical Communities: Cities, Erudition, and National Identity in Early Modern France*. Early Modern History and Modern History 32. Leiden: Brill, 2021.

Béroujon, Anne. "La satire de la passion généalogique: À propos d'une lettre de 1772." In *L'incorporation des ancêtres: Généalogie, construction du présent, du Moyen Âge à nos jours*, edited by Valérie Piétri and Isabelle Luciani, 143–57. Aix-en-Provence: Presses Universitaires de Provence, 2016.

Beuys, Barbara. *Maria Sibylla Merian: Künstlerin—Forscherin—Geschäftsfrau*. Berlin: Insel, 2017.

Bevilacqua, Alexander, and Frederic Clark, eds. *Thinking in the Past Tense: Eight Conversations*. Chicago: University of Chicago Press, 2019.

Biagioli, Mario. "Galileo's System of Patronage." *History of Science* 28 (1990): 1–62.

Birken, Sigmund von. *Der Briefwechsel zwischen Sigmund von Birken und Johann Michael Dilherr, Daniel Wülfer und Caspar von Lilien*, edited by Almut and Hartmut Laufhütte with Ralf Schuster. Werke und Korrespondenz 11. Berlin: de Gruyter, 2015.

———. *Tagebücher*. Edited by Joachim Kröll. Würzburg: Schöningh, 1971.

Bizzocchi, Roberto. *Généalogies fabuleuses: Inventer et faire croire dans l'Europe moderne*. Paris: Éditions Rue d'Ulm, 2010.

———. "Unglaubliche Genealogien: Eine Neubestimmung." *Quellen und Forschungen aus italienischen Archiven und Bibliotheken* 96 (2016): 245–63.

Blair, Ann. *Too Much to Know: Managing Scholarly Information before the Modern Age*. New Haven, CT: Yale University Press, 2010.

Blair, Ann, Anthony Grafton, Paul Duguid, and Anja-Silvia Goeing, eds. *Information: A Historical Companion*. Princeton, NJ: Princeton University Press, 2020.

Blom, Franciscus Joannes Maria. *Christoph and Andreas Arnold and England: The Travels and Book-Collections of Two Seventeenth-Century Nurembergers*. Nijmegen: n.p., 1981.

Bluche, François. *Les pages de la grand écurie*. Paris: n.p., 1968.

Bock, Hartmut. "Die Familiengeschichtsschreibung der Welser." *Mitteilungen des Vereins für Geschichte der Stadt Nürnberg* 95 (2008): 93–162.

Boislisle, Arthur de, ed. "Appendice VIII: Le Cardinal de Bouillon, Baluze et le procès des faussaire." In *Mémoires de Saint-Simon: Nouvelle édition collationnée sur le manuscrit autographe augmentée des additions de Saint-Simon au journal de Dangeau; Tome XIV*, 533–50. Paris: Hachette, 1899.

Boucher, Jacqueline. "Introduction." In *L'intendance de Lyonnais, Beaujolais, Forez en 1698 et en 1762: Édition critique du mémoire rédigé par Lambert d'Herbigny et des observations et compléments de la Michodière*, edited by Jacqueline Boucher, 9–37. Paris: Éditions du CTHS, 1992.

Bouchot, Henri. *Inventaire des dessins exécutés pour Roger de Gaignières et conservés aux départements des estampes et des manuscrits*. 2 vols. Paris: Hachette, 1891.

Bouquet, Mary. "Family Trees and Their Affinities: The Visual Imperative of the Genealogical Diagram." *Journal of the Royal Anthropological Institute* 2 (1996): 43–66.

Bouquet, Olivier. "Généalogies imperiales en republique: Le cas de la Turquie." *Revue d'Histoire Moderne et Contemporaine* 58 (2011): 146–78.

Bourcy, Hans von. "Nachweisungen und Inhalt der Werke des Regensburger Genealogen Johann Seifert." *Jahrbuch der Heraldischen Gesellschaft "Adler" in Wien* 10 (1928): 309–13.

Bourdeille, Pierre de, seigneur de Brantôme. *Oeuvres complètes.* Vol. 10, *Opuscules et pièces diverses.* Paris: Librairie Renouard, 1881.

Boutier, Jean. "'L'Accademia dei Nobili' di Firenze: Sociabilità ed educazione dei giovani nobili negli anni di Cosimo III." In *La Toscana nell'età di Cosimo III,* edited by Franco Angiolini, 205–24. Florence: Edifir, 1993.

Brennecke, Hanns Christof, ed. *Akademie und Universität Altdorf: Studien zur Hochschulgeschichte Nürnbergs.* Beihefte zum Archiv für Kulturgeschichte 69. Cologne: Böhlau, 2011.

Broadway, Jan. *"No historie so meete": Gentry Culture and the Development of Local History in Elizabethan and Early Stuart England.* New York: Manchester University Press, 2006.

Brown, Elizabeth A. R. *The Oxford Collection of the Drawings of Roger de Gaignières and the Royal Tombs of Saint-Denis.* Philadelphia: American Philosophical Society, 1988.

Bruni, Arnoldo. "Muratori epistolografo." *Muratoriana Online* (2017): 25–35.

Bruscoli, Francesco Guidi. "Der Handel mit Seidenstoffen und Leinengeweben zwischen Florenz und Nürnberg in der ersten Hälfte des 16. Jahrhunderts." *Mitteilungen des Vereins für Geschichte der Stadt Nürnberg* 86 (1999): 81–113.

Budde, Hendrik. *Die Kunstsammlung des Nürnberger Patriziers Willibald Imhoff unter besonderer Berücksichtigung der Werke Albrecht Dürers.* Uni-Press-Hochschulschriften 90. Münster: LIT, 1996.

Bulifon, Antonio. *Giornali di Napoli dal MDXLVII al MDCCVI.* Edited by Nino Cortese. Vol. 1 [the only one to appear], *MDXLVII–MDCXCI.* Naples: n.p., 1932.

Burguière, André. "L'état monarchique et la famille (XVIe–XVIIIe siècle)." *Annales: Histoire, Sciences Sociales* 56 (2001): 313–35.

———. "La mémoire familiale du bourgeois gentilhomme: Généalogies domestiques en France aux XVIIe et XVIIIe siècles." *Annales: Histoire, économie et société* 46 (1991): 771–88.

Burkardt, Johannes. *Die historischen Hilfswissenschaften in Marburg (17.–19. Jahrhundert).* Elementa diplomatica 7. Marburg: Institut für Historische Hilfswissenschaften, 1997.

Busch, Jörg W. *Die Mailänder Geschichtsschreibung zwischen Arnulf und Galvaneus Flamma: Die Beschäftigung mit der Vergangenheit im Umfeld einer oberitalienischen Kommune vom späten 11. bis zum frühen 14. Jahrhundert.* Munich: Fink, 1997.

Busino, Giovanni, and Alain Dufour. "Quattordici lettere inedite del Muratori a J. W. Imhof." *Rivista Storica Italiana* 64 (1957): 411–38.

Butaud, Germain, and Valérie Piétri. *Les enjeux de la généalogie, XIIe–XVIIIe siècle: Pouvoir et identité.* Paris: n.p., 2006.

Caiani, Ambrogio A. "Re-inventing the Ancien Régime in Post-Napoleonic Europe." *European History Quarterly* 47 (2017): 437–60.

Callard, Caroline. "La fabrication de la dynastie médicéenne." In *Florence et la Toscane, XIVe–XIXe siècles: Les dynamiques d'un état italien,* edited by Jean Boutier, Sandro Landi, and Olivier Rouchon, 399–418. Rennes: Presses Universitaires de Rennes, 2015.

———. *Le prince et la république: Histoire, pouvoir et société dans la Florence des Médicis au XVIIe siècle.* Collection Roland Mousnier 35. Paris: PUPS, 2007.

Canepari, Eleanora: "Généalogie et transmissions des biens dans le milieu artisanal (Rome, XVIe –XVIIe siècle)." In *L'entreprise généalogique: Pratiques sociales et imaginaires en Europe (XVe–XIXe siècles),* edited by Stéphane Jettot and Marie Lezowski, 157–70. Histoire des mondes modernes 2. Brussels: Peter Lang, 2016.

Carhart, Michael C. *Leibniz Discovers Asia: Social Networking in the Republic of Letters.* Baltimore, MD: Johns Hopkins University Press, 2019.

Castiglioni, Carlo. "I prefetti della Biblioteca Ambrosiana (Notizie bio-bibliografiche)." In *Miscellanea,* edited by Giovanni Galbiati, 399–429. Milan: Hoepli, 1951.

Chappey, Jean-Luc. *Ordres et désordres biographiques: Dictionnaires, listes de noms et réputation du XVIIIe siècle à Wikipédia.* Seyssel: Champ Vallon Editions, 2013.

Chartier, Roger. *The Order of Books: Readers, Authors, and Libraries in Europe between the Fourteenth and Eighteenth Centuries.* Stanford, CA: Stanford University Press, 1994.

Clark, Christopher M. *Time and Power: Visions of History in German Politics, from the Thirty Years' War to the Third Reich.* Princeton, NJ: Princeton University Press, 2019.

Claude-Gilbert Dubois. Review of *Artisans of Glory* by Orest Ranum. *Revue d'Histoire Littéraire de la France* 84 (1984): 598–600.

Claydon, Tony. "Daily News and the Construction of Time in Late Stuart England, 1695–1714." *Journal of British Studies* 52 (2013): 55–78.

Cooper, Alix. *Inventing the Indigenous: Local Knowledge and Natural History in Early Modern Europe.* Cambridge: Cambridge University Press, 2007.

Coote, Lesley. "Prophecy, Genealogy, and History in Medieval English Political Discourse." In *Broken Lines: Genealogical Literature in Late-Medieval Britain and France*, edited by Raluca Luria Radulescu and Edward Donald Kennedy, 27–44. Turnhout: Brepols, 2008.

Coulomb, Clarisse. "La publicité des archives de la noblesse sous l'ancien régime: L'exemple du Dauphiné, XVIIe–XVIIIe siècle." In *Archives familiales et noblesse provinciale: Hommages à Yves Soulingeas*, edited by René Favier, 89–102. Grenoble: Presses Universitaires de Grenoble, 2006.

Crick, Julia C., and Alexandra Walsham, eds. *The Uses of Script and Print, 1300–1700.* Cambridge: Cambridge University Press, 2004.

Crivelli, Tatiana. "The Print Tradition of Vittoria Colonna's Rime." In *A Companion to Vittoria Colonna*, edited by Abigail Brundin, Tatiana Crivelli, and Maria Serena Sapegno, 69–138. Leiden: Brill, 2016.

Croce, Benedetto: "I duelli a Napoli nel Seicento." *Archivio Storico per le Provincie Napolitane* 20 (1895): 543–58.

Crouch, David. *The Birth of Nobility: Constructing Aristocracy in England and France, 900–1300.* London: Routledge, 2005.

Csepregi, Zoltán, ed. *Pietas Danubiana: Pietismus im Donautal 1693–1755. 437 Schreiben zum Pietismus in Wien, Preßburg und Oberungarn.* Budapest: n.p., 2013.

Czech, Vinzenz, ed. *Fürsten ohne Land: Höfische Pracht in den sächsischen Sekundogenituren Weißenfels, Merseburg und Zeitz.* Schriften zur Residenzkultur 5. Berlin: Lukas-Verlag, 2009.

———. *Legitimation und Repräsentation: Zum Selbstverständnis thüringisch-sächsischer Reichsgrafen in der Frühen Neuzeit.* Schriften zur Residenzkultur 2. Berlin: Lukas-Verlag, 2003.

Darby, Paul, and Paul Clough. "Investigating the Information-Seeking Behaviour of Genealogists and Family Historians." *Journal of Information Science* 39 (2013): 73–84.

Daston, Lorraine. "The History of Science and the History of Knowledge." *Know* 1 (2017): 131–54.

Davis, Natalie Zemon. *Women on the Margins: Three Seventeenth-Century Lives.* Cambridge, MA: Harvard University Press, 1997.

Defauconpret, Benoit. *Les preuves de noblesse au XVIIIe siècle: La réaction aristocratique; Avec un recueil de tous les ordres, honneurs, fonctions, écoles, chapitres, réservés à la noblesse.* Paris: Intermédiaire des Chercheurs et Curieux, 1999.

Delille, Gérard. "Namen und Linien: Ein süditalienisches Dorf und seine Genealogen (1572–1730)." *L'Homme* (2009): 53–62.

Delisle, Leopold. *Le cabinet des manuscrits de la bibliothèque impériale: Étude sur la formation de ce dépôt comprenant les éléments d'une histoire de la calligraphie, de la miniature, de la reliure et du commerce des livre à Paris avant l'invention de l'imprimerie.* Histoire générale de Paris. Paris: Impr. Nat., 1868.

Del Tredici, Federico. *Comunità, nobili e gentiluomini nel contado di Milano del Quattrocento.* Milan: Unicopli, 2013.

Detering, Nicolas. *Krise und Kontinent: Die Entstehung der deutschen Europa-Literatur in der Frühen Neuzeit.* Cologne: Böhlau, 2017.

Dewald, Jonathan. *The European Nobility, 1400–1800.* New Approaches to European History. Cambridge: Cambridge University Press, 1996.

———. *Status, Power, and Identity in Early Modern France: The Rohan Family, 1550–1715.* University Park: Pennsylvania State University Press, 2015.

Diefenbacher, Michael, ed. *Das Nürnberger Buchgewerbe: Buch- und Zeitungsdrucker, Verleger und Druckhändler vom 16. bis zum 18. Jahrhundert.* Quellen und Forschungen zur Geschichte und Kultur der Stadt Nürnberg 31. Nuremberg: Stadtarchiv, 2003.

———. "Stadt und Adel: Das Beispiel Nürnbergs." *Zeitschrift für Geschichte des Oberrheins* 141 (1993): 51–70.

Diefenbacher, Michael, and Rudolf Endres, eds. *Stadtlexikon Nürnberg.* 2., verb. Aufl. Nuremberg: Tümmels, 2000.

Di Marco, Giampiero. "Librai, editori e tipografi a Napoli nel XVII secolo." *La Bibliofilía* 112 (2010): 21–62, 141–84.

Donati, Claudio. "Una famiglia lombarda tra XVI e XVIII secolo: Gli Estense di San Martino e i loro feudi." In *Archivi, territori, poteri in area estense,* edited by Euride Fregni, 435–54. Rome: Bulzoni, 1999.

———. *L'idea di nobiltà in Italia: Secoli XIV–XVIII.* Rome: Laterza, 1988.

Dooley, Brendan, ed. *The Dissemination of News and the Emergence of Contemporaneity in Early Modern Europe.* Aldershot: Ashgate, 2010.

———. *The Social History of Skepticism: Experience and Doubt in Early Modern Culture.* Baltimore, MD: Johns Hopkins University Press, 1999.

Dover, Paul Marcus. *The Information Revolution in Early Modern Europe.* New Approaches to European History. Cambridge, MA: Cambridge University Press, 2021.

Doyon, Julie, Isabelle Robin, and Nathalie Sage-Pranchère. "Les lois du sang." *Annales de Démographie Historique* 137 (2019): 5–19.

Drévillon, Hervé. *L'impôt du sang: Le métier des armes sous Louis XIV.* Paris: Tallandier, 2005.

Duby, Georges. *The Chivalrous Society.* Translated by Cynthia Postan. Berkeley: University of California Press, 1980.

Duff, Wendy, and Catherine Johnson. "Accidentally Found on Purpose: Information-Seeking Behavior of Historians in Archives." *Library Quarterly: Information, Community, Policy* 72 (2002): 472–96.

———. "Where Is the List with All the Names? Information-Seeking Behavior of Genealogists." *American Archivist* 66 (2003): 79–95.

Duindam, Jeroen. *Dynasties: A Global History of Power, 1300–1800.* Cambridge: Cambridge University Press, 2016.

———. *Dynasty: A Very Short Introduction.* Oxford: Oxford University Press, 2019.

Dülmen, Richard van. "Orthodoxie und Kirchenreform: Der Nürnberger Prediger Johannes Saubert (1592–1646)." *Zeitschrift für Bayerische Landesgeschichte* 33 (1970): 636–786.

Eberle, Wilhelm. *Das Reichsfürstentum Mindelheim unter Marlborough 1705–1715.* Munich: Technische Hochschule München, 1917.

Eddy, Matthew Daniel. "Diagrams." In *Information: A Historical Companion,* edited by Ann Blair, Anthony Grafton, Paul Duguid, and Anja-Silvia Goeing, 397–401. Princeton, NJ: Princeton University Press, 2020.

Egmond, Florike. *The World of Carolus Clusius: Natural History in the Making, 1550–1610.* Perspectives in Economic and Social History 6. London: Pickering & Chatto, 2010.

Eickmeyer, Jost, Markus Friedrich, and Volker Bauer, eds. *Genealogical Knowledge in the Making: Tools, Practices, and Evidence in Early Modern Europe.* Cultures and Practices of Knowledge in History 1. Berlin: de Gruyter Oldenbourg, 2019.

Endres, Rudolf. "Grundzüge der Verfassung der Reichsstadt Nürnberg." *Zeitschrift der Savigny-Stiftung für Rechtsgeschichte. Germanistische Abteilung* 111 (1994): 405–21.

———. "Nürnberg im 18. Jahrhundert." *Mitteilungen des Vereins für Geschichte der Stadt Nürnberg* 75 (1988): 133–53.

———. "Sozial- und Bildungsstrukturen fränkischer Reichsstädte im Spätmittelalter und in der Frühen Neuzeit." In *Literatur in der Stadt: Bedingungen und Beispiele städtischer Literatur des 15. bis 17. Jh*, edited by Horst Brunner, 37–72. Tübingen: Max Niemeyer, 1982.

———. "Verfassung und Verfassungswirklichkeit in Nürnberg im späten Mittelalter und in der Frühen Neuzeit." In *Verwaltung und Politik in Städten Mitteleuropas: Beiträge zu Verfassungsnorm und Verfassungswirklichkeit in altständischer Zeit*, edited by Wilfried Ehbrecht, 207–19. Städteforschung Reihe A, Darstellungen 34. Cologne: Böhlau, 1994.

Engel, Wilhelm. "Christian Junckers wissenschaftlicher Briefwechsel der Jahre 1700–1713." *Zeitschrift des Vereins für Thüringische Geschichte und Altertumskunde* 37 (1931): 54–90.

Erdner, Sven. "Leibniz und die braunschweigisch-lüneburgische Hausgeschichte. Leibniz' Suche nach den Vorfahren Azzos II. von Este zwischen 1685–1716 und sein Prioritätsstreit mit Ludovico Antonio Muratori." PhD diss., University of Hanover 2019.

———. "Plagiat an Leibniz' historiographischem Werk? Rekonstruktion frühmittelalterlicher Adelsgeschichte bei G. W. Leibniz und J. G. Eckhart." *Studia Leibnitiana* 35 and 36 (2003 and 2004): 194–224, 178–209.

Esteve, Cesc. "Censorship, Censure, and Historical Thought in Early Modern Spain." In *Disciplining History: Censorship, Theory and Historical Discourse in Early Modern Spain*, edited by Cesc Esteve, 126–66. London: Routledge, 2018.

———. "Contenerse en los límites de la prudencia: Autocensura y discurso histórico en la temprana modernidad." *Manuscrits* 35 (2017): 103–24.

Even-Ezra, Ayelet. *Lines of Thought: Branching Diagrams and the Medieval Mind*. Chicago: University of Chicago Press, 2021.

Fage, Émile. *Étienne Baluze: Sa vie, ses ouvrages, son exil, sa défense*. Reprint of Tulle edition, 1899. Geneva: Slatkine Reprints, 1971.

Fichtner, Paula Sutter. *Protestantism and Primogeniture in Early Modern Germany*. New Haven, CT: Yale University Press, 1989.

Figeac, Michel. *Les noblesses en France: Du XVIe au milieu du XIXe siècle*. Paris: Armand Colin, 2013.

Filangieri, Riccardo, ed. *Gli atti perduti della Cancelleria Angioina transuntati da Carlo de Lellis*. Rome: Nella Sede dell'Istituto, 1943.

———. "Notamenti e repertori delle cancellerie napoletane compilati da Carlo de Lellis ed altri eruditi tra secoli XVI e XVII." *Scritti di Paleografia e Diplomatica di Archivistica e di Erudizione* (1970): 177–200.

Filippini, Orietta. *Memoria della Chiesa, memoria dello stato: Il archivio di Carlo Cartari*. Percorsi. Bologna: Il Mulino, 2010.

Flandrin, Jean-Louis. *Families in Former Times: Kinship, Household and Sexuality*. Themes in the Social Sciences. Cambridge: Cambridge University Press, 1979.

Fleischmann, Peter. *Rat und Patriziat in Nürnberg: Die Herrschaft der Ratsgeschlechter vom 13. bis zum 18. Jahrhundert*. Nürnberger Forschungen, 31,2. Nuremberg: Verein für Geschichte der Stadt Nürnberg, 2008.

———. "Das Verhältnis des Nürnberger Rats zur Universität Altdorf." In *Akademie und Universität Altdorf: Studien zur Hochschulgeschichte Nürnbergs*, edited by Hanns Christof Brennecke, Dirk Niefanger, and Werner Wilhelm Schnabel, 7–28. Beihefte zum Archiv für Kulturgeschichte 69. Cologne: Böhlau, 2011.

Fleming, Juliet, ed. "The Renaissance Collage: Toward a New History of Reading." Special issue, *Journal of Medieval and Early Modern Studies* 45, no. 3 (2015): 443–612.

Fossier, François. "A propos du titre d'historiographe sous l'ancien régime." *Revue d'Histoire Moderne et Contemporaine* 32 (1985): 361–417.

Fouilleron, Thomas. "Français par le livre: Les princes de Monaco et l'incroyable longévité d'une généalogie fabuleuse (XVIIe–XIXe siècle)." *Revue Historique* 315 (2013): 601–36.

Fox, Adam. *Oral and Literate Culture in England 1500–1700.* Oxford: Oxford University Press, 2000.

Freytag, Rudolf. "Die Seifertschen Stammtafeln gelehrter Leute." *Familiengeschichtliche Blätter* 21 (1923): 7–12.

Friedrich, Markus. *The Birth of the Archive: A History of Knowledge.* Ann Arbor: University of Michigan Press, 2018.

———. "Genealogy and the History of Knowledge." In *Genealogical Knowledge in the Making: Tools, Practices, and Evidence in Early Modern Europe*, edited by Jost Eickmeyer, Markus Friedrich, and Volker Bauer, 1–22. Cultures and Practices of Knowledge in History 1. Berlin: de Gruyter Oldenbourg, 2019.

———. "Genealogy as Archive-Driven Research Enterprise in Early Modern Europe." *Osiris* 32 (2017): 65–84.

———. "The House of Schwarzenberg and the Genealogical Book Market ca. 1700." In *Genealogisches Wissen in Mittelalter und Früher Neuzeit: Konstruktion—Darstellung—Rezeption*, edited by Giovanni Cusa and Thomas Dorfner. Berlin: de Gruyter, forthcoming.

Friedrich, Susanne. *Drehscheibe Regensburg: Das Informations- und Kommunikationssystem des Immerwährenden Reichstags um 1700.* Berlin: Akademischer Verlag, 2007.

Frisch, Andrea. *The Invention of the Eyewitness: Witnessing and Testimony in Early Modern France.* Chapel Hill: University of North Carolina Press, 2004.

Fritsch, Thomas von. *Die Gothaischen Taschenbücher, Hofkalender und Almanach.* Limburg: Starke, 1968.

Frow, John. "Commodification." In *Information: A Historical Companion*, edited by Ann Blair, Anthony Grafton, Paul Duguid, and Anja-Silvia Goeing, 374–78. Princeton, NJ: Princeton University Press, 2020.

Fuchs, Franz, ed. *Medizin, Jurisprudenz und Humanismus in Nürnberg um 1500: Verein für Geschichte der Stadt Nürnberg.* Pirckheimer-Jahrbuch für Renaissance- und Humanismusforschung 24. Wiesbaden: Harrassowitz, 2010.

Fuchs, Thomas. *Traditionsstiftung und Erinnerungspolitik: Geschichtsschreibung in Hessen in der frühen Neuzeit.* Kassel: Verein für Hessische Geschichte und Landeskunde, 2002.

Gaab, Hans. "Zur Geschichte der Emmert-Sternwarte." Special issue, *Regiomontanus Bote* 18 (2005): 7–71.

García González, Francisco, and Michael Gasperoni. "La sangre y el parentesco: Representaciones y prácticas sociales de la consanguinidad (Edad Media y Época Moderna)." Special issue, *Revista de Demografía Histórica*, 2nd ser., 37 (2019): 17–24.

Gasperoni, Michael, and Jasmin Hauck. "La représentation graphique d'une asymétrie sociale et culturelle: Du genre dans les dispenses matrimoniales, entre Moyen Âge et époque moderne." *Genre et Histoire* 21 (2018), https://journals.openedition.org/genrehistoire/3020.

Gautsch, Karl. "Lebensbeschreibung des Dresdner Chronisten Anton Weck." *Archiv für die Sächsische Geschichte* NF 1 (1875): 349–68.

Gavinelli, Simona. "Il testamento di Attone di Vercelli." *Verbanus* 32 (2011): 322–48.

Gebhardt, Peter von, ed. "Ein Brief des Regensburger Genealogen Johann Seifert." *Familiengeschichtliche Blätter* 25 (1927): 185–86.

Gebhardt, Walter. "Patrizische Leidenschaften: Christoph Sigmund Holzschuher und seine Deductions-Bibliothek von Teutschland." *Mitteilungen des Vereins für Geschichte der Stadt Nürnberg* 91 (2004): 195–210.

Geevers, Liesbeth. "The Danish Habsburgs: Hans, Dorothea and Christina of Denmark as Part of the Habsburg Dynasty." In *Nordens historiker: En vänbok till Harald Gustafsson,* edited by Erik Bodensten, Kajsa Brilkman, David Larsson Heidenblad, and Hanne Sanders, with assistance of Harald Gustafsson, 273–86. Studia Historica Lundensia 28. Lund: Historiska Institutionen Lunds Universitet, 2018.

Geevers, Liesbeth, and Mariella Marini, eds. *Dynastic Identity in Early Modern Europe: Rulers, Aristocrats and the Formation of Identities.* Politics and Culture in Europe, 1650–1750. Farnham: Ashgate, 2015.

———. "Introduction: Aristocracy, Dynasty, and Identity in Early Modern Europe, 1520–1700." In *Dynastic Identity in Early Modern Europe: Rulers, Aristocrats and the Formation of Identities,* edited by Liesbeth Geevers and Mirella Marini, 1–22. Farnham: Ashgate, 2015.

Genicot, Léopold. *Les généalogies.* Typologie des Sources du Moyen Âge Occidental, A-I, 15. Turnhout: Brepols, 1975.

Gentile, Egildo. "I manoscritti di Carlo de Lellis." *Archivio Storico Italiano* 87 (1929): 309–13.

Germann, Wilhelm. *M. Christian Juncker und sein Hennebergisches Geschichtswerk.* Meiningen: Brückner und Renner, 1900.

Gersmann, Gudrun, ed. *Adlige Lebenswelten im Rheinland: Kommentierte Quellen der Frühen Neuzeit.* Cologne: Böhlau, 2009.

Godsey, William D. *Nobles and Nation in Central Europe: Free Imperial Knights in the Age of Revolution, 1750–1850.* New Studies in European History. Cambridge: Cambridge University Press, 2004.

Goldmann, Karlheinz. *Geschichte der Stadtbibliothek Nürnberg.* Nuremberg: n.p., 1957.

Golinelli, Paolo. *Benedetto Bacchini (1651–1721): L'uomo, lo storico, il maestro.* Biblioteca dell'edizione nazionale del carteggio di L.A. Muratori 12. Florence: L. S. Olschki, 2003.

Goody, Jack. *The Development of the Family and Marriage in Europe.* Cambridge: Cambridge University Press, 1983.

Gorman, Michael John. *The Scientific Counter-Revolution: The Jesuits and the Invention of Modern Science.* London: Bloomsbury Academic, 2020.

Gottschalk, Karin. *Eigentum, Geschlecht, Gerechtigkeit: Haushalten und Erben im frühneuzeitlichen Leipzig.* Frankfurt am Main: Campus, 2003.

Götz, Rolf. *Wege und Irrwege frühneuzeitlicher Historiographie: Genealogisches Sammeln zu einer Stammfolge der Herzöge von Teck im 16. und 17. Jahrhundert.* Tübinger Bausteine zur Landesgeschichte 8. Ostfildern: Thorbecke, 2007.

Grafton, Anthony. "Jean Hardouin: The Antiquary as Pariah." *Journal of the Warburg and Courtauld Institutes* 62 (1999): 241–67.

———. *What Was History? The Art of History in Early Modern Europe.* Cambridge: Cambridge University Press, 2007.

Grandmaison, Charles de. *Gaignières, ses correspondants et ses collections de portraits.* Niort: A. Clouzot, 1892.

Grolée-Virville, Alain de. *Les d'Hozier: Juges d'armes en France.* Paris: n.p., 1978.

Grulich, Oscar. *Geschichte der Bibliothek und Naturaliensammlung der kaiserlichen Leopoldinisch-carolinischen deutschen Akademie der Naturforscher.* Halle: W. Engelmann, 1894.

Guillén Berrendero, José A. "Valores nobiliarios, libros y linajes: Rodrigo Mendez de Silva, un nobilista portugués en la Corte de Felipe IV." *Mediterranea-Ricerche Storiche* 11 (2014): 35–60.

Gutiérrez-de-Armas, Judit. "Construir una identidad familiar a partir de antecedentes lejanos y remotos: El problemo de documentar los ancestros para la aristocracia de Canarias en el s. XVII." *Revista de Historia Canaria* 198 (2016): 11–40.

———. "Identidad nobiliaria y legitimación del poder a través de los archivios de familia." In *Monarquías en conflicto: Linajes y noblezas en la articulación de la monarquía hispánica,* edited by José Ignacio Fortea Pérez and Juan Eloy Gelabert González, 659–70. Madrid: Fundación Española de Historia Moderna, 2020.

Haddad, Élie. "Parenté et pratiques généalogiques dans la noblesse française à l'époque modern." In *L'entreprise généalogique: Pratiques sociales et imaginaires en Europe (XVe–XIXe siècles)*, edited by Stéphane Jettot and Marie Lezowski, 73–87. Histoire des mondes modernes 2. Brussels: Peter Lang, 2016.

———. "Qu'est-ce qu'une 'maison'?" *L'homme* 212 (2014): 109–38.

———. "The Question of the Imprescriptibility of Nobility in Early Modern France." In *Contested Spaces of Nobility in Early Modern Europe*, edited by Matthew P. Romaniello and Charles Lipp, 147–66. Farnham: Ashgate, 2011.

Haller von Hallerstein, Helmut. "Nürnberger Geschlechterbücher." *Mitteilungen des Vereins für Geschichte der Stadt Nürnberg* 65 (1978): 212–35.

Hamm, Bernd. "Die Reformation in Nürnberg." In *Deutschlands Auge & Ohr: Nürnberg als Medienzentrum der Reformationszeit*, edited by Thomas Schauerte, 13–33. Schriftenreihe der Museen der Stadt Nürnberg 8. Nuremberg: Tümmel, 2015.

Hampe, Theodor. "Kunstfreunde im alten Nürnberg und ihre Sammlungen (Nebst Beiträgen zur Nürnberger Handelsgeschichte)." *Mitteilungen des Vereins für Geschichte der Stadt Nürnberg* 16 (1904): 57–124.

Hansert, Andreas. *Geburtsaristokratie in Frankfurt am Main: Geschichte des reichsstädtischen Patriziats*. Vienna: Böhlau, 2014.

Harding, Elizabeth, and Michael Hecht, eds. *Die Ahnenprobe in der Vormoderne: Selektion—Initiation—Repräsentation*. Münster: Rhema-Verlag, 2011.

Hausenstein, Wilhelm. "Der Nürnberger Poet Sigmund von Birken in seinen historischen Schriften." *Mitteilungen des Vereins für Geschichte der Stadt Nürnberg* 18 (1908): 197–236.

Hecht, Michael. "The Production of Genealogical Knowledge and the Invention of Princely 'Dynasties.'" In *Genealogical Knowledge in the Making: Tools, Practices, and Evidence in Early Modern Europe*, edited by Jost Eickmeyer, Markus Friedrich, and Volker Bauer, 145–68. Cultures and Practices of Knowledge in History 1. Berlin: de Gruyter Oldenbourg, 2019.

Heck, Kilian. *Genealogie als Monument und Argument: Der Beitrag dynastischer Wappen zur politischen Raumbildung der Neuzeit*. Munich: Dt. Kunstverlag, 2002.

Heck, Kilian, and Bernhard Jahn, eds. *Genealogie als Denkform in Mittelalter und früher Neuzeit*. Tübingen: Niemeyer, 2000.

Heinemann, Olav. *Das Herkommen des Hauses Sachsen: Genealogisch-historiographische Arbeit der Wettiner im 16. Jahrhundert*. Schriften zur sächsischen Geschichte und Volkskunde 51. Leipzig: Leipziger Uni-Verlag, 2015.

Helgerson, Richard. *Forms of Nationhood: The Elizabethan Writing of England*. Chicago: University of Chicago Press 1992.

Hellström, Petter. *Trees of Knowledge: Science and the Shape of Genealogy*. Uppsala Studies in History of Ideas 51. Uppsala: Uppsala Universitet, 2019.

Hiller, Lotte. *Die Geschichtswissenschaft an der Universität Jena in der Zeit der Polyhistorie (1674–1763)*. Jena: Gustav Fischer, 1937.

Hirschmann, Gerhard. "Das Geschlechterbuch der Familie Holzschuher im Stadtarchiv Nürnberg." In *Aus sieben Jahrhunderten Nürnberger Stadtgeschichte: Ausgewählte Aufsätze; Festgabe zu seinem 70. Geburtstag*, edited by Gerhard Hirschmann, 95–107. Nürnberger Forschungen 25. Nuremberg: Selbstverl. d. Vereins für Geschichte d. Stadt Nürnberg, 1988.

———. "Johann Gottfried Biedermann und das Nürnberger Patriziat." *Mitteilungen des Vereins für Geschichte der Stadt Nürnberg* 54 (1966): 141–44.

———. "Das Nürnberger Patriziat." In *Aus sieben Jahrhunderten Nürnberger Stadtgeschichte: Ausgewählte Aufsätze; Festgabe zu seinem 70. Geburtstag*, edited by Gerhard Hirschmann, 123–42. Nürnberger Forschungen 25. Nuremberg: Selbstverl. d. Vereins für Geschichte d. Stadt Nürnberg, 1988.

———. "600 Jahre Genealogie in Nürnberg." In *Aus sieben Jahrhunderten Nürnberger Stadtge-*

schichte: Ausgewählte Aufsätze; Festgabe zu seinem 70. Geburtstag, edited by Gerhard Hirschmann, 155–66. Nürnberger Forschungen 25. Nuremberg: Selbstverl. d. Vereins für Geschichte d. Stadt Nürnberg, 1988.

Hofmann, Hanns Hubert. "Nobiles Norimbergenses: Beobachtungen zur Struktur der reichsstädtischen Oberschicht." In *Untersuchungen zur gesellschaftlichen Struktur der mittelalterlichen Städte in Europa,* edited by Konstanzer Arbeitskreis für Mittelalterliche Geschichte, 53–92. Vorträge und Forschungen 11. Sigmaringen: Patmos, 1966.

Holladay, Joan A. *Genealogy and the Politics of Representation in the High and Late Middle Ages.* Cambridge: Cambridge University Press, 2019.

Hoppen, K. Theodore. *The Common Scientist in the Seventeenth Century: A Study of the Dublin Philosophical Society (1683–1708).* London: Routledge & Paul, 1970.

Horn, Norbert. "Literaturgeschichtliche Aspekte der Rezeption in Spanien." *Tijdschrift voor Rechtsgeschiedenis* 37 (1969): 489–514.

Hummer, Hans. *Visions of Kinship in Medieval Europe.* Oxford: Oxford University Press, 2018.

Iasiello, Italo M. *Il collezionismo di antichità nella Napoli dei viceré.* Naples: Liguori, 2003.

Imhoff, Christoph Freiherr von. "Imhoff." In *Neue Deutsche Biographie,* 10:146–51. Berlin: Duncker und Humblot, 1974.

———. "Die Imhoffs, Handelsherren und Kunstliebhaber." *Mitteilungen des Vereins für Geschichte der Stadt Nürnberg* 62 (1975): 1–42.

Infelise, Mario. *Prima dei giornali: Alle origini della pubblica informazione (secoli XVI e XVII).* Rome: Laterza, 2002.

Ingledew, Francis. "The Book of Troy and the Genealogical Construction of History: The Case of Geoffrey of Monmouth's *Historia regum Britanniae.*" *Speculum* 69 (1994): 665–704.

Jacobs, Peter. *Werdener Annalen.* Düsseldorf: n.p., 1896.

Jahnel, Helga. "Die Imhoffs, eine Nürnberger Patrizier- und Großkaufmannsfamilie: Eine Studie zur reichsstädtischen Wirtschaftspolitik und Kulturgeschichte an der Wende vom Mittelalter zur Neuzeit (1351–1579)." PhD diss., University of Würzburg, 1950.

Jakob, Reinhard. "Der Skandal um einen Nürnberger Imhoff-Faktor im Lissabon der Renaissance: Der Fall Calixtus Schüler und der Bericht Sebald Kneussels (1512)." *Jahrbuch für Fränkische Landesforschung* 60 (2000): 83–112.

Jante, Peter Rudolf. *Willibald Imhoff: Kunstfreund und Sammler. Die Ausgaben Willibald Imhoffs für seine Kunstsammlung und der Versuch ihrer wertmäßigen Erfassung aufgrund handschriftlicher Aufzeichnungen Willibald Imhoffs unter besonderer Berücksichtigung des Haushaltsbuches.* Lüneburg: Stadtlexikon-Verlag, 1985.

Jettot, Stéphane. "Intelligible to the Mind and Pleasing to the Eye: Mapping Out Kinship in British Family Directories (1660–1830)." *History of the Human Sciences* (2022): forthcoming.

Jettot, Stéphane, and Marie Lezowski, eds. *L'entreprise généalogique: Pratiques sociales et imaginaires en Europe (XVe–XIXe siècles).* Histoire des mondes modernes 2. Brussels: Peter Lang, 2016.

Jöns, Dietrich Walter. "Sigmund von Birken: Zum Phänomen einer literarischen Existenz zwischen Hof und Stadt." In *Literatur in der Stadt: Bedingungen und Beispiele städtischer Literatur des 15. bis 17. Jahrhundert,* edited by Horst Brunner, 167–86. Göppinger Arbeiten zur Germanistik 343. Göppingen: Kümmerle Verlag, 1982.

Junge, Walter. *Leibniz und der Sachsen-Lauenburgische Erbfolgestreit.* Quellen und Darstellungen zur Geschichte Niedersachsens 65. Hildesheim: August Laux, 1965.

Jürgensen, Renate. *Bibliotheca Norica: Patrizier- und Gelehrtenbibliotheken in Nürnberg zwischen Mittelalter und Aufklärung.* 2 vols. Beiträge zum Buch- und Bibliothekswesen 43. Wiesbaden: Harrassowitz, 2002.

———. "Norimberga Literata." In *Stadt und Literatur im deutschen Sprachraum der Frühen Neuzeit,* edited by Klaus Garber, 425–90. Berlin: de Gruyter, 1998.

——. *Utile cum dulci: Mit Nutzen erfreulich. Die Blütezeit des Peginesischen Blumenordens in Nürnberg 1644 bis 1744*. Wiesbaden: Harrassowitz, 1994.

Jussen, Bernhard et al., eds. *Blood and Kinship: Matter for Metaphor from Ancient Rome to the Present*. New York: Berghahn Books, 2013.

Kagan, Richard L. *Clio and the Crown: The Politics of History in Medieval and Early Modern Spain*. Baltimore, MD: Johns Hopkins University Press, 2009.

Kagerer, Alexander. *Macht und Medien um 1500: Selbstinszenierungen und Legitimationsstrategien von Habsburgern und Fuggern*. Deutsche Literatur 23. Berlin: de Gruyter, 2017.

Kantzenbach, Friedrich Wilhelm. "Gottfried Arnolds Weg zur Kirchengeschichte 1699." *Jahrbuch der hessischen Kirchengeschichtlichen Vereinigung* 26 (1975): 207–41.

Kekulé von Stradonitz, Stephan. *Ausgewählte Aufsätze aus dem Bezirk des Staatsrechts und der Genealogie*. Berlin: Carl Heymann, 1905.

Keller, Katrin, and Paola Molino. *Die Fuggerzeitungen im Kontext. Zeitungssammlungen im Alten Reich und in Italien*. Vienna: Böhlau Verlag, 2015.

Kellner, Beate. *Ursprung und Kontinuität: Studien zum genealogischen Wissen im Mittelalter*. Munich: Fink, 2004.

Kenny, Neil. *Born to Write: Literary Families and Social Hierarchy in Early Modern France*. Oxford: Oxford University Press, 2020.

Ketelaar, Eric. "The Genealogical Gaze: Family Identities and Family Archives in the Fourteenth to Seventeenth Centuries." *Libraries & the Cultural Record* 44 (2009): 9–28.

Kirchhoff, Matthias. *Gedächtnis in Nürnberger Texten des 15. Jahrhunderts: Gedenkbücher, Brüderbücher, Städtelob, Chroniken*. Nürnberger Werkstücke zur Stadt- und Landesgeschichte 68. Nuremberg: Philipp Schmidt, 2009.

Klapisch-Zuber, Christiane. "The Genesis of the Family Tree." *I Tatti Studies: Essays in the Renaissance* 4 (1991): 105–29.

——. *Das Haus, der Name, der Brautschatz: Strategien und Rituale im gesellschaftlichen Leben der Renaissance*. Reihe Geschichte und Geschlechter 7. Frankfurt am Main: Campus-Verlag 1995.

——. *L'ombre des ancêtres: Essai sur l'imaginaire médiéval de la parenté*. L'esprit de la cité. Paris: Fayard, 2000.

Klapisch-Zuber, Christiane, and David Herlihy. *Les Toscans et leurs familles: Une étude du Catasto Florentin de 1427*. Paris: Presses de la Fondation Nationale des Sciences Politiques, 1978.

Klocke, Friedrich von. "Ein westfälischer Genealogenkreis des 18. Jahrhunderts (Steinen, Mellin, Michels, Rademacher, Roßkampf, Schmitz)." *Beiträge zur westfälischen Familienforschung* 13 (1954): 65–87.

Knight, Jeffrey Todd. *Bound to Read: Compilations, Collections, and the Making of Renaissance Literature*. Material Texts. Philadelphia: University of Pennsylvania Press, 2013.

Koch, Gerhard, ed. *Imhoff Indienfahrer: Ein Reisebericht aus dem 18. Jahrhundert in Briefen und Bildern*. Göttingen: Wallstein, 2001.

Köcher, Adolf. "Der preußisch-welfische Hoheitsstreit um die Harzgrafschaft Regenstein." *Zeitschrift des Harzvereins* 28 (1895): 542–58.

Koldau, Linda Maria. *Frauen—Musik—Kultur: Ein Handbuch zum deutschen Sprachgebiet der Frühen Neuzeit*. Cologne: Böhlau, 2005.

Korb, Gerhard. "Seifert und Hildebrand." *Familie und Volk* 10 (1961): 458–61.

Koschlig, Manfred. *Das Ingenium Grimmelshausens und das "Kollektiv": Studien zur Entstehungs- und Wirkungsgeschichte des Werkes*. Munich: Beck, 1977.

Kraus, Andreas. "Bürgerlicher Geist und Wissenschaft: Wissenschaftliches Leben im Zeitalter des Barocks und der Aufklärung in Augsburg, Regensburg und Nürnberg." *Archiv für Kulturgeschichte* 49 (1967): 340–90.

Kreiner, Artur. "Nürnbergs und Altdorfs Anteil an der Kaiserlichen Leopold-Carolinischen Deutschen Akademie der Naturforscher." *Mitteilungen des Vereins für Geschichte der Stadt Nürnberg* 37 (1940): 309–22.

Kuhn, Christian. *Generation als Grundbegriff einer historischen Geschichtskultur: Die Nürnberger Tucher im langen 16. Jahrhundert.* Formen der Erinnerung 45. Göttingen: V&R Unipress, 2010.

Laborderie, Olivier de. *Histoire, mémoire et pouvoir: Les généalogies en rouleau des rois d'Angleterre 1250–1422.* Bibliothèque d'histoire médiévale 7. Paris: Classiques Garnier, 2013.

Lafage, Franck. *Côme III de Médicis, grand-duc de Toscane: Un règne dans l'ombre de l'histoire (1670–1723).* Chemins de la mémoire: Série XVIIIe siècle. Paris: l'Harmattan, 2015.

Landois, Antonia. *Gelehrtentum und Patrizierstand: Wirkungskreise des Nürnberger Humanisten Sixtus Tucher (1459–1507).* Spätmittelalter, Humanismus, Reformation 77. Tübingen: Mohr Siebeck, 2014.

Landwehr, Achim. *Geburt der Gegenwart: Eine Geschichte der Zeit im 17. Jahrhundert.* Frankfurt: Fischer, 2014.

Laufhütte, Hartmut, and Ralf Schuster, eds. *Sigmund von Birken: Werke und Korrespondenz.* Vol. 13.2/I, *Der Briefwechsel zwischen Sigmund von Birken und Magnus Daniel Omeis, Joachim Heinrich Hagen, Sebastian Seelmann und Georg Wende.* Berlin: de Gruyter, 2018.

Lecoq, Anne-Marie, Marc Fumaroli, and Jean Robert Armogathe, eds. *La querelle des Anciens et des Modernes, XVIIe–XVIIIe siècles.* Collection folio classique 3414. Paris: Gallimard, 2001.

Levine, Joseph M. *Between the Ancients and the Moderns: Baroque Culture in Restoration England.* New Haven, CT: Yale University Press, 1999.

Lezowski, Marie. "Ramificazioni della storia: Verdi ramoscelli, innesti e gemme delle indagini genealogiche (1995–2017)." *Quaderni Storici* 53 (2018): 229–46.

Lhotsky, Alphons. "Apis Colonna: Fabeln und Theorien über die Abkunft der Habsburger." *Mitteilungen des Instituts für Österreichische Geschichtsforschung* 55 (1944): 171–246.

Liddy, Christian D., and Christian Steer. "John Lord Lumley and the Creation and Commemoration of Lineage in Early Modern England." *Archaeological Journal* 167 (2010): 197–227.

Lieber, Maria. "Jacob Wilhelm von Imhofs Korrespondenz mit Muratori (1696–1705)." In *Die Glückseligkeit des gemeinen Wesens: Wege der Ideen zwischen Italien und Deutschland im Zeitalter der Aufklärung,* edited by Fabio Marri, 17–66. Italien in Geschichte und Gegenwart 14. Frankfurt am Main: Lang, 1999.

Lingnau, Anna. *Lektürekanon eines Fürstendieners: Die Privatbibliothek des Friedrich Rudolf von Canitz (1654–1699).* Bibliothek Altes Reich 32. Berlin: de Gruyter, 2021.

Lohrer, Liselotte. *Cotta: Geschichte eines Verlags, 1659–1959.* Stuttgart: Cotta, 1959.

Lohsträter, Kai. "Zeitgeschichte, Empirie und Pragmatismus: Zeitungskollegs als Versuchsfelder (früh-)moderner historischer Methoden (1660–1820)." In *Praktiken frühneuzeitlicher Historiografie,* edited by Markus Friedrich and Jacob Schilling, 387–435. Cultures and Practices of Knowledge in History 2. Berlin: de Gruyter, 2019.

Loiseau, Jérôme. "Much Ado about Nothing? The Intendant, the Gentilshommes and the Investigations into Nobility in Burgundy (1664–1670)." *French History* 22 (2008): 275–94.

Lopes, Paulo Catarino, ed. *Portugal e a Europa nos séculos XV e XVI: Olhares, relações, identidade(s).* Estudios 21. Lisbon: IEM, 2019.

Ludwig, Heidrun. *Nürnberger naturgeschichtliche Malerei im 17. und 18. Jahrhundert.* Acta biohistorica 2. Marburg an der Lahn: Basilisken-Presse, 1998.

Lux, David S., and Harold John Cook. "Closed Circles or Open Networks? Communicating at a Distance during the Scientific Revolution." *History of Science* 36 (1998): 179–211.

Magnanini, Suzanne. "Postulated Routes from Naples to Paris: The Printer Antonio Bulifon and Giambattista Basile's Fairy Tales in Seventeenth-Century France." *Marvels & Tales* 21 (2007): 78–92.

Mährle, Wolfgang. *Academia Norica: Wissenschaft und Bildung an der Nürnberger Hohen Schule in Altdorf (1575–1623)*. Contubernium 54. Stuttgart: Steiner, 2000.

Mancino, Maria Rosaria. "I 'Panni ricamati' delle storie della vita di S. Tommaso d'Aquino." *Antichità Viva* 32 and 34 (1993 and 1995): 47–55 and 52–59.

Mareš, Karel. "Augustiniánský historiograf Valentin Weidner a jeho dvě klášterní kroniky." *Folia Historica Bohemica* 30 (2015): 387–409 (English summary, 407–9).

Maťa, Petr. "The False Orsini from over the Alps: Negotiating Aristocratic Identity in Late Medieval and Early Modern Europe." *Römische historische Mitteilungen* 55 (2014): 155–218.

———. "Patres Patriae or Proditores Patriae? Legitimizing and De-legitimizing the Authority of the Provincial Estates in Seventeenth-Century Bohemia." In *Whose Love of Which Country? Composite States, National Histories and Patriotic Discourses in Early Modern East Central Europe*, edited by Balázs Trencsényi and Márton Zászkaliczky, 406–42. Boston: Brill, 2010.

Mazauric, Simone. *Savoirs et philosophie à Paris dans la première moitié du XVIIe siècle: Les conférences du bureau d'adresse de Théophraste Renaudot (1633–1642)*. Paris: Éditions de la Sorbonne, 1997.

McDowell, Paula. *The Invention of the Oral: Print Commerce and Fugitive Voices in Eighteenth-Century Britain*. Chicago: University of Chicago Press, 2017.

Melville, Gert. "Vorfahren und Vorgänger: Spätmittelalterliche Genealogien als dynastische Legitimation zur Herrschaft." In *Die Familie als sozialer und historischer Verband. Untersuchungen zum Spätmittelalter und zur frühen Neuzeit*, edited by Peter-Johannes Schuler, 203–310. Sigmaringen: Thorbecke, 1987.

Menant, François. "La connaissance du Moyen Âge en Lombardie aux XVIIe et XVIIIe siècles." *Le Moyen Âge* 87 (1981): 419–54.

Merode-Westerloo, Jean-Philippe-Eugène de. *Mémoires de Feld-Maréchal Comte de Merode-Westerloo, Chevalier de la Toison d'Or, Capitaine de Trabans de l'Empereur Charles VI*. 2 vols. Brussels: n.p., 1840.

Meyer, Carla. *Die Stadt als Thema: Nürnbergs Entdeckung in Texten um 1500*. Mittelalter-Forschungen 26. Ostfildern: Thorbecke, 2009.

Millstone, Noah. "Designed for Collection: Early Modern News and the Production of History." *Media History* 23 (2017): 177–98.

———. *Manuscript Circulation and the Invention of Politics in Early Stuart England*. Cambridge Studies in Early Modern British History. Cambridge: Cambridge University Press, 2016.

Minuzzi, Sabrina. "Sul filo dei segreti medicinali: Praticanti e preofessionisti del mercato della cura a Venezia (secoli XVI–XVIII)." PhD diss., Università degli Studi di Verona, 2012.

Mirto, Alfonso. "Lettere di Antonio Magliabechi a Bernardo Benvenuti." *Studi secenteschi* 39 (1998): 205–42.

———. "Lettere di Giovanni Battista Bidelli junior a Bernardo Benvenuti." *Culture del Testo* 10–11 (1998): 71–84.

———. *Stampatori, editori, librai nella seconda metà del Seicento: Parte prima [ed unica]*. Florence: Centro Editoriale Toscano, 1984.

Mix, York-Gothart. "Genealogische Kalender als Medium europäischer Identität." In *Französische Almanachkultur im deutschen Sprachraum (1700–1815): Gattungsstrukturen, komparatistische Aspekte, Diskursformen*, edited by Hans-Jürgen Lüsebrink and York-Gothart Mix, 35–50. Göttingen: V&R Unipress 2014.

Montaigne, Michel de. *Essais*. Edited by P. Villey and V.-L. Saulnier. Online edition by P. Desan, University of Chicago. https://artflsrv03.uchicago.edu/philologic4/montessaisvilley/navigate/1/5/6.

Mori, Elisabetta. *L'archivio Orsini: La famiglia, la storia, l'inventario*. Carte scoperte 4. Rome: Viella, 2016.

Morsel, Joseph. *Noblesse, parenté et reproduction sociale à la fin du Moyen Âge*. Paris: Picard, 2017.

———. "Sociogenèse d'un patriciat: La culture de l'écrit et la construction du social à Nuremberg vers 1500." *Histoire Urbaine* 35 (2012): 83–106.

Müller, Andreas. *Die Ritterschaft im Herzogtum Westfalen 1651–1803: Aufschwörung, innere Struktur und Prosopographie*. Münster: Aschendorff, 2017.

Müller, Johannes. "Die Geschäftsreisen und die Gewinnanteile Endres Imhofs des Älteren als Teilhaber der Handelsgesellschaft "Peter Imhof und Gebrüder" von 1508–1525." *Vierteljahrschrift für Sozial- und Wirtschaftsgeschichte* 13 (1916): 153–79.

Müller, Uwe. "Die Leopoldina unter den Präsidenten Bausch, Fehr und Volckamer (1652–1693)." In *350 Jahre Leopoldina—Anspruch und Wirklichkeit: Festschrift der Deutschen Akademie der Naturforscher Leopoldina, 1652–2002*, edited by Benno Parthier and Dietrich von Engelhardt, 45–94. Halle: Druck-Zuck, 2002.

Mulsow, Martin. "Freethinking in Early-Eighteenth-Century Protestant Germany: Peter Friedrich Arpe and the *Traité des trois imposteurs*." In *Heterodoxy, Spinozism, and Free Thought in Early-Eighteenth-Century Europe: Studies on the "Traité des trois imposteurs,"* edited by Silvia Berti, Françoise Charles-Daubert, and Richard H. Popkin, 193–240. International Archives of the History of Ideas 148. Dordrecht: Springer, 1996.

———. "History of Knowledge." In *Debating New Approaches to History*, edited by Marek Tamm and Peter Burke, 159–87. London: Bloomsbury Academic, 2019.

Münch Miranda, Susana, and João Paolo Salvado. "Justice, Politics, and Diplomacy: The Marquis of Arronches's Criminal Case (Vienna and Lisbon 1696–1700)." *Atalanta* 5 (2017): 103–31.

Munck, Bert de, and Antonella Romano. "Introduction." In *Knowledge and the Early Modern City: A History of Entanglements*, edited by Bert de Munck and Antonella Romano, 1–30. Knowledge Societies in History. London: Routledge, 2020.

Muratori, Lodovico Antonio. *Carteggi con Mabillon* [. . .] *Maittaire*. Edited by Viola Corrado. Edizione nazionale del carteggio di L. A. Muratori 26. Florence: Leo S. Olschki, 2016.

Muto, Giovanni. "I trattati napoletani cinqucenteschi in tema di nobiltà." In *Sapere e/è potere: Discipline, dispute e professioni nell'università medievale e moderna; Il caso bolognese a confront*, edited by Luisa Avellini, 321–43. Collana convegni e colloqui, n.s., 14. Bologna: n.p., 1990.

Natale, Alfio Rosario. "Falsari milanesi del Seicento." In *Raccolta di studi in memoria di Sergio Mochi Onory*, 457–506. Contributi dell'Istituto di Storia Medioevale 2. Milan: Vita e Pensiero, 1972.

Nevitt, Marcus. "Ben Jonson and the Serial Publication of News." *Media History* 11 (2005): 53–68.

Niccolini, F. "La fine del dominio spagnuolo sull'Italia meridionale nelle biografie di due generali napolitani." *Bolletino dell'Archivio Storico del Banco di Napoli* 8 (1954): 122–88.

Niefanger, Dirk, and Werner Wilhelm Schnabel. "Literarische Gruppenbildung an der Universität Altdorf." In *Akademie und Universität Altdorf: Studien zur Hochschulgeschichte Nürnbergs*, edited by Hanns Christof Brennecke, Dirk Niefanger, and Werner Wilhelm Schnabel, 245–322. Cologne: Böhlau, 2011.

Nozza Bielli, Chiara. "Il collezionismo di Gaignières e la sua epoca: La produzione del sapere genealogico nella Francia di Luigi XIV." PhD diss., University of Padua, 2022.

Orlandi, Antonella. "Biografia e bibliografia nell'opera di Prospero Mandosio." *Esperienze Letterarie* 29:55–98.

Ostenfeld-Suske, Kira. "Official Historiography, Political Legitimacy, Historical Methodology, and Royal and Imperial Authority in Spain under Phillip II, 1580–99." PhD diss., Columbia University, 2014.

Östling, Johan. "Circulation, Arenas, and the Quest for Public Knowledge: Historiographical Currents and Analytical Frameworks." *History and Theory* 59 (2020): 111–26.

Palmieri, Stefano. *Degli archivi napolitani: Storia e tradizione*. Bologna: Il Mulino, 2002.

Paravicini, Werner. "Colonna und Orsini: Römische Ursprungslegenden im europäischen Adel am Ende des Mittelalters." In *Adelslandschaft Mitteldeutschland: Die Rolle des landsässigen Adels in der mitteldeutschen Geschichte (15.–18. Jahrhundert),* edited by Enno Bünz, Ulrike Höroldt, and Christoph Volkmar, 19–110. Veröffentlichungen der Staatlichen Archivverwaltung des Landes Sachsen-Anhalt 22. Leipzig: Leipziger Universitätsverlag, 2016.

Paul, Markus. *Reichsstadt und Schauspiel: Theatrale Kunst im Nürnberg des 17. Jahrhunderts.* Frühe Neuzeit 69. Berlin: de Gruyter, 2002.

Pečar, Andreas. "Dynastien—Träger der Staatsbildung? Überlegungen zu Herrschaft und Staatsbildung in kulturvergleichender Perspektive anlässlich einer prominenten Neuerscheinung." *Zeitschrift für Historische Forschung* 44 (2017): 51–67.

Pedìo, Tommaso. *Storia della storiografia del regno di Napoli.* Naples: Edizioni Frama's, 1973.

Peronnet, Michel. "L'invention de l'ancien régime en France." *History of European Ideas* 14 (1992): 49–58.

Pettegree, Andrew. *The Invention of News: How the World Came to Know about Itself.* New Haven, CT: Yale University Press, 2014.

Petto, Christine Marie. *When France Was King of Cartography: The Patronage and Production of Maps in Early Modern France.* Toposophia. Lanham, MD: Lexington Books, 2007.

Pfeiffer, Gerhard, ed. *Nürnberg: Geschichte einer europäischen Stadt.* Munich: Beck, 1971.

———. "Nürnberger Patriziat und fränkische Reichsritterschaft." In *Norica: Beiträge zur Nürnberger Geschichte,* edited by Karlheinz Goldmann, 35–55. Nuremberg: Stadtbibliothek, 1961.

———. "Westfälische Aufschwörungstafeln als familiengeschichtliche Quelle." *Westfalen* 18 (1933): 153–63.

Pieper, Lennart. *Einheit im Konflikt: Dynastiebildung in den Grafenhäusern Lippe und Waldeck in Spätmittelalter und Früher Neuzeit.* Norm und Struktur 49. Cologne: Böhlau, 2019.

Piétri, Valérie. "Bonne renommée ou actes authentiques: La noblesse doit faire ses preuves (Provence, XVIIe–XVIIIe siècles)." *Genèses* 74 (2009): 5–24.

———. "Une offrande à la vanité des puissants: L'écriture généalogique comme service littéraire (France méridionale, XVIIe–XVIIIe siècle)." In *L'entreprise généalogique: Pratiques sociales et imaginaires en Europe (XVe–XIXe siècles),* edited by Stéphane Jettot and Marie Lezowski, 171–89. Brussels: P.I.E.-Peter Lang, 2016.

Piétri, Valérie, and Isabelle Luciani. "Généalogie, construction du présent, récit de soi." In *L'incorporation des ancêtres: Généalogie, construction du présent, du Moyen Âge à nos jours,* edited by Valérie Piétri and Isabelle Luciani, 5–22. Aix-en-Provence: Presses Universitaires de Provence, 2016.

———, eds. *L'incorporation des ancêtres: Généalogie, construction du présent, du Moyen Âge à nos jours.* Corps & âmes. Aix-en-Provence: Presses Universitaires de Provence, 2016.

Pípalová, Anna-Marie. "Bohuslav Balbín and the Patriotic Reconceptualization of Bohemia, c. 1650–1675." *Historical Journal* 65 (2022): 1–23.

Pironti, Pasquale. *Bulifon—Raillard—Gravier: Editori francesi in Napoli.* Naples: Lucio Pironti Editore, 1982.

Plumb, J. H. *The Death of the Past.* Harmondsworth: Penguin Books, 1973.

Pohle, Jürgen. "Rivalidade e cooperação: Algumas notas sobre las casas comerciais alemãs em Lisboa no início de Quinhentos." *Cadernos do Arquivo Municipal,* 2a series 3 (2015): 19–38.

Poli, Oscar de. "Introduction aux Mémoires de Claude de Guénegaud (1675–1681)." *Revue des Questions Héraldiques, Archeologiques et Historiques* (1903–4): 337–50, 412–28, 495–507, 538–51, 602–14, 665–76.

Poncet, Olivier. "Cercles savants et pratique généalogique en France (fin XVIe siècle–milieu du XVIIe siècle)." In *L'opération généalogique: Cultures et pratiques européennes entre XVe et XVIIIe siècle,* edited by Olivier Rouchon, 101–39. Histoire. Rennes: Presses Universitaires de Rennes, 2014.

———. "L'usage des chartriers seigneuriaux par les érudits et généalogistes en France dans la première moitié du XVIIe siècle." In *Les chartriers seigneuriaux: Défendre ses droits, construire sa mémoire (XIIIe–XXIe siècle),* edited by Philippe Contamine, 247–65. Société de l'histoire de France, 538. Paris: Société de l'Histoire de France, 2010.

Pumfrey, Stephen, and Frances Dawbarn. "Science and Patronage in England, 1570–1625: A Preliminary Study." *History of Science* 42 (2004): 137–88.

Quondam, Amedeo, and Michele Rak, eds. *Lettere dal regno ad Antonio Magliabechi.* Fonti e documenti per la storia del Mezzogiorno d'Italia 4. Naples: Guida, 1978.

Rabinovitch, Oded. *The Perraults: A Family of Letters in Early Modern France.* Ithaca, NY: Cornell University Press, 2018.

Radulescu, Raluca Luria, and Edward Donald Kennedy, eds. *Broken Lines: Genealogical Literature in Late-Medieval Britain and France.* Turnhout: Brepols, 2008.

Raines, Dorit. *L'invention du mythe aristocratique: L'image de soi du patriciat vénitien au temps de la Sérénissime.* Venice: Ist. Veneto di Scienze Lettere ed Arti, 2006.

Ranum, Orest. *Artisans of Glory: Writers and Historical Thought in Seventeenth-Century France.* Chapel Hill: University of North Carolina Press, 1980.

Rascher, Helmut. "Die Kleiderstiftung des Wolfgang Münzer von 1577." *Mitteilungen des Vereins für Geschichte der Stadt Nürnberg* 57 (1970): 1–123.

Raymond, Joad. *The Invention of the Newspaper: English Newsbooks 1641–1649.* Oxford: Oxford University Press, 1996.

Rejzek, Antonín. *P. Bohuslav Balbín T. J. Jeho život a práce.* Dědictví sv. Prokopa v Praze 48. Prague: Dědictví sv. Prokopa, 1908.

Ressel, Magnus. *Protestantische Händlernetze im langen 18. Jahrhundert: Die deutschen Kaufmannsgruppierungen und ihre Korporationen in Venedig und Livorno von 1648 bis 1806.* Schriftenreihe der Historischen Kommission bei der Bayerischen Akademie der Wissenschaften 107. Göttingen: Vandenhoeck & Ruprecht, 2020.

Richter, Ludwig, ed. *Der Briefwechsel zwischen Bohuslav Balbín und Christian Weise 1678–1688: Lateinisch-Deutsche Ausgabe.* Stuttgart: Franz Steiner, 2010.

Ritz-Guilbert, Anne. "La collection Gaignières: Méthodes et finalités." *Bulletin Monumental* 166 (2008): 315–38.

———. *La collection Gaignières: Un inventaire du royaume au XVII*e *siècle.* Paris: CNRS Éditions, 2016.

Rodríguez Villa, Antonio. *Don Diego Hurtado de Mendoza y Sandoval, Conde de la corzana (1650–1720): Estudio histórico.* Madrid: Fortanet, 1907.

Rohmer, Ernst. "Die Hirten in der Grotte: Zur Funktion genealogischen Wissens in den Schriften des Sigmund von Birken." In *Der Franken Rom: Nürnbergs Blütezeit in der zweiten Hälfte des 17. Jahrhunderts,* edited by John Roger Paas, 276–88. Wiesbaden: Harrassowitz, 1995.

———. "Literatur und Theologie in Nürnberg: Johann Michael Dilherr und der peginesische Blumenorden." In *Aedificatio: Erbauung im interkulturellen Kontext in der Frühen Neuzeit,* edited by Andreas Solbach, 267–83. Tübingen: Niemeyer, 2005.

Römer, Christof. "Die Grafen von Regenstein-Blankenburg als Stand des Reiches und des Niedersächsischen Reichskreises." In *Zwischen Herrschaftsanspruch und Schuldendienst: Beiträge zur Geschichte der Grafschaft Regenstein,* edited by Heinz A. Behrens, 73–90. Jena: Bussert & Stadeler, 2004.

Rosa, Maria de Lurdes. *O morgadio em Portugal, sécs. XIV–XV: Modelos e práticas de comportamento linhagístico.* Histórias de Portugal 16. Lisbon: Editorial Estampa, 1995.

Rosa, Mario. "Un médiateur dans la république des lettres: Le bibliothécaire." In *Commercium Litterarium: La communication dans la république des lettres: Forms of Communication in the*

Republic of Letters, edited by Hans Bots and Françoise Wacquet, 81–99. Amsterdam: APA Holland University Press, 1994.

Rosenthal, Erwin. "Dürers Buchmalereien für Pirckheimers Bibliothek." *Jahrbuch der Preußischen Kunstsammlungen* 49 (1928): 1–54.

Rouchon, Olivier. "L'enquête généalogique et ses usages dans la Toscane des Médicis: Un exemple pisan de 1558." *Annales: Histoire, Sciences Sociales* 54 (1999): 705–37.

———, ed. *L'opération généalogique: Cultures et pratiques européennes entre XVe et XVIIIe siècle.* Rennes: Presses Universitaires de Rennes, 2014.

Roumy, Franck. "La naissance de la notion canonique de consanguinitas et sa réception dans le droit civil." In *L'hérédité entre Moyen Âge et époque moderne: Perspectives historiques,* edited by Maaike van der Lugt, 41–66. Florence: SISMEL Edizioni del Galluzzo, 2008.

Rovito, Pier Luigi. *Respublica dei Togati: Giuristi e società nella Napoli del Seicento.* Naples: Jovene Editore, 1982.

Rücker, Elisabeth, and William Stearn. *Maria Sibylla Merian in Surinam: Kommentar zur Faksimile-Ausgabe der Metamorphosis Insectorum Surinamensium (Amsterdam 1705) nach den Aquarellen in der Royal Library, Windsor Castle.* London: Pion, 1982.

Sabean, David Warren, and Simon Teuscher. "A New Approach to Long-Term Development." In *Kinship in Europe: Approaches to Long-Term Development (1300–1900),* edited by David Warren Sabean and Simon Teuscher, 1–32. New York: Berghahn Books, 2007.

———, eds. *Kinship in Europe: Approaches to Long-Term Development (1300–1900).* New York: Berghahn Books, 2007.

Sahlins, Marshall David. *What Kinship Is—and Is Not.* Chicago: University of Chicago Press, 2013.

Sannino, Anna Lisa. "Le storie genealogiche." In *Il libro e la piazza: Le storie locali dei regni di Napoli e di Sicilia in eta modern,* edited by Antonio Lerra, 109–55. Manduria: Lacaita, 2004.

Santi, Venceslao. *La precedenza tra gli Estensi e i Medici e "L'historia de' Prencipi d'Este" di G. Battista Pigna.* Ferrara: n.p., 1897.

Sauer, Christine. "Reformation der Bücher: Die Gründung der Stadtbibliothek als Folge des Anschlusses Nürnbergs an die neue Glaubenslehre." *Mitteilungen des Vereins für Geschichte der Stadt Nürnberg* 104 (2017): 101–36.

———, ed. *642 Jahre Stadtbibliothek Nürnberg: Von der Ratsbibliothek zum Bildungscampus.* Beiträge zur Geschichte und Kultur der Stadt Nürnberg 26. Wiesbaden: Harrassowitz, 2013.

Scarano, Paolo. "Un principe napoletano vicerè di Perù, Carmine Niccoló Caracciolo di Santo Buono." *Atti dell'Accademia di Scienze Morali e Politiche* 75 (1964): 309–59.

Schalk, Ellery. *From Valor to Pedigree: Ideas of Nobility in France in the Sixteenth and Seventeenth Centuries.* Princeton, NJ: Princeton University Press, 1986.

Schicketanz, Peter, ed. *Der Briefwechsel Carl Hildebrand von Cansteins mit August Hermann Francke.* Texte zur Geschichte des Pietismus, Abt. III; 1. Berlin: de Gruyter, 1972.

Schiele, Walter. *Johann Joseph Vöhlins genealogische Sammlung. Studien zur Person, zum Werk und zur Arbeitsweise des Historikers und Genealogen Johann Joseph Vöhlin von Frickenhausen (1709–1785).* Göppinger akademische Beiträge 16. Göppingen: Kümmerle, 1971.

Schilling, Jacob. "Caspar Sagittarius und die Numismatik seiner Schüler." *Neues Archiv für sächsische Geschichte* 89 (2018): 143–57.

Schmeisser, Martin, and Klaus Birnstiel. "Gelehrtenkultur und antitrinitarische Häresie an der Nürnberger Akademie zu Altdorf." *Daphnis* 39 (2010): 221–85.

Schmid, Barbara. *Schreiben für Status und Herrschaft: Deutsche Autobiographik in Spätmittelalter und früher Neuzeit.* Zurich: Chronos Verlag, 2006.

Schmitt-Maass, Christoph. *Fénelons "Télémaque" in der deutschsprachigen Aufklärung (1700–1832).* Berlin: de Gruyter, 2018.

Schnabel, Werner Wilhelm. *Österreichische Exulanten in oberdeutschen Reichsstädten: Zur Mi-*

gration von Führungsschichten im 17. Jahrhundert. Schriftenreihe zur bayerischen Landesgeschichte 101. Munich: Beck, 1992.

Scholz Williams, Gerhild. *Mediating Culture in the Seventeenth-Century German Novel: Eberhard Werner Happel, 1647–1690.* Ann Arbor: University of Michigan Press, 2014.

——. *Ways of Knowing in Early Modern Germany: Johannes Praetorius as a Witness to His Time.* Aldershot: Routledge, 2006.

Schömbs, Erwin. *Das Staatsrecht Johann Jakob Mosers (1701–1785): Zur Entstehung des historischen Positivismus in der deutschen Reichspublizistik des 18. Jahrhunderts.* Schriften zur Verfassungsgeschichte 8. Berlin: Duncker & Humblot, 1968.

Schrenck von Notzing, Niklas Freiherr von. "Die 'Genealogiae' des Nicolaus Rittershausen." In *Totum Me Libris Dedo: Festschrift zum 80. Geburtstag von Adolf Seebass,* edited by Alain Moirandat, Heide Spilker, and Verena Tammann, 177–86. Basel: Haus der Bücher, 1979.

Schröder, Thomas. *Die ersten Zeitungen: Textgestaltung und Nachrichtenauswahl.* Tübingen: Narr, 1995.

Schröttel, Gerhard. *Johann Michael Dilherr und die vorpietistische Kirchenreform in Nürnberg.* Einzelarbeiten aus der Kirchengeschichte Bayerns 34. Nuremberg: Selbstverlag des Vereins für bayerische Kirchengeschichte, 1962.

Schultheiss, Werner. "Der Nürnberger Großkaufmann und Diplomat Andreas I. Imhoff." *Mitteilungen aus der Stadtbibliothek Nürnberg* 6 (1957): 3–12.

Schunka, Alexander. *Ein neuer Blick nach Westen: Deutsche Protestanten und Großbritannien (1688–1740).* Jabloniana 10. Wiesbaden: Harrassowitz, 2019.

Schuster, Peter. "Familien- und Geschlechtsbewusstsein im spätmittelalterlichen Adel." *Geschichte und Region* 11 (2002): 13–36.

Schuwey, Christophe. *Un entrepreneur des lettres au XVIIe siècle: Donneau de Visé, de Molière au Mercure galant.* Lire le XVIIe siècle: Discours critique 2. Paris: Classiques Garnier, 2020.

——. "Naissance d'une figure déviante: La genèse des 'nouvellistes.'" In *L'errance au XVIIe siècle: 45e colloque de la North American Society for Seventeenth-Century French Literature,* edited by Lucie Desjardins, Marie-Christine Pioffet, and Roxanne Roy, 303–19. Biblio 17, 216. Tübingen: Narr, 2017.

——. "Les périodiques dans l'écriture de l'histoire au XVIIe siècle." In *Histoire de l'écriture et écriture de l'H(h)istoire,* edited by Bernadette Rey Mimoso-Ruiz, 121–44. Toulouse: Institut Catholique de Toulouse, 2016.

Schwemmer, Wilhelm. "Das Mäzenatentum der Nürnberger Patrizierfamilie Tucher vom 14.–18. Jahrhundert." *Mitteilungen des Vereins für Geschichte der Stadt Nürnberg* 51 (1962): 18–59.

Seibold, Gerhard. "Die Imhoffsche Handelsgesellschaft in den Jahren 1579 bis 1635." *Mitteilungen des Vereins für Geschichte der Stadt Nürnberg* 64 (1977): 201–14.

Seitz, Winfried. "Johann Paul Wurffbain (1655–1711): Leben und Werk." PhD diss., University of Marburg, 1982.

Sforza, Michele. *Giovanni Carafa Duca die Noja umanista-scienziato del secolo dei lumi: Appendice: I Carafa di Noja contro gli Acquaviva d'Aragona di Conversano.* Bari: Centro Studi Niccolaiani, 2005.

Shapin, Steven. *A Social History of Truth: Civility and Science in Seventeenth-Century England.* Chicago: University of Chicago Press, 1999.

Shapin, Steven, and Adi Ophir. "The Place of Knowledge: A Methodological Survey." *Science in Context* 4 (1991): 3–21.

Shovlin, John. "The Nobility." In *The Oxford Handbook of the Ancien Régime,* edited by William Doyle, 111–26. Oxford: Oxford University Press, 2012.

Silver, Larry. *Marketing Maximilian: The Visual Ideology of a Holy Roman Emperor.* Princeton, NJ: Princeton University Press, 2008.

Simon, Mattias. "Der Prediger Tobias Winkler in Nürnberg." *Mitteilungen des Vereins für Geschichte der Stadt Nürnberg* 42 (1951): 198–235.

Simunotti, Luisa. "Prime osservazioni sulle curiositates magliabechiane d'Oltremanica." In *Antonio Magliabechi nell'Europa dei saperi*, edited by Jean Boutier, Maria Pia Paoli, and Corrado Viola, 395–426. Seminari e convegni, 50,1. Pisa: Edizioni della Normale, 2017.

Slauter, Will. "Periodicals and the Commercialization of Information in the Early Modern Era." In *Information: A Historical Companion*, edited by Ann Blair, Anthony Grafton, Paul Duguid, and Anja-Silvia Goeing, 128–51. Princeton, NJ: Princeton University Press, 2020.

Smith, Jay M. *The Culture of Merit: Nobility, Royal Service, and the Making of Absolute Monarchy in France, 1600–1789*. Studies in Medieval and Early Modern Civilization. Ann Arbor: University of Michigan Press, 1996.

Smyth, Adam. *Material Texts in Early Modern England*. Cambridge: Cambridge University Press, 2018.

Solignat, Anne-Valérie. "Perpétuer la maison: La place des ancêtres dans la transmission de l'identité nobiliaire." In *L'incorporation des ancêtres: Généalogie, construction du présent, du Moyen Âge à nos jours*, edited by Valérie Piétri and Isabelle Luciani, 175–95. Aix-en-Provence: Presses Universitaires de Provence, 2016.

Solms-Laubach, Rudolf Graf von. *Geschichte des Grafen- und Fürstenhauses Solms*. Frankfurt: C. Adelmann, 1865.

Sommer, Wolfgang. "Das Wirken Johann Michael Dilherrs in der Reichsstadt Nürnberg in der Mitte des 17. Jahrhunderts." *Mitteilungen des Vereins für Geschichte der Stadt Nürnberg* 91 (2004): 181–93.

Somsen, Quinten. "Prussia's Franconian Undertaking: Dynasty, Law, and Politics in the Holy Roman Empire (1703–1726)." *Virtus* 24 (2017): 75–104.

Soria Mesa, Enrique, ed. *La biblioteca genealógica de don Luis de Salazar y Castro*. Córdoba: Universidad de Córdoba, Servicio de Publicaciones, 1997.

———. "Genealogía y poder: Invención de la memoria y ascenso social en la España moderna." *Estudis: Revista de Historia Moderna de la Universidad de Valencia* 30 (2004): 21–56.

Spagnoletti, Pierantonio. "The Naples Elites between City and Kingdom." In *A Companion to Early Modern Naples*, edited by Tommaso Astarita, 197–214. Brill's Companions to European History. Leiden: Brill, 2013.

Spener, Philipp Jakob. *Briefe aus der Dresdner Zeit*. Edited by Johannes Wallmann, 4 vols. so far. Tübingen: Mohr Siebeck, 2003–17.

———. *Briefe aus der Frankfurter Zeit: 1666–1686*. Edited by Johannes Wallmann, 7 vols. so far. Tübingen: Mohr Siebeck, 1992–2019.

———. *Briefwechsel mit Adam Rechenberg 1686–1704*. Edited by Johannes Wallmann. 1 vol. so far. Tübingen: Mohr Siebeck, 2019.

Spiegel, Gabrielle M. *The Past as Text: The Theory and Practice of Medieval Historiography*. Baltimore, MD: Johns Hopkins University Press, 1997.

Sporhan-Krempel, Lore. *Nürnberg als Nachrichtenzentrum zwischen 1400 und 1700*. Nürnberger Forschungen 10. Nuremberg: Selbstverlag des Vereins für Geschichte der Stadt Nürnberg, 1968.

Stauffer, Hermann. *Sigmund von Birken (1626–1681): Morphologie seines Werkes*. 2 vols. Berlin: de Gruyter, 2007.

Steiger, Johann Anselm. *Ikonographie und Meditation des Hohenliedes in der Barockzeit zwischen Konfessionalität und Transkonfessionalität: Die "Göttliche Liebesflamme" (1651); Johann Michael Dilherrs und Georg Philipp Harsdörffers sowie das Bildprogramm an der Patronatsempore in Steinhagen (Vorpommern)*. Leipzig: Evangelische Verlagsanstalt, 2016.

Stein, Wolfgang Hans, ed. *Johann Jakob Mosers Mömpelgardisches Staatsrecht*. Veröffentlichungen der Kommission für Geschichtliche Landeskunde in Baden-Württemberg, Reihe A, 35. Stuttgart: Kohlhammer, 1983.

Stollberg-Rilinger, Barbara. *Maria Theresia: Die Kaiserin in ihrer Zeit. Eine Biographie*. Munich: C. H. Beck, 2017.

Stromer, Wolfgang von. "Reichtum und Ratswürde: Die wirtschaftliche Führungsschicht der Reichsstadt Nürnberg 1348–1648." In *Führungskräfte der Wirtschaft in Mittelalter und Neuzeit: 1350–1850*, edited by Herbert Helbig, 1–50. Büdinger Vorträge 1968/69. Limburg: Starke, 1973.

Suppa, Silvio. *L'accademia di Medinacoeli: Fra tradizione investigante e nuova scienza civile*. Naples: Istituto Italiano per gli Studi Storici—Il Mulino, 1971.

Testa, Simone. *Italian Academies and Their Networks, 1500–1700: From Local to Global*. Italian and Italian American Studies. Houndmills: Palgrave Macmillan, 2015.

Teuscher, Simon. "Bilatéralité vs. conceptions androcentriques de la parenté en Europe: Quelques réflexions à partir des arbores consanguinitatis de la fin du Moyen Âge." *Genre et Histoire* 21 (2018): 1–18.

Teuscher, Simon, and Margareth Lanzinger: "Editorial: Trennende Verwandtschaft." *Historische Anthropologie* 22 (2014): 1–3.

Tiersonnier, Philippe. "Compte-rendu de 'Mémoires inédites sur le mariage de l'Infante Isabelle de Portugal (1675–1681), par M. de Guénegaud.'" *Bulletin de la Société d'Emulation du Bourbonnais* (1904): 207–8.

Timm, Elisabeth. "Grounding the Family: Locality and Its Discontents in Popular Genealogy." *Ethnologia Europaea: Journal of European Ethnology* 42 (2012): 36–50.

Tölle, Tom. "Generation, Theft, and Dynasty in Eighteenth-Century Württemberg: Preacher Johann Ulrich Pregitzer's Genealogies." In *Genealogical Knowledge in the Making: Tools, Practices, and Evidence in Early Modern Europe*, edited by Jost Eickmeyer, Markus Friedrich, and Volker Bauer, 291–317. Cultures and Practices of Knowledge in History 1. Berlin: de Gruyter Oldenbourg, 2019.

———. *Heirs of Flesh and Paper: A European History of Dynastic Knowledge around 1700*. Berlin: de Gruyter, 2022.

Torrell, Jean-Pierre. *Saint Thomas Aquinas*. Vol. 1, *The Person and His Work*. Translated by Robert Royal. Washington, DC: Catholic University of America Press, 1996.

Trevisan, Sara. "Genealogy and Royal Representation: Edmund Brudenell's Pedigree Roll for Elizabeth I (1558–60)." *Huntington Library Quarterly* 81 (2018): 257–75.

———, ed. *Mythical Ancestry in World Cultures, 1400–1800*. Turnhout: Brepols, 2018.

———. *Royal Genealogy in the Age of Shakespeare*. Cambridge: D. S. Brewer, 2020.

Trombetta, Vincenzo. "La Libraria di S. Angelo a Nido." *Accademie e Biblioteche d'Italia* 62 (1994): 11–43.

Tucker, Susan. *City of Remembering: A History of Genealogy in New Orleans*. Jackson: University Press of Mississippi, 2016.

Turner, James. *Philology: The Forgotten Origins of the Modern Humanities*. Princeton, NJ: Princeton University Press, 2014.

Tutino, Stefania. *A Fake Saint and the True Church: The Story of a Forgery in Seventeenth-Century Naples*. New York: Oxford University Press, 2021.

van Damme, Stéphane. *Le temple de la sagesse: Savoirs, écriture et sociabilité urbaine (Lyon, 17–18e siècles)*. Civilisations et sociétés, 119. Paris: Éditions de l'École des Hautes Études en Sciences Sociales, 2004.

van der Lugt, Maaike, ed. *L'hérédité entre Moyen Âge et époque moderne: Perspectives historiques*. Micrologus' Library, 27. Florence: SISMEL Edizioni del Galluzzo, 2008.

van Vugt, Ingeborg. "Geografia e storia di una rete epistolare: Contatti e mediazioni nell'epistolario di Magliabechi." In *Antonio Magliabechi nell'Europa dei saperi*, edited by Jean Boutier,

Maria Pia Paoli, and Viola Corrado, 259–92. Seminari e convegni, 50,1. Pisa: Edizioni della Normale, 2017.

Viola, Corrado. "Vecchia e nuova erudizione: Muratori e Magliabechi." *Studi Secenteschi* 54 (2013): 97–115.

Virieux, Maurice. "Guy Allard: Un provincial ami des lettres (1635–1716)." *Bulletin de l'Académie Delphinale* 8, no. 2 (1973): 60–81.

Vitale, Giuliana. "Modelli culturali nobiliari a Napoli tra quattro e cinquecento." *Archivio Storico per le Province Napoletane* 105 (1987): 27–104.

Vivo, Filippo de. "Pharmacies as Centres of Communication in Early Modern Venice." *Renaissance Studies* 21 (2007): 505–21.

Vogtherr, Thomas. "Archivtheorie und Archivpraxis im ausgehenden 17. Jahrhundert: Ahasver Fritsch, Jacob Bernhard Multz von Oberschönfeld und Georg Aebbtlin." In *Fundamenta Historia:. Geschichte im Spiegel der Numismatik und ihrer Nachbarwissenschaften*, edited by Reiner Cunz, Rainer Polley, and Andreas Röpcke, 403–9. Hanover: Niedersächsisches Landesmuseum, 2004.

Völkel, Markus. *"Pyrrhonismus historicus" und "fides historica": Die Entwicklung der deutschen historischen Methodologie unter dem Gesichtspunkt der historischen Skepsis.* Frankfurt am Main: Lang, 1987.

Voss, Jürgen. "Christian Juncker, ein Gelehrter und Schulmann der Leibnizzeit." *Jahrbuch für die Geschichte Mittel- und Ostdeutschlands* 16/17 (1968): 303–18.

Wallnig, Thomas, and Ines Peper, eds. *Central European Pasts: Old and New in the Intellectual Culture of Habsburg Europe, 1700–1750.* Cultures and Practices of Knowledge in History 6. Berlin: de Gruyter, 2022.

Weber, Camilla. "Der Briefwechsel des Johann Hiernoymus Imhoff: Networking aus der Provinz." *Morgen-Glantz* 23 (2013): 293–306.

———. "Die Korrespondenzen des Johann Hieronymus Imhoff: Religiöse, geographische und soziale Grenzgänge im konfessionellen Zeitalter." *Morgen-Glantz* 15 (2005): 271–98.

Weduwen, Arthur de, and Andrew Pettegree. *The Dutch Republic and the Birth of Modern Advertising.* The Handpress World 59. Leiden: Brill, 2020.

Wegele, Franz Xaver von. *Geschichte der deutschen Historiographie seit dem Auftreten des Humanismus.* Munich: Oldenbourg, 1885.

Weigel, Sigrid. *Genea-Logik: Generation, Tradition und Evolution zwischen Kultur- und Naturwissenschaften.* Munich: Fink, 2006.

Weigelt, Horst. *Geschichte des Pietismus in Bayern: Anfänge, Entwicklung, Bedeutung.* Arbeiten zur Geschichte des Pietismus 40. Göttingen: Vandenhoeck & Ruprecht, 2001.

Weil, François: *Family Trees: A History of Genealogy in America.* Cambridge, MA: Harvard University Press, 2013.

Weis, Joëlle. *Johann Friedrich Schannat (1683–1739): Praktiken historisch-kritischer Gelehrsamkeit im frühen 18. Jahrhundert.* Berlin: de Gruyter, 2022.

Werner, Theodor Gustav. "Repräsentanten der Augsburger Fugger und Nürnberger Imhoff als Urheber der wichtigsten Handschriften des Paumgartner-Archivs über Welthandelsbräuche im Spätmittelalter und am Beginn der Neuzeit." *Vierteljahrschrift für Sozial- und Wirtschaftsgeschichte* 52 (1965): 1–41.

Wickham, Chris. *Framing the Early Middle Ages: Europe and the Mediterranean, 400–800.* Oxford: Oxford University Press, 2006.

Willemetz, Geneviève. *Jean Anisson, 1642–1721: Un homme d'affaires et de culture au Grand Siècle.* Paris: Éditions des Cendres, 2004.

Williams, Kelsey Jackson. *The First Scottish Enlightenment: Rebels, Priests, and History.* Oxford: Oxford University Press, 2020.

Winter, Ursula. *Die europäischen Handschriften der Bibliothek Diez in der Deutschen Staatsbibliothek Berlin: Teil 3, Abschlussband.* Leipzig: Zentralantiquariat der DDR, 1986.

Wood, James B. *The Nobility of the Election of Bayeux, 1463–1666: Continuity through Change.* Princeton, NJ: Princeton University Press, 1994.

Woolf, Daniel R. *A Global History of History.* Cambridge: Cambridge University Press, 2011.

———. *The Social Circulation of the Past: English Historical Culture 1500–1730.* Oxford: Oxford University Press, 2003.

Woolf, Daniel R., et al., eds. *The Oxford History of Historical Writing.* 5 vols. New York: Oxford University Press, 2011–15.

Wrede, Martin. "Zwischen Mythen, Genealogie und der Krone: Rivalisierende Familiengedächtnisse im französischen Hochadel des 17. Jahrhunderts; Die Häuser Bouillon, Noailles und Bourbon." *Zeitschrift für Historische Forschung* 32 (2005): 17–43.

Wurster, Herbert Wilhelm. "Die Regensburger Geschichtsschreibung im 17. Jahrhundert: Historiographie im Übergang vom Humanismus zum Barock." *Verhandlungen des Historischen Vereins für Oberpfalz und Regensburg* 119 and 120 (1979 and 1980): 7–75, 69–210.

Wüst, Wolfgang. "Citronen, Pomeranzen, Spargel, Tabak: Exotik im Acker und Garten - Anbau und Konsum in frühen Quellen." *Mitteilungen des Vereins für Geschichte der Stadt Nürnberg* 104 (2017): 165–202.

———. "Patrizier: Zum Selbstverständnis reichsstädtischer Oligarchen in Süddeutschland." *Blätter für fränkische Familienkunde* 39 (2016): 9–32.

———. "Die Schwarzenberg in Franken und Böhmen: Freiherren—Grafen—Fürsten." *Jahrbuch für fränkische Landesforschung* 74 (2014): 115–30.

Wuttke, Dieter. "Humanismus in Nürnberg um 1500." *Zeitschrift für Bayerische Landesgeschichte* 48 (1985): 677–84.

Yakel, Elizabeth. "Seeking Information, Seeking Connections, Seeking Meaning: Genealogists and Family Historians." *Information Research* 10 (2004).

Yakel, Elizabeth, and Deborah Torres. "Genealogists as a 'Community of Records.'" *American Archivist* 70 (2007): 93–113.

Zerubavel, Eviatar. *Ancestors and Relatives: Genealogy, Identity, and Community.* Oxford: Oxford University Press, 2013.

Ziegler, Hannes. *Trauen und Glauben: Vertrauen in der politischen Kultur des Alten Reiches im Konfessionellen Zeitalter.* Affalterbach: Didymos-Verlag, 2017.

INDEX